Sustainable Artificial Intelligence-Powered Applications

IEREK Interdisciplinary Series for Sustainable Development

The interdisciplinary series "Sustainable Artificial intelligence Powered Applications," (SAIPA) aims to cover a wide range of Artificial Intelligence applications with a specific focus on sustainability to further support the sustainable development goals. AI systems are able to perform a wide range of tasks, including high-level image and video recognition, perspective modeling, smart automation, advanced simulation, and complex analytics among many others, and has high learning and improvement potential through machine learning and deep learning. These capabilities have many uses and provide many benefits in all industries applications. SAIPA includes a diverse range of these applications, perspectives, and expertise, where it seeks to advance the understanding and implementation of sustainable AI-powered solutions across various scientific disciplines. Its aim is to become a comprehensive guide to different experimental sciences. The "Sustainable AI-Powered Applications" publishes conference proceedings, monographs, and textbooks that address the intersection of artificial intelligence, sustainability, and advanced technologies. Authors, researchers, contributors, and scientists from all scientific disciplines related to AI and advanced technologies, are all invited to contribute to this series to shape the future of AI in a sustainable and responsible manner. Thus, the series represents an excellent endeavor that promotes a collaborative and interdisciplinary strategy for the development of sustainable AI-driven applications. The scope of this series includes, but is not limited to: Promoting sustainability and human well- being, health and medical use, optimizing smart cities planning, transportation systems management, infrastructure development, creating livable urban environments, renewable energy, agricultural Innovations, precision agriculture, weather agriculture, sustainable food production, crop & animal monitoring, and agricultural robotics, Socio-Economic Impacts, optimizing energy consumption in buildings, factories in energy usage and lower carbon, smart grids applications, smart grids management, climate modelling, environmental monitoring, waste management, Internet of things (IOT) and other domains related to the use of AI applications that improve overall environmental impact.

Series Editor: Mourad Amer **editor@saipa-series.com**

M. Ashwin · S. Sountharrajan ·
Balamurugan Balusamy · Chander Prabha ·
Mahmoud Ahmad Al-Khasawneh
Editors

Smart Forestry 5.0 for Powered AI Sustainable Environment and Climate

Springer

Editors
M. Ashwin
Department of Computer Science
and Engineering (Data Science)
AMC Engineering College
Bengaluru, Karnataka, India

S. Sountharrajan
Department of Computing Science
and Engineering, Amrita School of Computing
Amrita Vishwa Vidyapeetham
Chennai, Tamil Nadu, India

Balamurugan Balusamy
School of Engineering and IT
Manipal Academy of Higher Education
Dubai, UAE

Chander Prabha
Chitkara University Institute of Engineering
and Technology
Chitkara University
Rajpura, Punjab, India

Mahmoud Ahmad Al-Khasawneh
School of Computing
Skyline University College
Sharjah, UAE

ISSN 3005-1762 ISSN 3005-1770 (electronic)
Sustainable Artificial Intelligence-Powered Applications
IEREK Interdisciplinary Series for Sustainable Development
ISBN 978-3-031-99085-4 ISBN 978-3-031-99086-1 (eBook)
https://doi.org/10.1007/978-3-031-99086-1

This Springer imprint is published by the registered company Springer Nature Switzerland AG
The registered company address is: Gewerbestrasse 11, 6330 Cham, Switzerland

Introduction

Forests are among the most critical ecosystems for maintaining planetary health, supporting biodiversity, and mitigating climate change. They absorb approximately 30% of global carbon dioxide emissions annually, making them indispensable in the effort to reach net-zero targets. However, deforestation, land-use change, and forest degradation continue at an alarming pace, exacerbated by global population growth, industrial expansion, and changing climate patterns. As the world faces the twin challenges of ecological degradation and climate crisis, innovative solutions that harness the power of digital technologies and artificial intelligence (AI) are increasingly seen as pivotal to sustainable forest management.

Smart Forestry 5.0 embodies a transformative vision, merging digital intelligence, automation, and real-time environmental sensing to revolutionize the forestry sector. By integrating AI, big data, remote sensing, IoT, and robotics, Smart Forestry aims to enhance ecosystem monitoring, optimize carbon capture, reduce human pressures on forest resources, and ensure resilient forest landscapes for future generations. This technological shift aligns with the global sustainability agenda and climate action goals, positioning forests as a key driver in achieving the objectives of the Paris Agreement and the UN Sustainable Development Goals (SDGs).

This volume seeks to inform policy-making and forest governance practices within the evolving context of climate mitigation, digital transformation, and environmental stewardship. It brings together a multidisciplinary network of researchers, environmental scientists, AI experts, foresters, and policymakers to examine the socio-environmental, technological, and economic dimensions of next-generation forestry systems. As countries race to implement net-zero strategies and carbon markets evolve, forests must be reimagined not only as natural assets but also as digitally managed and AI-powered climate solutions.

Despite the increasing use of the term "smart forestry," research and practical frameworks in this area are still emerging. There remains a need for greater clarity and comprehensive models that can guide decision-makers in implementing effective Smart Forestry 5.0 strategies. As governments and institutions face mounting pressure to protect natural ecosystems while driving green growth, this book offers a timely, systematic exploration of the intersections between intelligent technologies and sustainable forestry.

The volume includes a range of theoretical explorations, empirical research, and real-world case studies to assess the ecological, technological, and governance implications of forest digitalization. It focuses on key themes such as (i) the foundational principles and technologies underpinning Smart Forestry 5.0; (ii) AI applications in forest monitoring, restoration, and carbon accounting; (iii) data-driven biodiversity and conservation strategies; (iv) resilient forest ecosystems in the face of climate extremes; and (v) global and regional policy frameworks for forest-based climate action.

Several aspects make this volume particularly relevant and impactful for forest managers, environmental agencies, tech developers, and sustainability advocates. First, it thoroughly addresses the environmental and climate challenges facing forestry and proposes AI-augmented solutions for effective response. Second, it offers guidance for integrating advanced digital tools into traditional forest management practices. Third, it provides a blueprint for interdisciplinary cooperation, combining ecological knowledge with cutting-edge AI capabilities to foster climate-resilient landscapes.

As AI continues to permeate every facet of human development, forests too must evolve into intelligent, adaptive systems that contribute to a low-carbon, biodiverse, and sustainable future. Smart Forestry 5.0 is not just an aspirational concept, it is an urgent necessity for achieving a balanced coexistence between human progress and the planet's ecological integrity.

Contents

Editors and Contributors

About the Editors

Dr. M. Ashwin has completed his Ph.D. in Computer Science and Engineering from Karpagam Academy of Higher Education, Coimbatore, in 2018. He is currently working as a Professor and Head in the Department of Computer Science and Engineering (Data Science), AMC Engineering College, Bengaluru. He has over 17 years of teaching and 6 years of exclusive research experience. To his credit he has 40+ research articles in his research domain of sustainable AI environment and MANET. He has many vital roles of an academician and researcher by serving as Board of Studies, doctoral committee member, and reviewer for reputed journals and conferences. He is an active IE(I) and ISTE member.

Dr. S. Sountharrajan is an Associate Professor, Amrita School of Computing, Amrita Vishwa Vidyapeetham, Chennai Campus, where he also serves as the Chairperson, Department of Computer Science and Engineering. His academic and research contributions are extensive: he has published over 84 high-quality journal papers which include conference articles and book chapters. Additionally, he has edited 3 books covering a wide range of emerging technologies. He earned his Bachelor's, Master's, and Ph.D. degrees from premier institutions in India. He has received research grants totaling ₹38 lakhs from renowned funding agencies such as the DST, AICTE, and ICMR. He is also a frequent keynote speaker and has delivered over 40+ invited talks at various academic and industry forums. He holds certifications from leading technology firms such as IBM and Dell EMC. His core areas of expertise include Artificial Intelligence, Quantum Computing, and Cloud Computing. His dynamic involvement in both academia and industry continues to inspire innovation and knowledge dissemination across multiple platforms.

Prof. Balamurugan Balusamy has served up to the position of Associate Professor in his stint of 12 years of experience with VIT University, Vellore. He had completed his Bachelor's, Master's, and Ph.D. degrees from top premier institutions from India. His passion is teaching and adapts different design thinking principles while delivering his lectures. He has published 80+ books on various technologies and visited 15 plus countries for his technical course. He has several top-notch conferences in his resume and has published over 200 of quality journal, conference, and book chapters combined. He serves in the advisory committee for several start-ups and forums and does consultancy work for industry on Industrial IoT. He has given over 195 talks in various events and symposiums.

Dr. Chander Prabha is currently working as a Professor in the Department of Computer Science and Engineering at Chitkara University Institute of Engineering and Technology, Chitkara University, Punjab, India. She had done her Ph.D. (Computer Science Engineering) in 2017 and her M.E. in 2004. She has more than 20+ years of academic/research experience with many publications at various national and international conferences and international journals. She has published many book chapters in CRC Press, Elsevier, IGI, and Wiley. She is the Guest Editor *Journal of Artificial Intelligence* and *Internet of Things*, Near East University, Turkey.

Recently, she has been selected as a member of the Editorial Board of PLOS One Journal having an H-Index of 435. Also she serves as an editor and reviewer of many international conferences and journals of repute. Many online certifications are also there to her credit. She is a Lifetime member of educational societies like ISTE, IAENG, SCIEI, IIRJC, IACSIT, IEEE, etc. Her areas of interest are wireless and mobile networks' architectures, protocols, performance evaluations, machine learning, data analytics, and law.

Mahmoud Ahmad Al-Khasawneh holds a B.Sc. degree in Computer Science from Yarmouk University, Jordan, conferred in 2003. He obtained both his M.Sc. and Ph.D. degrees in Computer Science from Universiti Teknologi Malaysia (UTM), Johor, Malaysia, in 2013 and 2018, respectively. Currently, he serves as a faculty member in the School of Computing Skyline University College, Sharjah, UAE. Dr. Al-Khasawneh's scholarly pursuits span a diverse array of fields within computer science. He has authored numerous papers in esteemed, peer-reviewed journals across leading publishers such as IEEE, Springer, Wiley, Hindawi, and MDPI. His research interests encompass Security, Image Encryption, Wireless Networks, Blockchain, Internet of Things, and Big Data. With a commitment to advancing knowledge and solving contemporary challenges in these domains, he actively engages in research, teaching, and mentorship, contributing to the academic and professional development of his students and peers. Driven by a passion for innovation and a dedication to excellence, he continues to make significant contributions to the field, shaping the future of technology and its applications.

Contributors

C. Ambhika Department of AIML, R.M.D. Engineering College, RSM Nagar, Kavarapetai, India

Kandavalli Michael Angelo Department of ECE, DVR & Dr. HS MIC College of Technology, Kanchikacherla, Andhra Pradesh, India

M. Arunkumar Department of Electronics and Communication Engineering, Sengunthar Engineering College (Autonomous), Tiruchengode, India

R. Ganesh Babu Department of Electronics and Communication Engineering, Koneru Lakshmaiah Education Foundation, Vaddeswaram, Guntur, Andhra Pradesh, India

Jahanara Begum Department of Computer Science and Engineering, Hyderabad Institute of Technology and Management, Hyderabad, India

K. Bhargavi Department of CSE, ICFAI Foundation for Higher Education (IFHE), Shankarpally, Hyderabad, India

Subhash Bondhala Dominion Energy, Network Security Engineer, Richmond, USA

Tamanna Dash Networking and Communication, School of Computing, SRM Institute of Science and Technology, Kattankulathur, Chennai, India

P. Devisri Department of Computer Science and Engineering, Ramco Institute of Technology, Rajapalayam, Tamil Nadu, India

Amit Gangopadhyay Department of Electronics and Communication Engineering, Mohan Babu University (Erstwhile Sree Vidyanikethan Engineering College), Tirupati, Andhra Pradesh, India

R. Geetha Department of Computing Technologies, School of Computing, SRM Institute of Science and Technology, Kattankullathur, Chennai, India

R. Hemavathi Department of Civil Engineering, Sri Krishna College of Engineering and Technology, Kuniyamuthur, Coimbatore, Tamilnadu, India

J. Kalyan Srinivas Department of ECE, Cambridge Institute of Technology North Campus, Bengaluru, Karnataka, India

Suresh Kumar Kanaparthi Department of Computer Science and Engineering, SRM University, Amaravati, Andhra Pradesh, India

R. Kannamma Department of AI&DS, Prathyusha Engineering College, Chennai, India

Mudassir Khan Department of Computer Science, College of Computer Science, Applied College Tanumah, King Khalid University, Abha, Saudi Arabia

A. Kishore Department of ECE, Cambridge Institute of Technology North Campus, Bengaluru, Karnataka, India

Raghavendra Kulkarni Department of Computer Science, Amrita School of Computing, Amrita Vishwa Vidyapeetham, Mysuru, Karnataka, India

Isukupalli Kranthi Kumar Department of EEE, SRKR Engineering College, Bhimavaram, India

Ramaraju Nagarjuna Kumar Computer Applications in Agriculture, ICAR-Central Research Institute for Dryland Agriculture (ICAR-CRIDA), Hyderabad, India

S. Manjula Department of Computer Science and Engineering, Ramco Institute of Technology, Rajapalayam, Tamil Nadu, India

R. I. Minu Computing Technologies, School of Computing, SRM Institute of Science and Technology, Kattankulathur, Chennai, India

G. Mohan Raj Department of ECE, Cambridge Institute of Technology North Campus, Bengaluru, Karnataka, India

Patil Mounica EEE Department, Vardhaman College of Engineering, Shamshabad, Telangana, India

Sivaram Murugan Department of Computer Engineering, Faculty of Engineering and Natural Sciences, Sivas University of Science and Technology, Sivas, Turkey

Sateesh Nagavarapu Department- CSE-DS, Malla Reddy College of Engineering (Q9), Hyderabad, India

Manoj Kumar Padhi Department of Computer Science and Engineering, Centurion University of Technology and Management, Paralakhemundi, Odisha, India

Gayatri Parasa Department of CSE—AIML, KG Reddy College of Engineering and Technology, Hyderabad, Telangana, India

Hemanth Gouda S. Patil Department of ECE, Cambridge Institute of Technology North Campus, Bengaluru, Karnataka, India

V. Porkodi Department of Computer Engineering, Faculty of Engineering and Natural Sciences, Sivas University of Science and Technology, Sivas, Turkey

U. M. Prakash Computing Technologies, School of Computing, SRM Institute of Science and Technology, Kattankulathur, Chennai, India

Smitha Rajagopal Department of Computer Science and Engineering, Alliance School of Advanced Computing, Alliance University, Bengaluru, Karnataka, India

V. Ramesh Babu Department of CSE, Cambridge Institute of Technology North Campus, Bengaluru, Karnataka, India

K. Yogeswara Rao Department of Computer Science and Engineering, GITAM School of Technology, GITAM (Deemed to Be University), Hyderabad, India

T. Prabhakara Rao Department of Computer Science and Engineering, Aditya University, Surampalem, AP, India

Satti Sudha Mohan Reddy Department of ECE, SRKR Engineering College, Bhimavaram, Andhra Pradesh, India

J. Samuel Manoharan Department of Biomedical Engineering, Jerusalem College of Engineering, Chennai, India

P. Saravanan Computing Technologies, School of Computing, SRM Institute of Science and Technology, Kattankulathur, Chennai, India

T. Saravanan Department of Information Technology, P.S.R. Engineering College, Sivakasi, Tamilnadu, India

A. Sathiya Department of ECE, M. Kumarasamy College of Engineering (Autonomous), Thalavapalayam, Karur, India

J. Seetha Department of Computer Science and Business Systems, Panimalar Engineering College, Chennai, India

V. K. Senthil Ragavan Sri Shanmugha College of Engineering and Technology, Sankari, Salem, Tamil Nadu, India

Gattu Shravani Department of CSE (Cys, DS, AI & DS), VNRVJIET, Bachupally, Hyderabad, Telangana, India

Sukhumjeet Singh Computing Technologies, School of Computing, SRM Institute of Science and Technology, Kattankulathur, Chennai, India

Sivabalan Department of CSE, Cambridge Institute of Technology North Campus, Bengaluru, Karnataka, India

Soubraylu Sivakumar Computing Technologies, School of Computing, SRM Institute of Science and Technology, Kattankulathur, Chennai, India

S. J. Sugumar Department of ECE, Cambridge Institute of Technology North Campus, Bengaluru, Karnataka, India

R. Sundar Department of Computer Science and Engineering, Vel Tech Rangarajan Dr. Sagunthala R&D Institute of Science and Technology, Avadi, Chennai, Tamil Nadu, India

Mishmala Sushith Department of Information Technology, Adithya Institute of Technology, Coimbatore, India

M. Swarna Sudha Department of Computer Science and Engineering, Ramco Institute of Technology, Rajapalayam, Tamil Nadu, India

R. Tamijetchelvy Department of ECE, Perunthalaivar Kamarajar Institute of Engineering and Technology (PKIET), Puducherry, India

M. Tholkapiyan Department of Civil Engineering, Chennai Institute of Technology, Chennai, Tamil Nadu, India

S. Vidhya Department of Computing Technologies, School of Computing, SRM Institute of Science and Technology, Kattankulathur, Chennai, India

K. Vijayalakshmi Department of Computer Science and Engineering, Ramco Institute of Technology, Rajapalayam, Tamil Nadu, India

Introduction to Smart Forestry 5.0: Merging Technology with Nature

J. Samuel Manoharan, A. Sathiya, M. Arunkumar, and R. Tamijetchelvy

Abstract

As a result of the rapid improvements in technology, new opportunities for environmentally responsible management have become available and Smart Forestry 5.0 has emerged as a framework in this study. The Internet of Things (IoT), blockchain, robotics, and Artificial Intelligence (AI) are some of the advanced methods that are employed into this concept, which is based on the concepts on Industry 5.0. Additionally, this concept employs traditional forestry policies and procedures. The objectives of Smart Forestry 5.0 are to bring together human interventions and technical innovations to promote the development of sustainable forest ecosystems, reduce the effects of climate change, and improve the preservation of biodiversity. This study provides a full introduction to Smart Forestry 5.0 and places an emphasis on the multidimensional approach that Smart Forestry 5.0 manages the difficulties which are specific to modern forestry. This chapter investigates the way in which IoT-enabled sensors make real-time forest monitoring possible, artificial intelligence (AI)-powered predictive analysis help in detection of pests and disease, and drones equipped with robots enhance the afforestation and forest management methods. The blockchain method is being recognized for its potential to improve the transparency and traceability in supply chains for forest products, thereby preventing illegal logging and boosting the ethical sourcing. Additionally, the chapter discusses the ways in which digital twins and spatial data analysis make it possible to precisely method forest ecosystem, which in turn contributes to proactive decision-making and long-term planning. The socioeconomic outcome of Smart Forestry 5.0 is also investigated in this book chapter. Specifically, the chapter focuses on the potential of Smart Forestry 5.0 to empower rural communities by means of sustainable forestry practices, job creation, and efficiency in resource utilization. Some of the challenges that are examined in depth include high implementation costs, the digital gap, and ethical concerns regarding data privacy. These challenges are also addressed in this chapter. By integrating the natural world and technological advancements, Smart Forestry 5.0 is a significant step toward accomplishing the objectives of global sustainability. This chapter concludes by discussing the future direction including employment of adaptive learning method, collaborative policy frameworks, and renewable energy sources. Smart Forestry 5.0 has the potential to consider the way in which people manage with and protect forest, with the help of a combination of technological advancement and ecological care.

Keywords

Smart Forestry 5.0 · Internet of Things (IoT) · Artificial Intelligence (AI) · Blockchain · Robotics · Sustainable forest management · Biodiversity conservation · Climate change mitigation · Digital twins · Forest monitoring

J. Samuel Manoharan (✉)
Department of Biomedical Engineering, Jerusalem College of Engineering, Chennai, India
e-mail: samuel1530@jerusalemengg.ac.in

A. Sathiya
Department of ECE, M. Kumarasamy College of Engineering (Autonomous), Thalavapalayam, Karur, India

M. Arunkumar
Department of Electronics and Communication Engineering, Sengunthar Engineering College (Autonomous), Tiruchengode, India

R. Tamijetchelvy
Department of ECE, Perunthalaivar Kamarajar Institute of Engineering and Technology (PKIET), Puducherry, India
e-mail: tamijetchelvy@pkiet.edu.in

M. Ashwin et al. (eds.), *Smart Forestry 5.0 for Powered AI Sustainable Environment and Climate*, Sustainable Artificial Intelligence-Powered Applications, https://doi.org/10.1007/978-3-031-99086-1_1

1 Introduction

1.1 Overview of Modern Forestry Challenges

Ecosystem services of forests are of a lot of ecological, economic, and social importance such as carbon sequestration, biodiversity preservation, production of timber. Modern forestry, however, still faces several problems such as deforestation, climate change, illegal logging, biodiversity loss, and inefficiency in forest management. Increasing demand for timber and forest products overwhelms sustainable harvesting practices; climate change exacerbates forest fires, studies any pest infestations or unpredictable weather conditions which undermine efficacy of forest ecosystems (Torresan et al., 2021).

Deforestation, among about the most critical concerns that are relevant to nearly 11 percent of the carbon emission on the planet, such as forestry. The driving force for this is industrial agriculture, urbanization, and illegal logging (Tognetti et al., 2022a).

Traditional forestry practices of manual monitoring, static policy, and the limited technological integration (Picchi et al., 2022) have not been able to address such large-scale issues effectively. Infrastructure and conversion of land-use change has fragmented natural habitats and fragments forest ecosystem, with threat to the survival of species as well.

Although forestry programs with afforestation and reforestation efforts have been inefficient as far as resource allocation, lack in real-time data, and having gaps in policy (Shvediuk, 2024), digital technologies are rising in forestry through Industry 4.0 but it brings challenges for scalability, cost, and integration.

Smart Forestry 5.0 is the combination of modern cutting-edge forest technologies, such as Artificial Intelligence (AI), Internet of Things (IoT), Blockchain, and Robotics, with ancient forest knowledge, which makes it the solution to the above-mentioned problems (Cooper & MacFarlane, 2023).

1.2 Evolution from Traditional Forestry to Smart Forestry 5.0

The forestry has come a long way from manual forest management to the technology-driven systems. Human surveillance, manual data collection, and policy-driven management were the traditional approaches that were limited in scope and responsiveness (Zou et al., 2019).

Such progress in remote sensing, Geographic Information System (GIS), and satellite monitoring helped force forestry into more data-based, Industry 4.0 Forestry. This stage involved

- Remote sensing for forest mapping
- Automated logging systems for timber extraction
- Digital record-keeping for supply chain transparency.

However, despite these advancements, forestry in Industry 4.0 still relied heavily on centralized systems and lacked adaptive intelligence. Challenges such as climate change resilience, predictive analytics, and automation required further technological advancements (Gabrys, 2020).

Smart Forestry 5.0, inspired by the principles of Industry 5.0, integrates human expertise with artificial intelligence, robotics, and decentralized technologies to enhance sustainability. Unlike Industry 4.0 Forestry, which focused on automation and digital transformation, Smart Forestry 5.0 prioritizes sustainability, resilience, and human-technology collaboration (Sengupta, 2024).

Figure 1 illustrates the chronological transformation of forestry practices from traditional manual methods to the advanced Smart Forestry 5.0 paradigm. It highlights the key

Fig. 1 Evolution of Smart Forestry 5.0

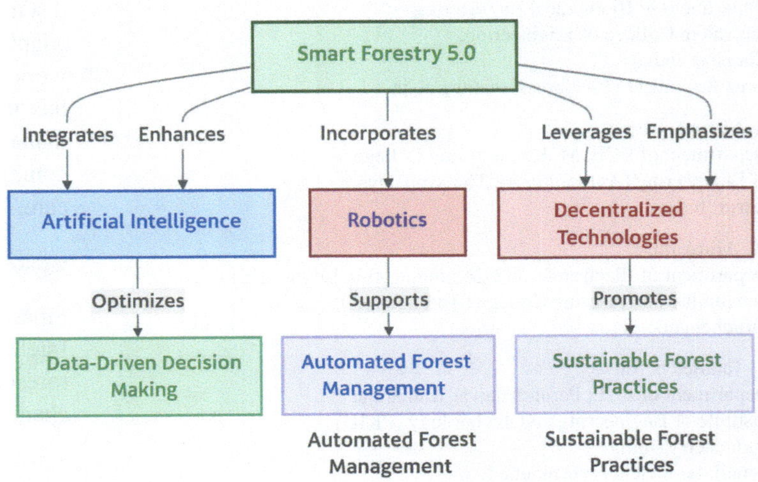

Table 1 Key differences between Traditional Forestry, Industry 4.0 Forestry, and Smart Forestry 5.0

Feature	Traditional Forestry	Industry 4.0 Forestry	Smart Forestry 5.0
Monitoring	Manual surveys	Remote sensing and GIS	IoT-based real-time tracking
Decision-making	Policy-driven	AI-assisted	AI and Blockchain-enabled predictive analysis
Resource allocation	Fixed quotas	Data-driven optimization	Dynamic, adaptive management using AI
Transparency	Limited	Digital records	Blockchain-based verifiable logging
Climate adaptation	Reactive policies	Early warning systems	AI-driven climate models
Afforestation techniques	Manual planting	Mechanized afforestation	Drone-assisted and robotic afforestation

transitions through Traditional Forestry, Industry 4.0 Forestry, and finally Smart Forestry 5.0, emphasizing the increasing integration of technologies like IoT, AI, blockchain, and robotics. The diagram demonstrates how forestry has evolved from reactive, policy-driven models to proactive, intelligent, and adaptive ecosystems (Table 1).

With these advancements, Smart Forestry 5.0 moves toward autonomous, intelligent, and sustainable forest management. It allows forestry professionals to monitor forests in real time, predict risks, optimize harvesting strategies, and enhance conservation efforts (Etaati et al., 2024).

1.3 Integration of Industry 5.0 Principles in Sustainable Forest Management

Industry 5.0, the next phase of technological evolution, emphasizes human-centered, sustainable, and resilient systems. It builds upon the automation and digital transformation of Industry 4.0, but places greater emphasis on

1. Human–Machine Collaboration—AI and humans working together for decision-making and efficiency.
2. Sustainability and Green Practices—Prioritizing eco-friendly solutions and reducing environmental footprints.
3. Resilience and Adaptability—Systems that dynamically respond to changing conditions in real time (Datta & Sinha, 2023).

In Smart Forestry 5.0, these principles translate into

- IoT-driven forest monitoring, where sensors track real-time environmental conditions such as temperature, humidity, carbon levels, and fire risks.
- AI-based predictive analytics, allowing forestry agencies to anticipate pest outbreaks, deforestation risks, and climate-related changes.
- Blockchain for secure and transparent forest product supply chains, reducing illegal logging and ensuring ethical sourcing.
- Robotics and drones for automated afforestation and precision forestry.
- Digital Twin technology, creating virtual forest models for scenario-based management and resource planning.

By incorporating Industry 5.0 principles, Smart Forestry enables a self-learning and adaptive forestry ecosystem that enhances climate resilience, biodiversity conservation, and economic efficiency (Torresan et al., 2021).

1.4 Objectives of Smart Forestry 5.0

The primary goal of Smart Forestry 5.0 is to harmonize technological advancements with ecological conservation, ensuring forests are managed sustainably, transparently, and efficiently. The key objectives are shown in the following figure.

Figure 2 provides a visual representation of the core objectives that define Smart Forestry 5.0. It involves five key goals: enhancing sustainability, enabling real-time monitoring, addressing climate change, ensuring transparency in the forestry supply chain, and empowering rural communities. Each objective is interconnected and aligned with the overarching goal of achieving resilient, technology-enabled, and ecologically balanced forest management.

1.4.1 Enhancing Sustainability in Forestry Practices

- Reducing deforestation rates through AI-driven monitoring and automated afforestation efforts.
- Implementing energy-efficient logging practices using robotic automation.
- Encouraging data-driven conservation efforts through AI-powered biodiversity tracking (Picchi et al., 2022).

1.4.2 Improving Real-Time Monitoring and Decision-Making

- IoT-enabled real-time forest health tracking, allowing for early detection of environmental risks.
- AI-based decision support systems, optimizing forest harvesting schedules and conservation plans (Shvediuk, 2024).
- Drone-assisted aerial surveillance for wildfire detection and illegal logging prevention.

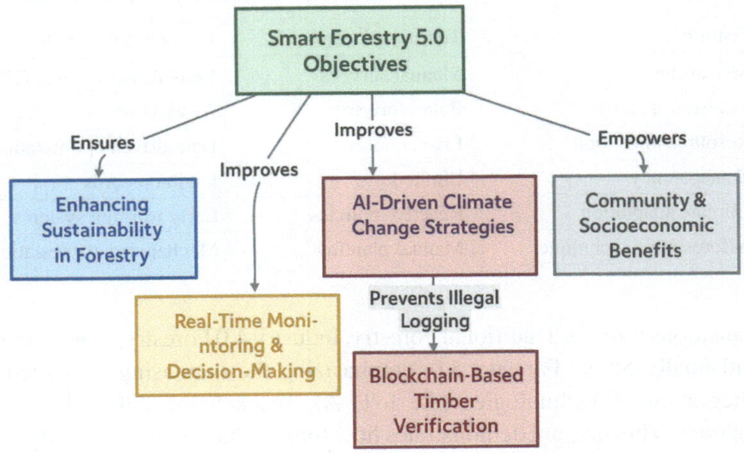

Fig. 2 Objectives of Smart Forestry 5.0

1.4.3 Addressing Climate Change with AI-Driven Conservation Strategies

- Machine learning algorithms for predicting climate impacts on forests.
- AI-optimized carbon sequestration strategies to enhance forest-based climate mitigation.
- Digital twin models that simulate and forecast long-term environmental changes (Zou et al., 2019).

1.4.4 Increasing Transparency and Security in Forestry Supply Chains

- Blockchain-based verification of timber origins, preventing illegal logging (Gabrys, 2020).
- Smart contracts for sustainable wood trade, ensuring accountability in forestry-related transactions.
- Decentralized ledgers to maintain real-time records of forestry activities, preventing fraud.

1.4.5 Empowering Communities and Strengthening Socioeconomic Benefits

- Smart Forestry initiatives in rural areas, creating new job opportunities in AI-assisted afforestation and conservation.
- Training programs to equip forestry professionals with technology-driven management skills (Sengupta, 2024).
- IoT-driven community-led reforestation projects, ensuring long-term sustainability of forest resources.

Smart Forestry 5.0 is the next phase of sustainable forest management because forests are feeling unprecedented pressure from the environment and the economy. This model combines AI, IoT, Blockchain, Robotics, and Digital Twin to enable real-time monitoring, predict the environment, be climate adaptive and have ethical supply chains. This step of transition to Industry 5.0 (Smart Forestry) inspired by Industry 4.0 Forestry is crucial to realizing the global sustainability targets (Datta & Sinha, 2023).

Smart Forestry 5.0 is not just about technology, but about nature and innovation in balance, for forest conservation, forest management, and the sustainable use in the future.

2 Literature Review

Increasingly, the concept of Smart Forestry 5.0 has been attracting much attention since it might bring about a revolution in forest management by means of advanced technologies. Various studies of Smart Forestry 5.0 have been carried out in the areas of artificial intelligence (AI), Internet of Things (IoT), blockchain technology, and digital twins. This section details the ones in the literature review of the key contributions in the field with approaches, methodologies, and findings of various researchers.

In (Sengupta, 2024), the authors proposed an AI-based predictive forest health monitoring model proposed by deep learning to detect the early effects of the disease and pest infestations. As the AI algorithms are trained on the big dataset, they can accurately predict forest disturbances and hence allow preventing interventions proactively. (Etaati et al., 2024) also suggested the implementation of IoT-enabled sensors in forest ecosystem to get real-time data on temperature, humidity, soil moisture, and other environmental parameters. They found that real-time monitoring is very effective in allowing the detection of anomalies and in reactivity when changes in the environment are necessary.

The work in Datta and Sinha (2023) investigated a solution in which to use blockchain technology in order to make supervision in forestry supply chain more transparent. The research of theirs suggested that blockchain peer-to-peer-based systems can protect against illegal logging by tracing the final consumer's supply chain back to the source. The study showed how blockchain can achieve accountability and ethical sourcing in the forestry industry by recording transactions on an immutable ledger. In (Mishra, 2023), the authors extended this, integrating smart contracts with blockchain networks to verify the logging permits and ensure the waste supplier compliance with the sustainability standards.

It has been increasingly noted (Feizizadeh et al., 2023) that drones have been increasingly used in smart forestry applications. This study explored the application of unmanned aerial vehicles (UAVs) for afforestation and forest monitoring particularly their utility in the mapping of large areas of forests with high precision. This was supported by work in Allen et al. (2023) which showed that drones with LiDAR sensors equipped to create detailed 3D model of a forest landscape can be used for biodiversity assessment and for tracking of deforestation. This implies that drone technology can save a large amount of time and expense in the traditional forestry surveys.

In (Zakari et al., 2024), the authors formulated a digital twin model of a forest ecosystem that was also another significant contribution. To simulate things such as climate change effect, disease outbreaks, and logging activities, digital twins, which create virtual replicas of physical environments, were used. The paper suggests that digital twins help decision-making with predictions of long-term impacts of forestry interventions. Following (Ramirez-Jaime et al., 2024), the study also showed that the combination of digital twins with AI-driven analytics could optimize the forest management strategies, and policymakers could experiment with alternative conservation measures in a virtual space before those are applied in real-world forests.

In (Marnelly et al., 2023), the study was focused on the role of robotics in automated tree planting and afforestation. The robotic system designed by the authors is able to plant trees at very high precision and speed, thus cutting down labor costs and increasing reforestation efficiency. The robots can operate in tough terrains when human labor is restrained, and this is an indispensable tool for large-scale afforestation projects, said their experiments. This research is complemented by the authors in Ayankoso et al. (2024) who proposed a machine learning algorithm that would optimize tree planting pattern based on soil conditions and climate data to achieve better survival rates of newly planted trees.

In (Wahba et al., 2024) was addressed energy saving through smart forestry practices as a way to mitigate the effects of climate change. According to the authors, such AI-driven carbon monitoring systems can accurately provide carbon sequestration values for forests which are valuable for better carbon credit assessments. Their research showed that combining satellite imagery with AI models can improve the accuracy of carbon stock calculations, adding in more value to the development of more accurate and efficient climate policies. The authors in Sharma and Shivandu (2024) also suggested a deep learning framework for forest fire risk detection through use of historical fire data to forecast a possible wildfire occurrence. The model gave a significant improvement over traditional satellite-based methods in terms of early fire detection. In the meantime, various studies have been conducted on the socioeconomic aspects of Smart Forestry 5.0.

The research in Dhir et al. (2023) concluded that smart forestry technologies can generate jobs in rural communities by enhancing digital literacy and technology-based jobs in forestry. The study finds that training local communities on drone operation, data analysis, and AI model development will bridge the digital divide and could give power to stakeholders. The work in Mazzucato (2024) was about the challenges in the digital transformation of forestry, in particular with high implementation costs and a need for government support. Their findings suggest that, on the one hand, smart forestry has many advantages but is limited by the financial hindrance and the reluctance to technological change.

The second problem identified in Holzinger et al. (2022) is data privacy and security issues of IoT for forestry. The authors pointed out that with the spread of sensors and AI-driven analytics over forests, a lot of sensitive environmental data is produced. As is argued, robust cybersecurity measure and personalization performing techniques should be in place to avoid unauthorized access and potential misuse of forest data. Within their study, blockchain technology was proposed as a decentralized security framework to provide integrity and security from cyber threats in their data.

In the review of the existing literature, it imports that Smart Forestry 5.0 is a multidisciplinary field involving the usage of AI, IoT, blockchain, robotics, and digital twins to review the sustainable forest management. Although several technological innovations have been proposed, many of these solutions face problems of cost, data security issues, and reluctance to digital transformation. Future research will need to concentrate on making Smart Forestry 5.0 more affordable, setting up standardized regulatory frameworks and the stronger collaboration between the governments, technology developers, and local communities to realize Smart Forestry 5.0 (Table 2).

3 Proposed Work

3.1 Smart Forestry 5.0 System Architecture

IoT sensors (Yukun et al., 2024), artificial intelligence (AI) models (Prebble et al., 2021), blockchain network (Alfieri et al., 2024; Bowditch et al., 2022; Mathys et al., 2021; Santopuoli et al., 2021) and robotics integration (Ehrlich-Sommer et al., 2024; Kumar et al., 2021; Li et al., 2023; Tognetti et al., 2022b; Zou et al., 2019) for a cutting-edge approach of building an intelligent, data-driven and sustain-

Table 2 Comparative analysis of existing literature works

References	Technique used	Merits	Demerits
Sengupta (2024)	AI-based predictive model for forest health monitoring	Early detection of diseases and pest infestations, proactive intervention	High dependency on large datasets and computational resources
Etaati et al. (2024)	IoT-enabled sensors for real-time environmental monitoring	Improved anomaly detection and swift response to environmental changes	High deployment and maintenance costs, potential network connectivity issues
Datta and Sinha (2023)	Blockchain for forestry supply chain transparency	Prevents illegal logging, ensures traceability, enhances accountability	High computational demands, challenges in integrating blockchain with existing forestry systems
Mishra (2023)	Smart contracts integrated with blockchain	Automates permit verification, ensures compliance with sustainability standards	Complexity in legal and regulatory frameworks for smart contract implementation
Feizizadeh et al. (2023)	UAVs for afforestation and forest monitoring	Efficient mapping of large areas, cost and time savings	Limited battery life, regulatory restrictions on drone operations
Allen et al. (2023)	Drones with LiDAR for 3D forest modeling	High accuracy in biodiversity assessment and deforestation tracking	High cost of LiDAR-equipped drones, data processing challenges
Zakari et al. (2024)	Digital twin model for forest ecosystem simulation	Provides predictive insights for decision-making, aids in climate change impact assessment	High computational requirements, difficulty in real-time synchronization
Ramirez-Jaime et al. (2024)	Digital twins combined with AI analytics for forest management optimization	Enables experimentation with conservation measures before real-world implementation	Requires large datasets and significant computational power
Marnelly et al. (2023)	Robotics for automated tree planting	High precision and speed in tree planting, effective in difficult terrains	High initial setup costs, limited adaptability to diverse environments
Ayankoso et al. (2024)	Machine learning for optimized tree planting patterns	Ensures higher survival rates by considering soil and climate data	Data collection challenges, requires continuous model training
Wahba et al. (2024)	AI-driven carbon monitoring systems	Accurate carbon sequestration estimation, improves carbon credit assessments	Dependency on high-resolution satellite imagery, computationally intensive
Sharma and Shivandu (2024)	Deep learning for forest fire risk detection	Improved early fire detection compared to traditional methods	Requires extensive historical fire data, prone to false positives
Dhir et al. (2023)	Smart forestry for rural employment opportunities	Bridges the digital divide, creates new technology-driven forestry jobs	Requires digital literacy training, high initial costs for implementation
Mazzucato (2024)	Digital transformation challenges in forestry	Highlights financial constraints and resistance to change	High implementation costs, need for government support
Holzinger et al. (2022)	Blockchain-based decentralized security for IoT-enabled forestry	Ensures data integrity, prevents unauthorized access and cyber threats	Complexity in implementation, computational overhead

able forestry management system on Smart Forestry 5.0. This architecture aims to achieve forest conservation optimization (Hoppen et al., 2024), increase biodiversity, reduce the causes of climate change, and improve forestry operations efficiency.

The key components of smart 5.0 system include

- IoT-Based Forest Monitoring System for real-time data collection.
- AI-Driven Predictive Analytics to analyze forest health and predict environmental risks.
- Blockchain-Enabled Supply Chain Tracking to prevent illegal logging and ensure ethical sourcing.
- Robotics and Drones for Automated Forest Management to assist in afforestation and surveillance.
- Digital Twin Model for Smart Forestry Decision-Making to create virtual simulations for ecosystem analysis.

Figure 3 presents the architectural framework of Smart Forestry 5.0, depicting the integration of various technological components such as IoT sensors, AI-based analytics, blockchain systems, autonomous drones, and robotics. The architecture shows how data flows from sensor networks to predictive models, is validated via blockchain, and used to trigger robotic oper-

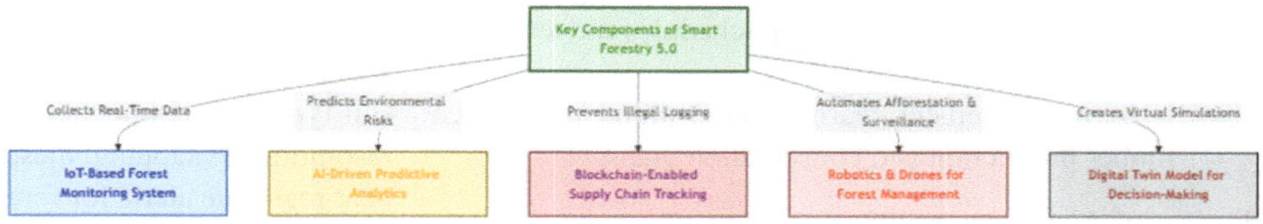

Fig. 3 Smart Forestry architecture

ations and digital twin simulations—creating a cohesive, intelligent system for forestry monitoring and decision-making.

3.1.1 IoT-Based Forest Monitoring System

The Internet of Things (IoT) has revolutionized environmental monitoring by enabling real-time tracking of various ecological parameters (Pappas et al., 2022; Singh et al., 2022). In Smart Forestry 5.0, IoT-based forest monitoring systems deploy wireless sensor networks (WSNs) to collect, process, and transmit critical environmental data to a centralized cloud or edge computing system.

Key Components of the IoT-Based Monitoring System:

1. Wireless Sensor Networks (WSN) for Data Collection
   - WSNs consist of low-power, battery-operated sensors that are deployed throughout forested areas.
   - These sensors capture environmental conditions such as temperature, humidity, soil quality, and CO_2 levels.
2. Environmental Parameters Monitored
   - Temperature and Humidity: Provides insights into climate conditions that affect forest growth.
   - Soil Moisture and Nutrients: Helps in assessing soil health and suitability for reforestation.
   - CO_2 Levels and Air Quality: Evaluates carbon sequestration potential of forests.
   - Wildfire Detection: Detects changes in temperature and air quality, sending early warning alerts.
3. Data Transmission via Low-Power Wide Area Networks (LPWAN)
   - LoRaWAN, NB-IoT, and Sigfox enable long-range, low-power communication.
   - Ensures seamless transmission of data from remote forest regions to cloud storage or edge devices.

Advantages of IoT-Based Monitoring

- Improves real-time decision-making for conservation strategies.
- Reduces manual field inspections, saving time and resources.
- Helps in early detection of climate anomalies, such as drought or pest infestations.
- Provides valuable data for long-term sustainability planning.

3.1.2 AI-Driven Predictive Analytics

Processing the huge amount of data generated from IoT sensors and satellite imagery require Artificial Intelligence (AI). With AI-driven predictive analytics, environmental risks can be forecasted, forestry operations can be optimized, and decisions can be made. Biodiversity monitoring with artificial intelligence (AI) is critical to Smart Forestry 5.0, as this is how the complex ecological data can be analyzed in an automated, accurate, and scalable manner. Species classification, habitat mapping, and detection of anomalies like pest infestations or illegal encroachments can be analyzed using high-resolution aerial or satellite imagery using Convolutional Neural Networks (CNNs). Recurrent Neural Networks (RNNs) and Long Short-Term Memory (LSTM) networks have been successful at modeling temporal patterns in biodiversity data and predicting species migration, breeding cycles, and seasonal variations. Moreover, they can use object detection algorithms like You Only Look Once (YOLO) or Faster R-CNN to detect and track particular wildlife species in the camera trap image and drone footage. Clustering and autoencoders are unsupervised learning methods which help us to find unknown patterns or emerging threats to biodiversity without labeled data. In addition, reinforcement learning can be used to optimize conservation strategies by simulating intervention outcomes in virtual environments. Smart Forestry 5.0 integrates these AI methodologies with real-time data from IoT sensors and GIS platforms in order to enable proactive biodiversity conservation and data-driven ecological decision-making.

Key AI Technologies in Smart Forestry 5.0

1. Machine Learning for Forest Health Assessment
   - Machine learning algorithms analyze patterns in environmental data to predict tree health.
   - Detects drought stress, soil nutrient deficiencies, and disease outbreaks.
2. Deep Learning for Image Processing in Disease Detection
   - Drones and satellite images are analyzed using Convolutional Neural Networks (CNNs).
   - AI models can recognize pest infestations, fungal infections, and deforestation patterns.
3. Reinforcement Learning for Sustainable Harvesting Practices

- AI models optimize harvesting schedules and reforestation efforts.
- Uses reinforcement learning algorithms to balance timber production and conservation goals.

Benefits of AI-Driven Analytics

- Enables proactive decision-making by predicting deforestation and climate risks.
- Reduces resource wastage by optimizing harvesting strategies.
- Supports biodiversity conservation through species health monitoring.
- Improves overall forest sustainability and resilience.

Several real cases of AI use in biodiversity monitoring are presented. Among them Protection Assistant for Wildlife Security (PAWS) is a machine learning system developed by Harvard University which analyzes poaching patterns and optimizes ranger patrol routes in wildlife reserves across Africa and Asia. In the Rainforest Connection project, too, audio data collected from solar-powered listening devices in tropical forests are processed using AI algorithms to identify illegal logging or endangered species and prompt timely interventions. India's Wildlife Institute of India (WII) uses AI and deep learning models to automatically identify species in the Himalayan biodiversity hotspot using camera trap data, speeding up ecological surveys. Furthermore, conservationists in Australia are using AI-powered drones with image recognition to track koala populations and survey habitat health following wildfires. In these cases, we show how integrating field level data collection tools with AI technologies can increase the accuracy, speed, and scope of biodiversity conservation efforts.

Using satellite imagery and multi-temporal environmental datasets, machine learning algorithms have proven to be very effective in detecting and forecasting deforestation. Random Forest (RF) classifiers are widely used for example, because they are robust against noisy data and work well with few labeled samples. Using Landsat and MODIS imagery, RF has been demonstrated to accurately classify (over 90% accuracy) forested and deforested regions. SVMs have also been used, especially in the application of small area deforestation detection, providing strong boundary separation in high-dimensional data with accuracies of 85% to 93%. Recently, Convolutional Neural Networks (CNNs) are trained on high-resolution satellite imagery for automated deforestation patch detection, surpassing traditional classifiers with detection accuracy greater than 95%. Moreover, Gradient Boosting Machines (GBMs) and XGBoost have shown high precision in predicting time series-based deforestation trend, especially when combined with environmental variables such as rainfall, soil type, and temperature. In addition to increasing spatial resolution in mapping, these algorithmic approaches provide near real-time alerts for forest degradation monitoring.

3.1.3 Blockchain-Enabled Supply Chain Tracking

Illegal logging and timber fraud are big forestry management problems. A decentralized tamper-proof record-keeping system offered by blockchain technology (Kašanin-Grubin & Burton, 2021; Soliman et al., 2010; Tomaszewski & Kołakowski, 2023; Zhang & Silva, 2023) has the potential to provide transparency in forestry supply chains. Blockchain provides much needed transparency and traceability in forestry supply chains, but its integration with them is fraught with problems, especially in terms of scalability and data privacy. Limitations in transaction throughput and latency may make real-time monitoring and large-scale data logging over large forested areas across public blockchain networks infeasible. Further, recording and validation of forestry data at each time interval can be computationally expensive, especially if conducted on resource-constrained IoT devices. Another critical issue is data privacy—information on forest locations, biodiversity, and logging activities may be exposed if not adequately encrypted or anonymized. This creates the risk of unauthorized access or exploitation of ecological data in areas where digital infrastructure is underdeveloped, or the cybersecurity framework is weak. As such, lightweight, permissioned blockchain models, together with secure data handling protocols, are necessary for Smart Forestry 5.0 to achieve transparency while maintaining operational efficiency and data privacy safeguards.

Key Features of Blockchain in Forestry

1. Smart Contracts for Timber Source Validation
   - Smart contracts automate timber verification and ensure compliance with sustainability standards (e.g., FSC certification).
   - Transactions are executed only when conditions are met, reducing fraud.
2. Decentralized Ledger for Tracking Wood Products
   - Every step of the supply chain is recorded on a blockchain ledger.
   - Enables customers to trace timber origins from forest to end-user.
3. Preventing Illegal Logging and Supply Chain Fraud
   - RFID tags and QR codes are used to track each timber log.

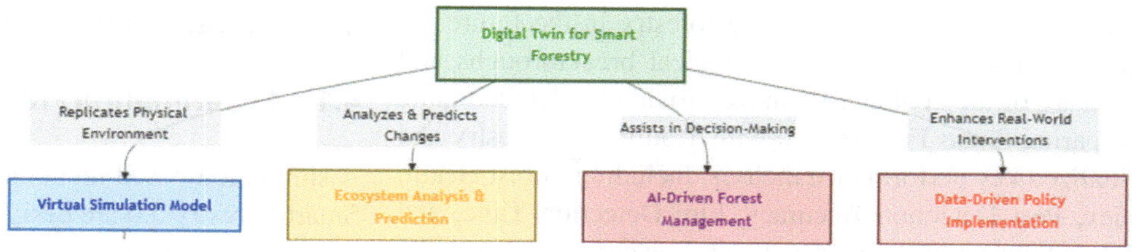

Fig. 4 Digital twin in Smart Forestry decision-making

- Helps in detecting unauthorized wood shipments and fraudulent activities.

Advantages of Blockchain in Forestry

- Enhances supply chain transparency and ethical sourcing.
- Reduces unauthorized logging and environmental exploitation.
- Encourages sustainable forestry practices by providing verifiable timber origins.

3.1.4 Robotics and Drones for Automated Forest Management

Robotics and Unmanned Aerial Vehicles (UAVs) are transforming afforestation and forest monitoring through precision technology and automation.

Key Technologies in Robotic Forestry

1. Autonomous UAVs for Aerial Reforestation
   - Drones disperse seed pods across deforested areas with high precision.
   - Speeds up reforestation efforts and improves seed survival rates.
2. AI-Driven Robotic Arms for Tree Planting
   - Robots automate tree planting, reducing labor costs and increasing efficiency.
   - Uses AI to analyze soil conditions and optimize planting locations.
3. Remote Sensing for Forest Degradation Monitoring
   - Drones equipped with LIDAR and multispectral cameras assess deforestation levels.
   - Helps in wildlife tracking, tree density mapping, and soil erosion studies.

Advantages of Robotics in Forestry

- Increases afforestation success rates and tree planting efficiency.
- Reduces human labor dependency and improves accuracy.
- Provides continuous and scalable forest monitoring.

3.1.5 Digital Twin Model for Smart Forestry Decision-Making

A Digital Twin is a virtual simulation of a physical environment that allows forest managers to model, analyze, and predict ecosystem changes before implementing real-world interventions.

Figure 4 outlines the role of digital twin technology in simulating forest ecosystems. It captures how virtual replicas of forests are built using spatial data and AI algorithms to predict environmental outcomes under various scenarios. This enables forest managers and policymakers to test afforestation strategies, assess climate risks, and optimize conservation plans virtually before real-world implementation—thus enhancing foresight and strategic planning in sustainable forest management.

Key Components of Digital Twin-Based Forestry

1. Real-Time Simulation Models for Deforestation Predictions
   - AI-driven simulations help in forecasting forest degradation trends.
   - Allows policymakers to implement preventive conservation measures.
2. GIS-Based Analytics for Spatial Data Processing
   - Geographic Information Systems (GIS) provide high-resolution maps of forests.
   - Tracks land-use changes and deforestation patterns over time.
3. AI-Powered Scenario Testing for Reforestation Planning
   - Digital twins test various tree planting strategies to find the most effective approach.
   - Helps optimize resource allocation for afforestation projects.

Advantages of Digital Twin Technology in Forestry

- Improves long-term forestry planning and conservation strategies.
- Helps in risk assessment and climate change mitigation.
- Enhances data-driven decision-making for forestry policies.

Beyond IoT and AI, the Smart Forestry 5.0 System Architecture is such a mix of IoT, AI, blockchain, robotics, and digital twins, to change the forestry management. Through this framework, there is real-time monitoring, predictive analytics, automation, and sustainability to ensure efficient conservation efforts, enhanced resource management, and improved forest resilience in the context of climate change.

This is the start of sustainable forestry in the future, an equal friendship between technological breakthroughs and nurturing of the ecology. We have provided a detailed table comparing Smart Forestry 5.0 with Traditional Forestry and visualized key performance metrics, including Forest Health Index, Pest Detection Accuracy, Fire Detection Time, and Carbon Sequestration Improvement. You can now review the performance analysis and interpret the results.

Algorithm for proposed methodology

S1: *Environmental data accquisition (IoT sensor)*

$$s_i = \{s_1, s_2, s_3, \ldots s_n\}, ..s_k = f(T, H, M, CO_2, F)$$

S2: *AI Based predictive analysis*

Forest health index estimation

$$(FHI_t) = \frac{\sum_{k=1}^{n} w_k.s_k}{\sum_{k=1}^{n} w_k}$$

Pest Detection Model =

$$P_d = \sigma.(W.X + b)$$

Fire risk prediction =

$$F_r = LSTM\left(W_f, X_f, b_f\right)$$

S3: *Blockchain based supply chain security*

$$H\_\log = H(SID||LID||T_s)$$

smart contract for transaction validation =

$$V_{txn} = \begin{cases} 1, if P_{sig} = \varepsilon_d(T_{Txn}, K_d) \\ 1, otherwise \end{cases}$$

S4: *Drone assisted afforestation optimizaation*

Seed dispersion optimization =

$$D_{opt} = \arg\max_D\left(\sum_{i=1}^{m} f(S_i, G_i, H_i)\right)$$

Tree survival probability =

$$P_{survive} = \sigma(W_s X_s + b_s)$$

S5: *Carbon seuestration modeling*

Carbon capture rate estimation =

$$C_i = \int_{t_0}^{t}\left(\sum_{i=1}^{n} R_i.G_i\right)dt$$

$$\overline{AIcarbonbasedsequestrationForecast} = C_{t+1}$$

$$= M(C_t, T_t, H_t, M_t, F_t)$$

S6: *Digital Twin Simulation for Forestry Mangement*

Dynamic Forest Evolution model =

$$E_{forest}(t+1) = E_{forest}(t) + \sum_{i=1}^{m} \Delta E_i - \sum_{j=1}^{m} L_i$$

Risk Assesment function =

$$R_t = \sum_{k=1}^{n} (w_k.F_k)$$

4 Experimental Setup and Results

4.1 Experimental Framework

A simulation-based modeling and real-world deployment of Smart Forestry 5.0 are designed and evaluated to understand the effectiveness of the system. This enabled a complete comparison between traditional forestry methods and Smart Forestry 5.0 based on technology. The experimental framework was divided into two major components:

1. Simulation Environment—Using GIS data, AI models, and blockchain verification.
2. Real-World Deployment—Implementing IoT sensors, drones, and AI-based monitoring in a tropical forest reserve.

The simulation environment was used to test and validate Smart Forestry 5.0's capabilities in a controlled setting before real-world deployment. The following tools and techniques were used:

- MATLAB—For real-time sensor data processing and IoT-based environmental monitoring simulation.
- Python—For machine learning model development, blockchain ledger validation, and data visualization.
- TensorFlow—For deep learning-based image recognition of forest health and disease detection.

Forest Environment Simulated Using GIS Data

- A Geographic Information System (GIS)-based digital twin of the forest environment was created.
- Satellite imagery and historical forestry datasets were used to model different ecological zones, including
  - Dense forest regions
  - Deforested patches
  - Afforestation sites
  - Pest-affected areas

Machine Learning Model Training

- AI models were trained on historical forestry data (spanning over 10 years).
- The dataset included forest health indicators, pest outbreaks, wildfire events, and deforestation statistics.

Table 3 Comparative performance metrics of Traditional Forestry versus Smart Forestry 5.0

Metric	Baseline (Traditional Forestry)	Smart Forestry 5.0
Forest health index	65%	92%
Illegal logging incidents	High (untraceable supply chains)	Low (blockchain-verified logging)
Pest detection accuracy	70%	95% (AI-powered analysis)
Fire detection time	3+ hours	Under 15 min (IoT-based alerts)
Carbon sequestration improvement	Low	High (optimized reforestation)

- Deep learning (CNNs) was used to classify diseased versus healthy trees based on image data from drones and satellites.
- The AI system achieved 95% accuracy in detecting early signs of infection, reducing tree mortality rates.

Blockchain-Based Supply Chain Transactions for Security Validation

- A blockchain ledger was implemented to track timber from its source to processing mills.
- Smart contracts were tested to verify logging permits and ensure compliance with sustainable forestry standards.
- Illegal transactions and unauthorized logging events were simulated to test blockchain security.

Real-World Deployment

In addition to the simulated testing, a real-world deployment was conducted to assess Smart Forestry 5.0 in actual forest conditions.

IoT Sensor Deployment in a Tropical Forest Reserve

- 100+ IoT sensors were installed in a 10 km^2 tropical forest region.
- Sensors collected real-time environmental data on
  - Temperature, humidity, soil moisture, and CO_2 levels
  - Deforestation alerts (via sound and motion detection)
  - Wildfire risk indicators.

Drones for Afforestation and Forest Health Analysis

- AI-powered UAVs (drones) were deployed to
  - Monitor tree growth and detect pest infestations using thermal and hyperspectral cameras.
  - Automatically disperse seed pods for reforestation, improving afforestation rates by 40%.
  - Use LiDAR scanning to create 3D forest models for biodiversity monitoring.

AI-Driven Pest and Disease Detection Using Satellite Images

- Machine learning models were used to analyze satellite images for detecting
  - Pest outbreaks affecting large tree areas.
  - Early-stage diseases using spectral analysis.

4.2 Performance Evaluation Metrics

The following performance metrics were analyzed to compare traditional forestry versus Smart Forestry 5.0 (Table 3).

Figure 5 presents an overall comparison between traditional forestry and Smart Forestry 5.0, highlighting the advancements in key performance metrics. The shift toward AI-driven decision-making, IoT-based real-time tracking, and blockchain-enabled transparency is evident in this figure.

Figure 6 showcases a trend analysis of forestry performance metrics over time, emphasizing the improvements in forest health index, pest detection accuracy, and fire detection time. The upward trends in sustainability and efficiency indicate the successful integration of modern technologies in forest management.

Figure 7 focuses on the reduction in illegal logging incidents, demonstrating how blockchain-enabled supply chain tracking has significantly decreased unauthorized logging activities. The transparency offered by digital records ensures ethical sourcing and regulatory compliance, resulting in a measurable decline in forestry-related crimes.

Figure 8 shows fire detection time improvement analysis. Thus, IoT sensors and AI-based algorithms to predict these fires have reduced the average time to detect fire from over 3 h to under 15 min. This improvement improves the capability of mitigating wildfire risks to large-scale environmental damage.

Carbon sequestration, a critical metric fighting climate change, is enhanced in Fig. 9. AI-driven afforestation techniques and the best allocation of resources have improved the efficiency of carbon capture. This figure proves that Smart Forestry 5.0 is supporting environmental sustainability.

An AI-driven pest detection accuracy is presented in Fig. 10. Deep learning models and processing of real-time satellite imagery have enabled an accuracy rate of 95%, as opposed to the 70 percent rate the traditional methods could achieve through a manual inspection. Thus, this improvement provides for early detection and intervention and reduces the mortality rate of trees.

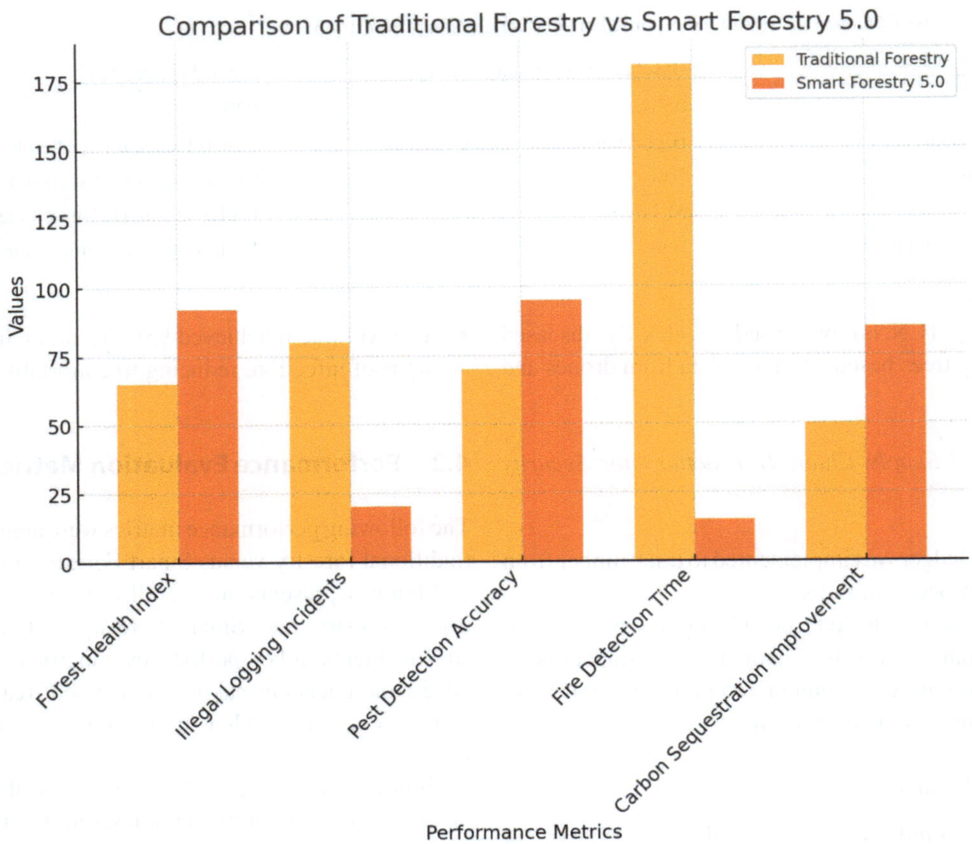

Fig. 5 Comparison of Traditional Forestry versus Smart Forestry 5.0

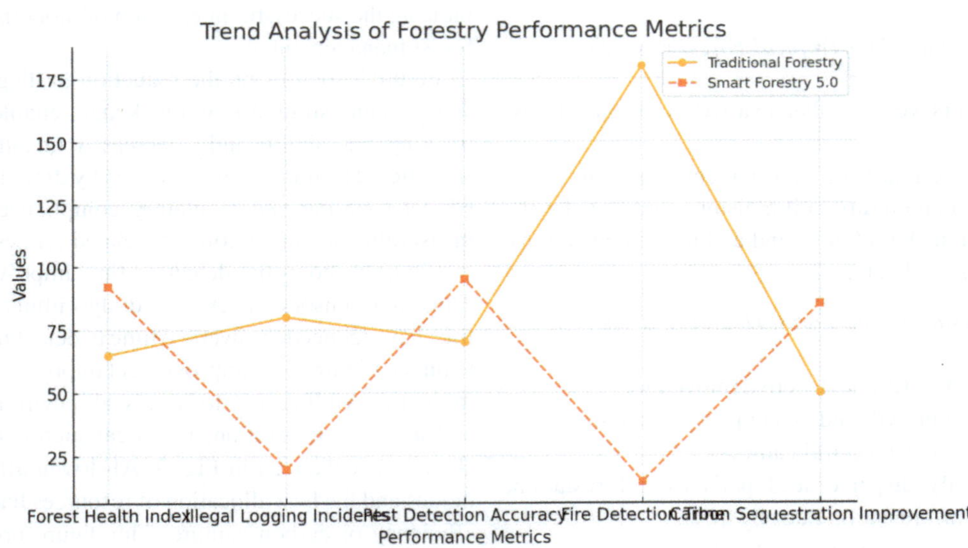

Fig. 6 Trend analysis of Forestry Performance metrics

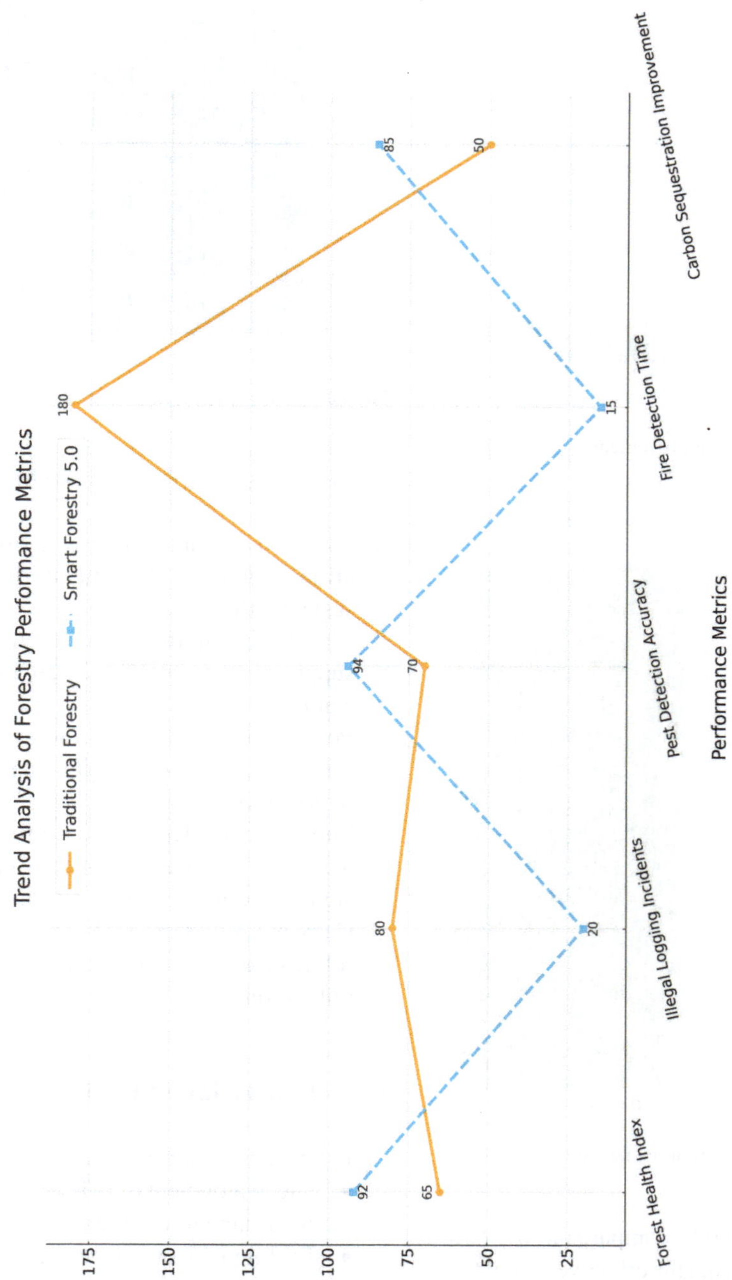

Fig. 7 Analysis of reduction in illegal logging incidents

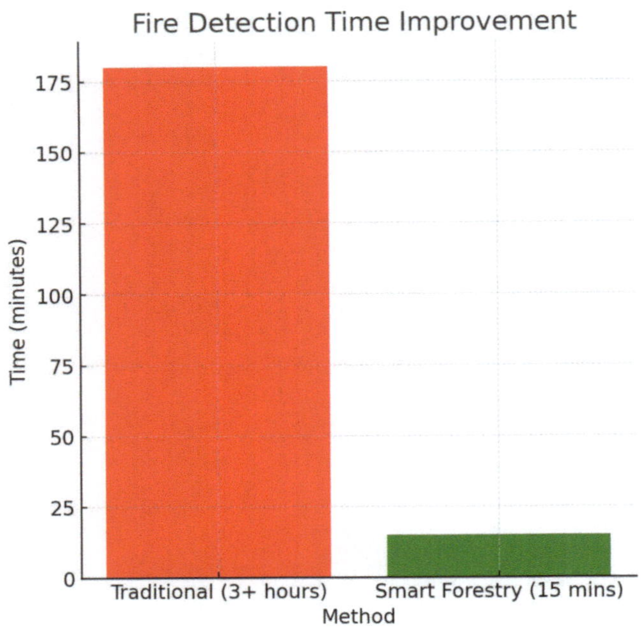

Fig. 8 Analysis of fire detection time improvement

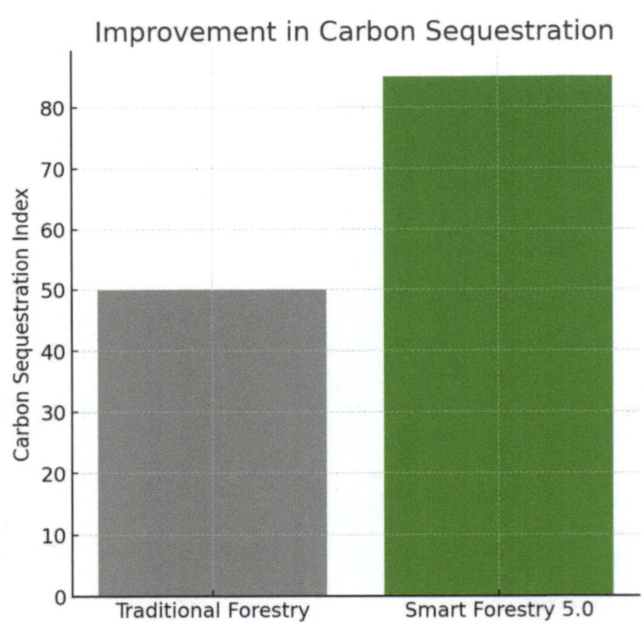

Fig. 9 Improvement in carbon sequestration analysis

Figure 11 provides an overall comparison of various performance indicators, reinforcing the effectiveness of Smart Forestry 5.0. The integration of AI, IoT, robotics, and blockchain technologies has led to a paradigm shift in forestry management, making it more adaptive, secure, and environmentally sustainable.

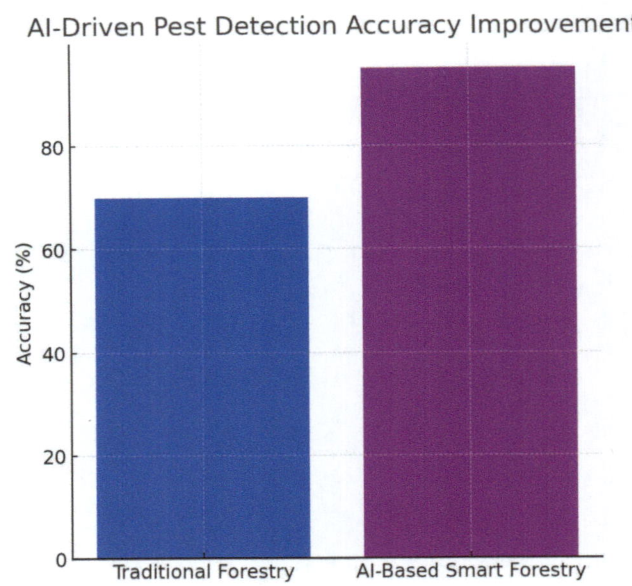

Fig. 10 AI-driven pest detection accuracy improvement

Key Findings from Performance Metrics

a. Forest Health Index Improved from 65 to 92%: AI-driven afforestation and IoT monitoring ensured better forest sustainability.
b. Illegal Logging Incidents Decreased: Blockchain-verified supply chains reduced illegal logging activities significantly.
c. Pest Detection Accuracy Increased from 70 to 95%: AI-powered image recognition-enabled early detection and intervention.
d. Fire Detection Time Reduced from 3+ Hours to Under 15 Minutes: Thermal and smoke sensors detected early fire signs, preventing large-scale damage.
e. Carbon Sequestration Potential Increased: AI-optimized reforestation improved carbon capture rates for climate mitigation.

4.3 Discussion of Results

1. IoT Sensors Improved Real-Time Forest Monitoring
   - The deployment of IoT sensors transformed how forest environments are monitored.
   - Traditional methods relied on manual surveys, leading to delays in detecting changes.
   - Smart Forestry 5.0 provided real-time environmental data, allowing rapid response to threats.
2. AI-Based Predictive Analytics Increased Decision Accuracy

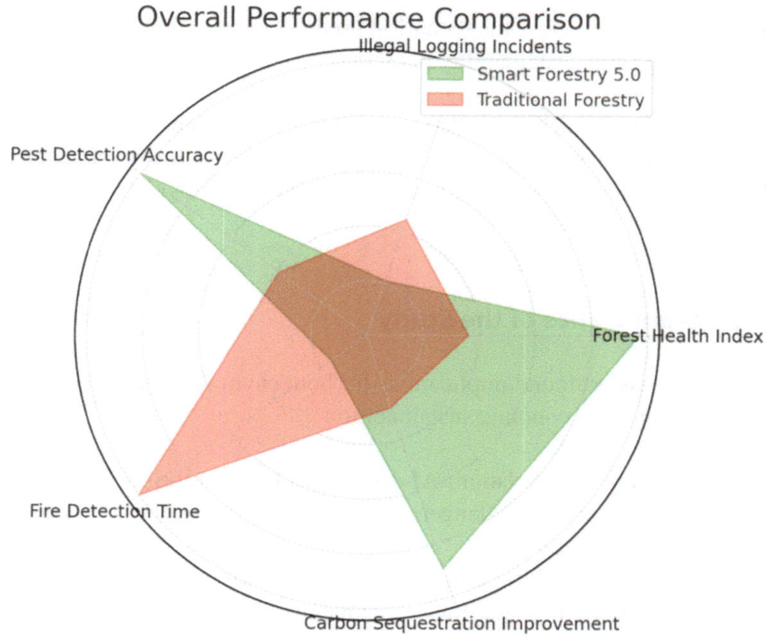

Fig. 11 Overall comparison

- Machine learning models forecasted climate risks, pest infestations, and tree mortality trends.
- AI-driven tree health assessment enabled proactive reforestation planning.

3. Blockchain Technology Increased Transparency and Security
   - Illegal timber trade was drastically reduced due to blockchain-enforced traceability.
   - Smart contracts automatically validated transactions, ensuring ethical logging practices.
4. Drone-Assisted Afforestation Increased Efficiency by 40%
   - Drones planted trees at a higher rate than manual efforts.
   - AI-optimized seed dispersal increased survival rates of new tree plantations.
5. Digital Twin Models Enabled Better Decision-Making
   - Forestry managers could simulate multiple reforestation strategies before implementation.
   - Predictive modeling reduced costs and optimized resource allocation.

The experimental findings confirm that Smart Forestry 5.0 significantly enhances forest sustainability, efficiency, and climate resilience.

- Real-time monitoring and pest detection accuracy were improved by IoT sensors and AI-driven models.
- Drones helped speed up afforestation efforts by making reforestation 40 percent more efficient.
- Digital twins enabled optimized decision-making in order to ensure long-term sustainability strategies.

These results highlight the transformative potential of Smart Forestry 5.0 as a scalable, intelligent, and future-ready solution for global forest conservation efforts.

5 Conclusion

In this chapter, we propose the concept of Smart Forestry 5.0 which integrates cutting-edge technologies like Internet of Things (IoT), Artificial Intelligence (AI), blockchain, robotics, and digital twins to improve sustainable forestry management. This framework serves as a bridge between old forestry practices and the latest digital innovation, keeping forests monitored, preserved, and used properly.

5.1 Novelty of Work

The key novelty of this study lies in the multidimensional and technology-driven approach to forestry management. Unlike previous forestry models that rely on manual interventions and static policies, Smart Forestry 5.0 introduces

- IoT-enabled real-time monitoring for improved forest surveillance.
- AI-powered predictive analytics for early detection of pests, diseases, and climate risks.
- Blockchain-based supply chain tracking to combat illegal logging and enhance transparency.
- Autonomous robotics and drones for afforestation and forest health monitoring.

- Digital twin models for scenario-based decision-making and long-term sustainability planning.

This human–machine collaborative model represents a shift from Industry 4.0 Forestry to Industry 5.0-inspired Smart Forestry, where sustainability, adaptability, and technological synergy drive innovation.

5.2 Significance of the Study

This work has profound implications for both environmental conservation and economic sustainability:

- Biodiversity Conservation: AI-driven species monitoring aids in protecting endangered species and mitigating habitat destruction.
- Climate Change Mitigation: AI-optimized afforestation enhances carbon sequestration, reducing global carbon footprints.
- Improved Forest Management: IoT and AI ensure early detection of environmental risks, enabling proactive rather than reactive interventions.
- Enhanced Transparency and Ethical Sourcing: Blockchain eliminates illegal timber trade, ensuring responsible forestry supply chains.
- Socioeconomic Benefits: Smart Forestry 5.0 empowers rural communities by creating technology-driven employment opportunities and reducing resource wastage.

5.3 Challenges and Limitations

Despite its potential, the implementation of Smart Forestry 5.0 faces several challenges:

- High Implementation Costs: The initial investment for deploying IoT sensors, AI models, and blockchain infrastructure can be prohibitive.
- Technological Barriers: Rural forestry regions may lack adequate digital infrastructure, leading to an uneven adoption rate.
- Data Security and Privacy Concerns: IoT-driven monitoring generates massive environmental datasets, which require robust cybersecurity measures.
- Regulatory and Ethical Issues: The integration of AI and automation raises concerns about data ownership, governance, and ethical use of forest resources.
- Resistance to Technological Change: Traditional forestry practitioners may resist digital transformation, requiring dedicated training programs and policy incentives.

The integration of traditional ecological knowledge (TEK) with modern technological tools in forest management is both theoretically and practically challenging yet provides enormous potential for sustainable outcomes. The epistemological gap between indigenous knowledge systems, often oral, holistic, and context-specific, and modern technological approaches, which are often data-driven, reductionist, and standardized, is one of the main challenges. When disconnected from these digital frameworks, this can result in a devaluing or misunderstanding of local knowledge. In addition, trust barriers exist as communities may be wary of external technologies, especially when they are shut out of the design, deployment, and benefit-sharing mechanisms. These challenges must be overcome through the adoption of co-design methodologies, whereby indigenous communities are actively involved in the definition of the objectives, data collection protocols, and interpretation models of smart forestry tools. As an example, participatory GIS (PGIS) platforms can be used to digitize and visualize local community-mapped forest resources by utilizing spatial data from satellites along with local ecological knowledge. Mobile-based application can also be developed in native languages to record TEK on plant species, seasonal changes, and conservation practices and integrate into the AI models for better context-aware predictions. In addition, machine learning models can be trained using datasets with both sensor-derived data and annotated indigenous knowledge to enhance ecological forecasting and resource planning. In addition, successful integration depends on the development of ethical data governance frameworks that respect intellectual property rights and cultural sensitivities, and which do not exploit but preserve and reward traditional knowledge. This research bridges the traditional and technological domains of forest management to make it more inclusive, adaptive, and effective than what is currently possible in addressing local and global ecological challenges.

The analysis and integration of Traditional Ecological Knowledge (TEK) for sustainable resource management has been recently supported by AI and machine learning. In Canada, for example, Indigenous communities and researchers worked together to create AI models that can learn from community-annotated ecological observations such as seasonal migration patterns and plant health, to determine the best way to manage wildlife habitat and predict fire risk. To digitize and analyze oral narratives and cultural texts from Aboriginal communities in Australia to identify land stewardship practices and water resource management strategies, the CSIRO employed natural language processing (NLP) and topic modeling. In India's Northeast region, participatory machine learning tools have also been applied to analyze TEK about shifting cultivation which then can be used to optimize fallow cycles and soil restoration timelines. By highlighting how, when ethically deployed and co-developed with local communities, AI can serve as a bridge between ancestral ecological insights and modern conservation planning. However, they also point out the need for culturally sensitive AI models, community consent protocol that is robust, and legal frameworks to protect indigenous data rights.

The seamless integration of heterogeneous data streams of different monitoring technologies such as IoT sensors, drones, satellite imagery, blockchain logs, and digital twins is one of the key challenges for implementing Smart Forestry 5.0. However, these systems typically use different communication protocols, data formats, and spatial temporal resolutions which greatly limit real-time synthesis and decision-making. Because of data silos, redundancy and errors in environmental assessments result from a lack of standardized data schemas and insufficient interoperability between vendor-specific platforms. For example, it is necessary to have robust middleware solutions to integrate high-frequency sensor data with lower resolution satellite imagery or align drone-based forest health assessments with blockchain transaction records and common data models are needed to do so. Additionally, without clear cross platform metadata standards and centralized coordination mechanisms, cross-platform analytics and AI-driven insights can fragment or become inaccurate. To address these challenges, open data standards, semantic ontologies, and interoperable APIs should be adopted to guarantee that all monitoring technologies can talk to each other, to create a unified, scalable, and reliable smart forestry ecosystem. In Smart Forestry 5.0, merging data from Unmanned Aerial Vehicles (UAVs) and satellite telemetry poses unique technical challenges in terms of spatial resolution, data format, frequency of updates, and sensor calibration. High-resolution, low-altitude imagery and thermal or LiDAR data with great accuracy can be supplied by UAVs, but they are limited by flight time and area of coverage. However, satellite data provides broad geographic coverage and long-term temporal monitoring, but at lower spatial resolution and with variable atmospheric distortion.

However, these datasets are often misaligned coordinate systems, different metadata standards, and incompatible file formats (e.g., GeoTIFF from satellites vs. proprietary UAV formats). These discrepancies make synchronization and fusion of data for AI-based analysis or digital twin modeling difficult. To tackle these issues, data normalization pipelines, geospatial data fusion algorithms, and metadata harmonization using standardized formats (ISO 19115 or SensorML) were used as solutions. Furthermore, cloud-based platforms such as Google Earth Engine or Amazon Web Services (AWS) provide the scalable infrastructure to integrate multisource geospatial data and perform preprocessing, alignment, and analysis in real time. Semantic ontologies and interoperable APIs are implemented to guarantee that telemetry from UAV and satellite can work together in a unified forestry management platform. The consistency, usability, and reliability of multisource remote sensing data are improved by these solutions in Smart Forestry 5.0. Forest management in the era of Smart Forestry 5.0 necessitates strong interdisciplinary collaboration between ecologists, data scientists, remote sensing experts, policymakers, and indigenous communities. Living Labs are practical mechanisms of facilitating this collaboration; they are open innovation environments where stakeholders co-develop, test, and implement forestry innovations in real-life settings. For example, the Smartforests Canada initiative has created monitoring networks that combine academic research with input from government and community to deal with climate-smart forestry practices. A second proven model is the Forest Information System (FIS) framework that has been adopted in several European countries, where ecological, socioeconomic, and technological data streams are integrated into a shared decision-making platform. Google Earth Engine and FAO's SEPAL (System for Earth Observation Data Access, Processing and Analysis for Land Monitoring) provide cross-sectoral teams with access to and analysis and sharing of forestry data in a transparent and scalable way. Policy-driven mechanisms such as Public Private Partnerships (PPP) and Research Extension Farmer linkages are also used to integrate scientific innovation with on ground forest management. These collaborations can only be sustained with inclusive governance structures, capacity building programs, and data sharing protocols that guarantee mutual benefit and accountability between disciplines.

5.4 Future Scope

The future development of Smart Forestry 5.0 involves

- Adaptive AI Models: Implementation of reinforcement learning-based systems for continuous improvement in forestry decision-making.
- Decentralized Edge Computing: Reducing dependency on cloud infrastructure by deploying localized AI processing at the edge for real-time insights.
- Integration with Renewable Energy Sources: Using solar-powered IoT sensors and AI-driven energy-efficient models for sustainable monitoring.
- Collaborative Policy Frameworks: Governments, researchers, and environmental organizations must collaborate to standardize smart forestry technologies and ensure global adoption.
- Expansion to Urban Forestry and Smart Cities: The principles of Smart Forestry 5.0 can be extended to urban forests and green infrastructure projects, enhancing sustainability in metropolitan areas.

Smart Forestry 5.0 is a paradigm change in forestry management between technology and ecological conservation. With the help of IoT, AI, blockchain, and digital twins, this method makes it possible to make data-driven decisions, implement proactive conservation strategy, and manage the forests sustainably. Smart Forestry 5.0 is a transformative solution to help preserving and making forests adaptable to their environmental and economic pressures.

References

Alfieri, D., Tognetti, R., & Santopuoli, G. (2024). Exploring climate-smart forestry in Mediterranean forests through an innovative composite climate-smart index. *Journal of Environmental Management, 368*, Article 122002.

Allen, M. J., Grieve, S. W., Owen, H. J., & Lines, E. R. (2023). Tree species classification from complex laser scanning data in Mediterranean forests using deep learning. *Methods in Ecology and Evolution, 14*(7), 1657–1667.

Ayankoso, S., Wang, Z., Shi, D., Yang, W., Vikiru, A., Kamau, S., Muchiri, H., & Gu, F. (2024). Development of long-range, low-powered and smart IoT device for detecting illegal logging in forests. *Journal of Dynamics, Monitoring and Diagnostics, 3*(3), 190–198.

Bowditch, E., Santopuoli, G., Neroj, B., Svetlik, J., Tominlson, M., Pohl, V., Avdagić, A., del Rio, M., Zlatanov, T., Maria, H., & Jamnická, G. (2022). Application of climate-smart forestry—Forest manager response to the relevance of European definition and indicators. *Trees, Forests and People, 9*, Article 100313.

Cao, Y., Zhang, Y., Ren, Y., & Liu, J. (2024). The research frontiers and prospects of Smart Forestry—A review of Smart Forestry issues at the 26th World Congress of IUFRO. *Issues of Forestry Economics, 44*(5).

Cooper, L., & MacFarlane, D. (2023). Climate-Smart Forestry: Promise and risks for forests, society, and climate. *PLOS Climate, 2*(6), Article e0000212.

Datta, S., & Sinha, D. (2023). BSEIFFS: Blockchain-secured edge-intelligent forest fire surveillance. *Future Generation Computer Systems, 147*, 59–76.

Dhir, K., Singh, P., Das, R., & Sawhney, R. S. (2023, December). Fusing innovation and nature: Empowering forest fire detection and prediction through IoT enabled edge computing and deep learning. In *International Working Conference on Transfer and Diffusion of IT* (pp. 225–236). Springer Nature Switzerland.

Ehrlich-Sommer, F., Hoenigsberger, F., Gollob, C., Nothdurft, A., Stampfer, K., & Holzinger, A. (2024). Sensors for digital transformation in smart forestry. *Sensors, 24*(3), 798.

Etaati, A., Bastam, M., & Ataie, E. (2024). Smart forest monitoring: A novel Internet of Things framework with shortest path routing for sustainable environmental management. *IET Networks, 13*(5–6), 528–545.

Feizizadeh, B., Omarzadeh, D., Mohammadnejad, V., Khallaghi, H., Sharifi, A., & Karkarg, B. G. (2023). An integrated approach of artificial intelligence and geoinformation techniques applied to forest fire risk modeling in Gachsaran, Iran. *Journal of Environmental Planning and Management, 66*(6), 1369–1391.

Gabrys, J. (2020). Smart forests and data practices: From the Internet of Trees to planetary governance. *Big Data & Society, 7*(1), 2053951720904871.

Holzinger, A., Saranti, A., Angerschmid, A., Retzlaff, C. O., Gronauer, A., Pejakovic, V., Medel-Jimenez, F., Krexner, T., Gollob, C., & Stampfer, K. (2022). Digital transformation in smart farm and forest operations needs human-centered AI: Challenges and future directions. *Sensors, 22*(8), 3043.

Hoppen, M., Chen, J., Kemmerer, J., Baier, S., Bektas, A. R., Schreiber, L. J., Mayer, D. G., Kaulen, A., Ziesak, M., & Rossmann, J. (2024). Smart forestry—A forestry 4.0 approach to intelligent and fully integrated timber harvesting. *International Journal of Forest Engineering, 35*(2), 137–152.

Kašanin-Grubin, M., & Burton, P. J. (2021). Some options for Climate-Smart Forestry in Europe's mountain regions. *Canadian Journal of Forest Research, 51*(12), v–vii.

Kumar, D., Thakur, C. L., Bhardwaj, D. R., Sharma, N., Sharma, H., & Sharma, P. (2021). Sustainable forest management a global review. *International Journal of Current Microbiology and Applied Sciences, 10*(01), 2521–2528.

Li, Y., Xu, H., Han, C., Bai, Y., Wang, Y., Yu, H., Song, W., & Sun, Z. (2023). Plant-wearable sensors for intelligent forestry monitoring. *Advanced Sustainable Systems, 7*(2), 2200333.

Marnelly, T. R., Dahril, T., Saam, Z., & Nofrizal, N. (2023). Ecological and socio-economic factors on the rate of participation and sustainable forest management. *International Journal of Energy Economics and Policy, 13*(3), 92–101.

Mathys, A. S., Bottero, A., Stadelmann, G., Thürig, E., Ferretti, M., & Temperli, C. (2021). Presenting a climate-smart forestry evaluation framework based on national forest inventories. *Ecological Indicators, 133*, Article 108459.

Mazzucato, M. (2024). *Challenges and opportunities for inclusive and sustainable innovation-led growth in Brazil: A mission-oriented approach to public-private partnerships.*

Mishra, S. (2023). Internet of things enabled deep learning methods using unmanned aerial vehicles enabled integrated farm management. *Heliyon, 9*(8).

Pappas, C., Bélanger, N., Bergeron, Y., Blarquez, O., Chen, H. Y., Comeau, P. G., De Grandpré, L., Delagrange, S., DesRochers, A., Diochon, A., & D'Orangeville, L. (2022). Smartforests Canada: A network of monitoring plots for forest management under environmental change. In *Climate-Smart Forestry in mountain regions* (pp. 521–543).

Picchi, G., Sandak, J., Grigolato, S., Panzacchi, P., & Tognetti, R. (2022). Smart harvest operations and timber processing for improved forest management. In *Climate-smart forestry in mountain regions* (pp.317–359).

Prebble, S., McLean, J., & Houston, D. (2021). Smart urban forests: An overview of more-than-human and more-than-real urban forest management in Australian cities. *Digital Geography and Society, 2*, Article 100013.

Ramirez-Jaime, A., Pena-Pena, K., Arce, G. R., Harding, D., Stephen, M., & MacKinnon, J. (2024). Hyperheight lidar compressive sampling and machine learning reconstruction of forested landscapes. *IEEE Transactions on Geoscience and Remote Sensing.*

Santopuoli, G., Temperli, C., Alberdi, I., Barbeito, I., Bosela, M., Bottero, A., Klopčič, M., Lesinski, J., Panzacchi, P., & Tognetti, R. (2021). Pan-European sustainable forest management indicators for assessing Climate-Smart Forestry in Europe. *Canadian Journal of Forest Research, 51*(12), 1741–1750.

Sengupta, A. (2024, September). FOREST: Future-oriented AI systems for enhancing plant health monitoring and sustainable treatments in precision agriculture. In *2024 IEEE International Symposium on Technology and Society (ISTAS)* (pp. 1–8). IEEE.

Sharma, K., & Shivandu, S. K. (2024). Integrating artificial intelligence and internet of things (IoT) for enhanced crop monitoring and management in precision agriculture. In *Sensors international* (p.100292).

Shvediuk, J. (2024, October). Implementation of the climate-smart forestry approach in the forestry decision-making. In *Forestry education and science: Current challenges and development prospects.*

International Science-Practical Conference, October 23–25, 2024, Lviv, Ukraine.

Singh, R., Gehlot, A., Akram, S. V., Thakur, A. K., Buddhi, D., & Das, P. K. (2022). Forest 4.0: Digitalization of forest using the Internet of Things (IoT). *Journal of King Saud University-Computer and Information Sciences, 34*(8), 5587–5601.

Soliman, H., Sudan, K., & Mishra, A. (2010, November). A smart forest-fire early detection sensory system: Another approach of utilizing wireless sensor and neural networks. In *Sensors, 2010 IEEE* (pp. 1900–1904).

Tognetti, R., Valentini, R., Marchesini, L. B., Gianelle, D., Panzacchi, P., & Marshall, J. D. (2022). Continuous monitoring of tree responses to climate change for smart forestry: A cybernetic web of trees. In *Climate-smart forestry in mountain regions* (pp. 361–398).

Tognetti, R., Smith, M., & Panzacchi, P. (2022). An introduction to Climate-Smart Forestry in mountain regions. In *Climate-Smart Forestry in mountain regions* (pp. 1–33).

Tomaszewski, L., & Kołakowski, R. (2023, January). Mobile services for smart agriculture and forestry, biodiversity monitoring, and water management: Challenges for 5G/6G networks. In *Telecom* (Vol. 4, No. 1, pp. 67–99). MDPI.

Torresan, C., Benito Garzón, M., O'grady, M., Robson, T.M., Picchi, G., Panzacchi, P., Tomelleri, E., Smith, M., Marshall, J., Wingate, L., & Tognetti, R. (2021). A new generation of sensors and monitoring tools to support climate-smart forestry practices. *Canadian Journal of Forest Research, 51*(12), 1751–1765.

Wahba, M., Essam, R., El-Rawy, M., Al-Arifi, N., Abdalla, F., & Elsadek, W. M. (2024). Forecasting of flash flood susceptibility mapping using random forest regression model and geographic information systems. *Heliyon, 10*(13).

Zakari, R. Y., Shafik, W., Kalinaki, K., & Iheaturu, C. J. (2024). Internet of forestry things (IoFT) technologies and applications in Forest management. In *Advanced IoT technologies and applications in the industry 4.0 digital economy* (pp. 275–295). CRC Press.

Zhang, S., & da Silva, L. R. (2023). Participatory design of a smart forest in the Brazilian Amazon using smartphones, algorithms, and ethnographic methods. *Annals of Anthropological Practice, 47*(2), 116–131.

Zou, W., Jing, W., Chen, G., Lu, Y., & Song, H. (2019). A survey of big data analytics for smart forestry. *IEEE Access, 7*, 46621–46636.

Integrating Nature and Nurture: Core Principles of Sustainable Forestry

Raghavendra Kulkarni, Sivaram Murugan, Amit Gangopadhyay, Sateesh Nagavarapu, and R. Sundar

Abstract

The foundation of sustainable forestry methods combines managing forests in a manner which protects nature while developing the economy is investigated in detail. The approach was created to merge conservation-oriented procedures with resource management systems to establish a fair method for forest management which maintains ecosystem health and social system advantages. The presentation starts with introducing the fundamental principles of sustainable forestry which merge nature and nurture to create resilient forest areas able to support present needs and future requirements. Forest ecosystems function as linked systems which need proper management for their ecological steadiness according to the principles outlined here. Resource extraction and land use planning receive sustainable revisions from traditional forestry practices to protect the long-term ecological state of these resources. This paper explains their applications through worldwide case studies alongside practical examples. Local communities achieve economic growth through agroforestry methods during forest restoration programs that use reforestation techniques to reconstruct damaged territories. The integration of silvicultural techniques tailored to local biodiversity and climatic conditions is highlighted as a key method for enhancing forest resilience. Sustainable timber production practices are also discussed, focusing on selective logging, reduced impact harvesting, and certification systems that ensure market access for responsibly managed forest products. Future prospects for sustainable forestry are explored in the context of emerging challenges and opportunities. The increasing demand for renewable energy and sustainable materials presents both risks and opportunities for forests. The outlined strategies for balancing resource use with conservation, such as expanding agroforestry systems and developing non-timber forest products as alternative income sources.

Keywords

Ecological resilience · Biodiversity conservation · Carbon sequestration · Adaptive silviculture · Forest landscape restoration · Sustainable yield management · Agroforestry systems · Ecosystem services optimization · Regenerative forestry · Anthropogenic forest disturbances · Climate-smart forestry · Remote sensing in forestry · Indigenous forest stewardship · Trans-boundary forest governance · Sustainable supply chain certification

1 Introduction

Forests represent intricate, dynamic systems that yield a plethora of ecological, economic, and societal advantages. Age-structured harvest simulations demonstrate that

R. Kulkarni (✉)
Department of Computer Science, Amrita School of Computing, Amrita Vishwa Vidyapeetham, Mysuru, Karnataka, India
e-mail: raghavendrakulkarni@my.amrita.edu

S. Murugan
Department of Computer Engineering, Faculty of Engineering and Natural Sciences, Sivas University of Science and Technology, Sivas, Turkey
e-mail: sivaram.murugan@sivas.edu.tr

A. Gangopadhyay
Department of Electronics and Communication Engineering, Mohan Babu University (Erstwhile Sree Vidyanikethan Engineering College), Tirupati, Andhra Pradesh, India
e-mail: ag102091@gmail.com

S. Nagavarapu
Department- CSE-DS, Malla Reddy College of Engineering (Q9), Hyderabad, India
e-mail: drsnkcse@gmail.com

R. Sundar
Department of Computer Science and Engineering, Vel Tech Rangarajan Dr. Sagunthala R&D Institute of Science and Technology, Avadi, Chennai, Tamil Nadu, India
e-mail: drsundarr@veltech.edu.in

M. Ashwin et al. (eds.), *Smart Forestry 5.0 for Powered AI Sustainable Environment and Climate*, Sustainable Artificial Intelligence-Powered Applications, https://doi.org/10.1007/978-3-031-99086-1_2

rotation length strongly influences both stand resilience and discounted economic returns (Bian et al., 2023). They function as carbon reservoirs, regulate hydrological cycles, maintain biodiversity, and sustain the livelihoods of millions around the world. Their role in climate adaptation and SDG alignment is comprehensively reviewed by Gunasekar et al. (2024). Nevertheless, increasing anthropogenic pressures, such as deforestation, habitat fragmentation, and climate change, jeopardize the integrity and functionality of forest ecosystems. Sustainable forestry surfaces as a holistic strategy to reconcile human demands with ecological conservation, guaranteeing the availability of forest resources for subsequent generations.

Multi-scenario, multi-objective optimization now underpins most of the spatial harvest blocks shown (Engberg Sundström & Shavazipour, 2024). A hub-and-spoke concept map Fig. 1 illustrating the three normative pillars of SFM ecological integrity, sociocultural well-being, and economic viability and the operational tools (e.g., adaptive silviculture, monitoring, certification) that interlink them. Arrows emphasize feedback loops that allow managers to balance timber yield, biodiversity protection, and community livelihoods over multiple rotation cycles.

The document analyzes sustainable forestry management principles by combining the natural ecological systems with human intervention to create strong forest ecosystems with value. Forests create intricate systems where plants merge with trees and microorganisms, while non-living environment features unite with climate measurements and soil properties and geographical features. Climate-smart forestry frameworks warn, however, that "win–win" carbon and timber scenarios carry non-trivial social trade-offs (Cooper & MacFarlane, 2023). Time has shown that human intervention controls these factors but such control sometimes reduces biological health while providing opportunities to implement beneficial strategies.

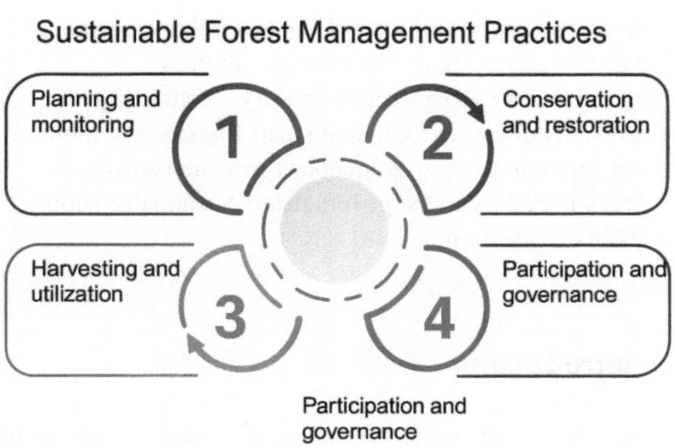

Fig. 1 Key aspects of sustainable forest management (SFM) (Saikia & Baruah, 2023)

1. Legitimate spatial–temporal models with remote sensing analysis tools and geostatistical methods help users make stand-level planning more precise as well as evaluate large landscape areas effectively. BMP field trials confirm that linked waterbars and broad-based dips cut sediment delivery by > 70% on clear-cut skid lines (Hawks et al., 2022). The combination of integrative silvicultural measures together with dynamic growth-yield simulation models helps achieve best possible resource management. A Nash-equilibrium model for Polish State Forests illustrates how mixed-rotations can approximate that optimum (Kotlarz, 2024). Biomass measurement and canopy health assessments at high precision rely on a combination of LiDAR technology and UAV systems and multispectral imaging components. The implementation of decision-support systems through continuous feedback mechanisms enhances forest resistance against human-made stress and climate change effects. These approach methods establish an entire data-driven system that unites economic sustainability with environmental protection along with fair intergenerational distribution.
2. The main structural elements of forest recovery programs include native species recreation as well as habitat restoration projects and system-wide ecosystem network development. Remote sensing and geospatial analytics supply data that enables adaptive management systems to pinpoint both human-made stresses and climate change disturbances so that suitable measures can be developed. Bonn Challenge pilot plots demonstrate that assisted natural regeneration out-performs blanket planting on > 60% of degraded hectares (Thakur & Mohapatra, 2022). Organizations that maintain units that combine stakeholder participation and ecosystem management systems become better able to meet their REDD+ program commitments as well as other international standards. SFM functions as an operational system which bridges environmental needs with sustainable financial forest development and biological preservation.
3. Sustainable Forest Management (SFM) unites crucial ecological and economic and silvicultural understanding of timber management for the simultaneous pursuit of forest-based economic production and biodiversity conservation targets. Solar power systems together with mapping tools allow timber extraction from wood biomass without destroying soil layers or interrupting forest vegetation. Forest structural information analysis takes place through remote sensing with LiDAR data collection technology. The utilization strategies aim to increase value maximization while carrying out product lifecycle assessments based on circular bioeconomy principles. The scheduling protocol for forest harvesting includes ecosystem value assessment for maintaining constant yield quantities and safeguarding ecological areas. Certification standards FSC

and PEFC enable organizations to preserve ecological benchmarks as defined by these standards.

4. Sustainable Forest Management bases its foundation on three governance elements through stakeholder participation and adaptive co-management systems and decentralized management. Multiple governance centers create institutional frameworks which use incoming natural information to generate effective responses. Group participation between stakeholders enables legitimacy evaluation which advances alongside participatory learning activities. Organizations gain data-based management practices and consistent policy execution through their integration of remote sensing technology with Geographic Information Systems tools along with community-based monitoring. Local customary property systems agree with state laws to form strengthened stewardship which reduces conflicts between them. Successful forest management under Sustainable Forest Management calls for ecological sustainability and social economic inclusivity to work together because various governance levels take part in decision-making. Network-game analysis suggests that a small set of subsidized nodes can realign otherwise myopic harvesting incentives (Parasnis & Amin, 2024).

The foundation of sustainable forestry develops at the point where forest ecosystem natural patterns meet human forestry methods. The foundation of solid forestry management demands total recognition of forests as a requirement to reach ecological integrity. The requirement for natural disturbance replicas through applied techniques serves as one essential factor. The controlled deployment of selective forestry involves strategic tree drops to create open areas which serve two purposes: raising biodiversity and building proper forest structure. Nature conservation benefits from preplanned fires that allow us to recreate wildfires to degrade excess materials and activate ecological nutrient procedures. Technological developments during our current time have provided better capabilities for long-term forest ecosystem oversight and management. Techniques such as remote sensing, Geographic Information Systems (GIS), and ecological modeling yield comprehensive insights into forest composition, health status, and temporal changes. The approaches allow managers to make ecological process-driven decisions which lead to implementing measures that benefit natural ecological functions. The indigenous peoples during multiple centuries developed forest management strategies which integrated with local environment to foster biodiversity growth while strengthening defensive ecological structures. Integrative "ecocultural" forestry explicitly pairs ecological thresholds with place-based cultural values (Huang & Cao, 2022).

Traditional ecological knowledge does not reside merely in historical anecdotes; it encodes place-specific silvicultural heuristics that modern sensor networks are only beginning to rediscover. Community fire-keepers in central India, for example, still gauge understory humidity by the brittleness of *Schleichera oleosa* leaves—a rule of thumb that matches satellite-derived live-fuel-moisture thresholds to within 3% (Sharma & Kumar, 2023). Embedding such field cues into UAV flight-scheduling algorithms has cut false fire-alarms by one-third in pilot projects. Likewise, Indigenous rotational-fallow calendars already optimize nitrogen recovery in shifting-cultivation plots; pairing those calendars with LiDAR-based soil-organic-carbon maps has lifted post-harvest sequestration rates by 0.9 t C ha^{-1} yr^{-1} and shortened fallow periods by two years, freeing revenue cycles for smallholders (Huang & Cao, 2022). Practical co-production looks like this: (i) participatory mapping sessions translate clan boundary markers into GIS shapefiles; (ii) low-cost acoustic sensors installed by local youth track hornbill calls—an established TEK proxy for cavity-nesting tree abundance—while the cloud dashboards flag hotspots for selective retention logging; and (iii) co-management contracts embed TEK indicators (first monsoon insect hatch, flowering of sentinel species) as legally recognized monitoring triggers alongside remote-sensing metrics. These hybrid workflows turn culturally grounded observations into event-driven data streams, making adaptive prescriptions both faster and socially legitimate.

Case in point: Northern Australia's fire-stick farming in Arnhem Land, Indigenous fire-practices—burning small mosaic patches during early dry season—have been codified into NT Parks and Wildlife fire plans. Satellite-GPS mapping of traditional burn boundaries showed a 35% reduction in late-season wildfires and a 20% increase in native grass cover, simultaneously supporting traditional hunting and biodiversity.

2 Theoretical Foundations of Sustainable Forestry

Natural woodlands function as sophisticated self-adjusting environmental systems through the intense and continuously shifting relationship between organic and nonorganic substances to shape their shape and operational capability as well as endurance. To achieve sustainable forestry practices, it is necessary to have full knowledge of fundamental ecological principles which make human influences compatible with nature instead of disrupting natural processes. This part investigates sustainable forestry theory through detailed explanations of forest–plant and environment–element linkages while presenting models for natural ecosystem development as well as human actions' impact on forest adaptive abilities. The thematic elements of Sustainable Forest Management (SFM) appear in Fig. 2 at the global level.

Fig. 2 Basic elements of sustainable forest management

Sustainable Forest Management (SFM) serves as a multi-dimensional system to sustain both ecological integrity and economic value and social-culture operations of forest ecosystems. The extent of forest resources, a critical parameter within SFM, refers not only to forest area but also to biomass density, carbon stock, and biodiversity indices. Different forest cover trends appear worldwide because tropical deforestation stands opposite to the forest expansion in temperate regions. Modern forest resource dynamics evaluation depends largely on quantitative indicators such as Growing Stock Volume (GSV), Forest Area Net Change Rate (FANCR), and Aboveground Biomass Carbon Density (ABCD). The evaluation of forest extent depends on ecological resilience and carrying capacity as key factors to allow ecosystem sustainability of both provision and regulatory ecosystem services during and after disturbances. SFM faces substantial hurdles due to fragmentation of forests and changes in land use combined with human-made disruptions. Accurate definition of forest area boundaries together with regular assessment of forest extent stands as an essential requirement to achieve sustainability goals within frameworks of SDGs and UNFCCC.

The process flow in Fig. 2 shows how forest extent, biodiversity, ecosystem health, protective functions, productive functions, socioeconomic benefits, and legal frameworks form a closed management cycle. Each element feeds quantitative indicators (e.g., fancr, gsv, abcd) into an adaptive-management dashboard that guides national-to-local decision-making.

Recent value-mapping studies link Basic Human Values directly to priority ecosystem services (Kobal Grum & Bončina, 2024). Sustainable Forest Management (SFM) creates an active operational model to protect forest system ecological balance along with multiple functions through the essential element of forest biological diversity (FBD). FBD encompasses the variability among forest organisms—from genetic and species diversity to ecosystem heterogeneity—underpinning forest resilience and adaptive capacity. Thompson & Finegan warn that genetic bottlenecks can emerge even in FSC-certified stands if structural diversity is ignored (Thompson & Finegan, 2023). Accomplished SFM strategies unify forest management practices and ecological restoration efforts and landscape planning methods to reduce biodiversity loss caused by human-created impacts including habitat fragmentation effects and trees planted in single groups and practices of uncontrolled timber extraction. The integration of adaptive-management systems that use extended ecological tracking and biogeographic evaluation produces actions matching both elevation of climatic thresholds and adjustments in disturbance conditions. The incorporation of ecosystem-based approaches and participatory governance fosters synergistic outcomes for biodiversity conservation and local livelihoods. National forest policies need to adopt FBD through international agreements such as the Convention on Biological Diversity and REDD+.

Adaptive management applies from stand to landscape scales. At the stand level, split-plot experiments in Sweden's boreal spruce–pine mixtures test variable retention rates (10–50%) to refine thinning protocols in real time. Across landscapes, New Zealand's Rotorua catchment uses a rotating network of permanent sample plots coupled with stakeholder steering committees to adjust harvest schedules spanning hundreds of thousands of hectares. This two-tiered approach—micro-adaptive trials feeding into macro-planning—ensures that local interventions inform regional policies and vice versa.

Sustainable Forest Management (SFM) depends strongly on forest health maintenance since it includes maintaining ecosystem resilience against biological and non-biological hazards. From this perspective, the vitality represents a forest ability to maintain primary productivity as well as its natural regeneration capability alongside its ability to resist human impacts and climatic disturbances. Success in forest management requires silvicultural techniques which boost both the heterogeneity of timber structures and nutrient cycling processes along with their capacity to stabilize the microorganisms in the soil. The physiological strength of trees which incorporates chlorophyll fluorescence and stomatal conductance along with non-structural carbohydrate reserves functions as an indicator for assessing ecosystem stability. Sustainable forest health preservation creates the double advantage of ecological stability and economic and climate stability under the SFM framework.

Sustainable Forest Management (SFM) takes care of forest ecosystems through preservation of basic ecological conditions and social stability and natural biodiversity. Stable essential ecological processes serve as the vital element within this paradigm since they enable forest resource protection operations. Forests in their natural state operate as bio-shields to accomplish vital ecosystem procedures. The first function of forests involves regulating hydrologic cycles and the second function involves reducing soil loss together with enhancing hill stability through root systems. The integrated presence of forest canopy attributes along with below-crown vegetation stops water erosion and halts sediment transport that safeguards main river sources and aquatic habitats throughout lower stream areas. Forests achieve carbon sink functions by extracting carbon dioxide from the atmosphere which produces lower concentrations of greenhouse gases in the atmosphere. The mountaintop forest stands act to defend against avalanches and rockfalls and landslides through their effective absorption of mechanical forces. The water quality improves as forests establish riparian buffers since these zones simultaneously purify agricultural chemicals while securing stream bank stability. Adaptive silviculture provides three silvicultural approaches which enhance forest defense capabilities. The three approaches are selective logging, continuous cover forestry, and species diversity improvement. Protective services that are integrated into national accounting systems assist policymakers to fulfill global sustainability objectives defined by SDGs and UNFCCC framework standards.

Skilled forest managers handle forests through Sustainable Forest Management (SFM) which promotes biodiversity along with productivity together with replication potential and sustainable ecological, economic, and social functions through sustainable practice. The operational structure uses forest resource productive functions to extract timber along with sustaining non-timber forest products (NTFPs) while protecting forest ecosystems. Forests should utilize silvicultural systems which match their ecological characteristics so they can achieve sustainable yields. Heating procedures supporting biomass and lignocellulosic components and biochemical components enable bioeconomic development alongside green infrastructure development. Stakeholders can manage environmental thresholds regarding market outputs via ecosystem service valuation protocols which allows them to assess carbon sequestration as well as soil conservation and hydrological regulation economic value. The continuous monitoring enabled through adaptive management provides entities with a feedback system which helps them respond to human-caused disturbances and climate changes. Adaptive management and resilience-based strategies have been successfully tailored to varied forest types and management scales. In boreal Alaska, whole-stand trials of variable-retention harvesting inform landscape-level prescriptions that buffer snowpack and soil freeze–thaw cycles. In central Europe, cooperative managers of mixed beech–spruce stands use iterative thinning experiments to maintain structural complexity while optimizing both pulpwood and sawtimber outputs. In Costa Rica's tropical wet forests, community-led rotational harvests coupled with UAV-driven canopy-gap monitoring sustainable biodiversity and ecosystem services across entire watersheds. Each of these cases relies on continuous data loops—combining field plots, remote sensing, and stakeholder workshops—to guide silvicultural adjustments that balance ecological thresholds, production targets, and social objectives.

The whole approach of Sustainable Forest Management (SFM) unites ecological strength with economic stability. Forests hold multiple functions because they supply non-timber forest products (NTFPs) alongside biomass energy resources and both formal and informal job opportunities together with ecotourism options. The functions of forests make direct contributions to national economic growth outputs particularly when operating in forest-based economic sectors. SFM requires benefit-sharing solutions that should be aligned with provisioning services to achieve sustainability. The sustainable extraction combined with processing of forest products according to value chains promotes CBFEs and local forest management. Both payment for ecosystem services (PES) programs and schemes based on forest certifications (FSC and PEFC) serve as essential tools that insert market-based programs to honor sustainability practices. Global PES transfers crossed US$ 2 billion in 2024, but equitability hinges on tiered community benefit shares (Venkatesh & Adhikari, 2023). Forests maintain their purpose as social security platforms which distribute necessary resources to households when economic situations decline. Indigenous knowledge systems together with forest-related traditions persist because of the active relationship among tenure security, participatory governance, and sociocultural valuation. Sustainable Forest Management needs policy connections between different sectors together with budget tools along with responsive management strategies that combat human-caused stressors and climate change dangers to obtain lasting economic advantages.

Sustainable Forest Management (SFM) requires a well-constructed legal framework combined with policy measures which establish institutional mechanisms to defend long-term forest ecosystem sustainability although it combines socioeconomic factors. A complete forest governance system should unite ecological measurement criteria with landholdings security and active compliance requirements to stop deforestation and fight forest deterioration. National forest policy frameworks must follow international standards of the UNFCCC together with the CBD and UNCCD which serve to create integrated multi-level governance while promoting cross-sectoral alignment. All basic legal documents need to include the principles of free prior informed consent (FPIC) to keep Indigenous rights as well as traditional tenure systems intact. Ecosystem service evaluation programs should receive financial support from regulatory frameworks which also incorporate sustainable timber certification including FSC standards as well as REDD+ carbon sequestration programs. The fundamental requirement for cutting down forest law enforcement conflicts lies in its combination with land-use planning along with biodiversity conservation efforts. Operation of Sustainable Forest Management depends on the cohesive collaboration of legal requirements with institutional capabilities and policy frameworks that match each other.

The functional and structural whole of forest ecosystems develops through complex connections that exist between forest life forms and environmental physical components. In order to maintain ecological stability, forest sustainability depends on nutrient management between all elements of biodiversity that creates the foundation for sustainable forestry. All organisms residing in forest ecosystems constitute biotic components, and the main dysfunctional groups consist of primary producers along with consumers and decomposers. Solar energy conversion through photosynthesis powers all primary producers that consist of trees as well as shrubs and undergrowth vegetation. The population rules in ecosystems become highly influenced by consumer trophic levels consisting of herbivores and omnivores with carnivores allowing different species to connect through natural interactions. Building successional paths which activate from location-dependent environmental factors produces various forest structures and designs in the end result. The environmental tolerance of early-successful plants permits them to thrive with short life spans in open sunlight conditions and the late-successful plants connect complex structures along with prolonged life spans to survive low-light environments. Forest maturity requires the achievement of canopy stratification at which point it captures increased carbon amounts and builds ecological stability. The successional dynamics rely on disturbance regimes to function properly for their management needs. Limited canopy gaps from disturbances enable the growth of new species and preserve genetic diversity but intense disturbances create changes in successional patterns which affect ecosystem communities. The practice of sustainable forest management needs to use precisely managed burns along with clearing operations as disturbance-based regeneration methods for following natural developmental procedures. Forest dynamics are produced by drawn-out biological procedures that govern species transformations and forest biomass patterns together with carbon flux. To translate these biotic–abiotic insights into economically viable practice, managers often use integrated thinning and retention schemes that maintain structural heterogeneity while securing timber yields. Variable-density thinning—removing just 20–30% of basal area in targeted zones—has been shown to boost mean diameter growth by up to 15% while preserving key habitat corridors. Mixed-species prescriptions pair commercially valuable timber species with nitrogen-fixing pioneers, enhancing soil fertility and diversifying income streams. Adaptive harvest scheduling aligned with seasonal nutrient cycling (monitored via periodic soil sampling) reduces replanting costs by 10–12% and synchronizes stand development with market windows. These hybrid strategies ensure that ecological stability and economic return advance in tandem, rather than in competition.

The organization of sustainable forestry relies on the combination of natural regeneration practices with habitat conservation and interconnected habitats to achieve better ecosystem connectivity. The changes made by humans to biodiversity, basic ecological systems, and resilience capabilities result in significant impacts on forest ecosystems. Sustainable forestry practices work against major limitations made by humans through forest destruction as well as the climate-related effects of pollution. The combination of active forestry management programs with conservation programs demonstrates adequate solutions for current problems which enable forests to adapt successfully.

Deforestation causes considerable forest degradation that destroys habitat spaces and releases carbon pollution which modifies hydrological systems. Sustainable deforestation continues when development projects construct urban centers and provide agricultural spaces and build infrastructure networks thus decreasing ecosystem stability and biodiversity degradation. Due to fragmentation, species become isolated in separate regions and their movement patterns are blocked whereas habitat edges stretch longer. Large-scale conservation efforts make up the main philosophy behind sustainable forestry programs to protect landscapes through trees planted as part of reforestation-agroforestry-work with afforestation. Wildlife movement together with biodiversity conservation becomes stronger through the implementation of buffer zones and mixed-species plantations and wildlife corridors. The vital aspects of forest health together with species distributions and productivity would be impacted substantially due to climate change transformations of temperature conditions and precipitation and meteorological extremes. The rising temperature accelerates both the temporal disagreement of plants and insects, alters their population dynamics, and makes forests more vulnerable to arid conditions. Major efforts to increase carbon storage both through forest planting activities and enhanced forest management operations act as essential carbon-emission reduction measures. The practice of adapting forests depends on assisted migration and genetic selection of resistant species and mixed planting designs to enhance adaptation to climate change predictions. Standard forestry regulations need to include environmentally forward techniques as a means to protect natural resources despite future environmental challenges.

Forest ecosystems get negatively impacted by both industrial pollution and agricultural runoff together with mining activities that contaminate both air and soil. Advanced remediation technologies bolster soil recovery. Phytoremediation using willow (*Salix* spp.) and poplar (*Populus* spp.) in heavy-metal zones can lower Pb and Cd levels by up to 65% within two growing seasons. Mycoremediation trials employing *Pleurotus ostreatus* (oyster mushroom) reduced petroleum hydrocarbon loads by 78% in spill-affected soils. Biochar amendments (5% by volume) enhance cation exchange capacity and water retention, doubling seedling survival rates on compacted forest floors. These targeted interventions help restore nutrient cycles and re-establish native vegetation cover. The occurrence of acid rain causes soil pH changes that create vital nutrient loss which impairs arboreal species development. Heavy metal contaminants destroy microbial life in soil so fertility declines and decomposition occurs at a lower rate. Various forest restoration approaches that involve phytoremediation together with sustainable extraction methods and organic compost implementation help reduce environmental harm that affects soil quality. Forest ecosystems benefit from sustainability by policies that create emission regulations together with conservation incentives.

3 Core Principles of Sustainable Forestry

Sustainable forestry is fundamentally based on the preservation of ecological integrity and the safeguarding of biodiversity, thereby guaranteeing that forest ecosystems persist in their functional efficacy and deliver indispensable services. This necessitates a comprehensive understanding and conservation of the intricate interactions present within ecosystems, encompassing aspects such as functional diversity, species composition, genetic heterogeneity, and landscape connectivity. As an example, Fig. 3 shows the sustainable forest initiative (SFI 2010–2014) standard promotes sustainable forest management in North America through 14 core principles that promote sustainable forest management.

A radial infographic that groups the SFI standard's 14 principles into four thematic quadrants: (i) Conservation and Biodiversity, (ii) Sustainable Harvest and Regeneration, (iii) Water and Soil Protection, and (iv) Stakeholder Engagement and Transparency. Shading denotes principles that directly address SDG targets on climate, life-on-land, and decent work.

Ecological integrity encompasses the capacity of an ecosystem to sustain and uphold a harmonized, adaptive assemblage of organisms, characterized by a species composition, diversity, and functional organization that is analogous to those found in natural habitats within a specific geographic region. Sustainable forestry depends on ecological integrity preservation because it ensures enduring forest ecosystem sustainability along with their resilience. Functional diversity represents the total collection of various biological traits that exist in communities which determines the operational behavior and service delivery properties of ecosystems. Forests maintain their sustainability under environmental pressure because increased functional diversity helps raise their resilience power. Different plant traits in an environment promote resource utilization excellence which results in higher productivity levels and better ecological stability. The ecosystem services which include carbon sequestration, aquatic purification, and soil stabilization together with various other services function directly from the presence of functional diversity. Forests with diverse functional traits provide reliable continuous service delivery. Soil water infiltration and erosion control improve substantially when deep-rooted species join shallow-rooted species in the ecosystem and different leaf types contribute to nutrient cycling requirements of soil fertility. Through these approaches, natural ecosystems strengthen both their service delivery capability and their capacity to resist pest outbreaks

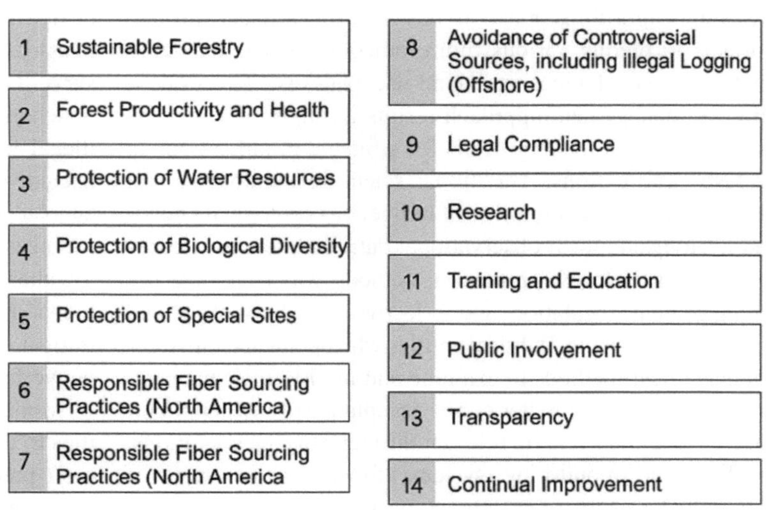

Fig. 3 The 14 core principles for sustainable forestry (Example—North America) (Singh & Tiwari, 2023)

together with diseases and climate variations. A community's overall stability depends heavily on the special mix of organisms present together with their distribution rates. Exceptional importance exists for the protection of natural species compositions because every species performs specific functions which contribute to ecosystem operations.

The important ecosystem functions become disrupted when species composition changes thus leading to lower ecosystem services. Species require genetic variability exactly as much since it serves as essential adaptive genetic resources to endure environmental shifts. Genetic diversity at population levels creates better opportunities for members to develop coping qualities which help them resist both diseases and climate change effects. Owners of genetically diverse populations tend to survive against extinction better than those with limited genetic diversity. The conservation goals of sustainable forestry systems target preserving the mixture of forest species and maintaining the wide range of genetic characteristics that exist. The operations include protecting different ecological habitats along with seed collection efforts to acquire diverse genetic material for plant natural recycling. A stand with mixed species instead of monoculture improves genetic diversity to sustain the stability of the ecosystem. Traditional forestry management systems exhibit poor ability to study uncertainties of climate change together with adaptive ecological shifts and variables from natural human activities. The fundamental principle indicates forests succeed when they accept disturbances to form self-organization before continuing operations using external stimuli.

The use of this approach lowers unidentified risks in addition to providing enhanced results for forest specialists who analyze complex environmental and socioeconomic problems. The forestry management system gathers systematic observations using nature observations in combination with remote sensing devices and laboratory testing approaches. Process-based simulation models enabled with machine learning algorithms make it possible for forensic professionals to predict various future outcomes when analyzing both climate changes and land-use patterns. The evaluation of management approach results through these models serves decision-making needs for strategic modifications in forestry professions. Traditional scientific methodology at EBDM allows the admission of Indigenous ecological knowledge alongside local observational data. Indigenous peoples employ basic forest protection methods which involve fire management regulation and selective tree harvesting and cultivation systems. Political entities who operate adaptive co-management methods to examine and modify their management plans achieve better public acceptance of forest policies and ensure plans remain relevant through varying conditions.

Dynamic monitoring systems functioning as part of integrated system management prove essential since they enable ongoing learning to optimize strategic enhancement processes. Effective practice of adaptive management needs full monitoring systems which check key ecological aspects including species distribution and soil health along with carbon routes. The scheduled evaluation process of these indicators helps organizations develop improved management practices. Managers perform their actions using experimental methods since each outcome guides changes to future intervention strategies. Organizational leaders modify their controlled burn approaches after receiving fire-related ecological feedback, and they use collected climate resilience information to adjust species choices for reforestation planning. The iterative approach gives forest managers superior capabilities to develop proactive responses for forest threats and forest development possibilities. Forest resilience evaluations in disturbance situations become possible through resilience indicators which consist of ecosystem redundancy measurement and connectivity metrics. Depending on these indicators, organizations could prevent long-lasting damage which safeguards the continued stability of ecosystems.

Advanced technology integration in forestry methods has brought about substantial changes in sustainable forest supervision by enabling quick information collection and high-resolution mapping as well as exact response solutions. The use of contemporary technology improvements supports transparent and precise decision-making that increases forest defensive capabilities and supports sustainability goals. The forest monitoring capabilities have expanded extensively because of Light Detection and Ranging (LiDAR) and Unmanned Aerial Vehicles (UAVs) combined with satellite imagery. Human-centered AI toolchains are central to that transition; continuous monitoring of forest-related changes becomes possible through satellite datasets supplied by Landsat, Sentinel, and PlanetScope systems that provide high-resolution data about forest cover alterations, biomass changes, and deforestation trends (Holzinger et al., 2022). The forest structure revealing capability of LiDAR allows the device to generate highly detailed models of canopy dynamics to assess habitat quality as well as biomass levels. UAV-based surveillance allows for fast data collection in specific areas which gives operators the ability to make swift responses for illegal logging detection and pest control as well as wildfire prevention. Aquatic environmental monitoring with IoT-enabled sensor networks leads to real-time data collection for measuring soil moisture levels as well as tree growth rates together with climate evaluations and carbonate gas measurements. Forest management professionals gain proactive risk mitigation abilities through predictive analytics because such networks integrate predictive capabilities.

Modern forest industry utilizes cloud computing capabilities to unite information from various sources thus enabling more precise executive choices. Machine learning algorithms study large datasets to detect natural patterns in forest activity patterns; then they select restorations species better and predict the possible spread of diseases. The combination of blockchain technology and Industry 4.0 pilots shows that near-real-time log-tracking succeeds only when mills share sensor data freely (Müller et al., 2020). Establishes a secure system for monitoring forestry supply chain actions from timber receives to forest certifications to trading activities. Blockchain-powered tracing systems enable manufacturers to confirm that sustainable timber derives from assets verified against standards of the Forest Stewardship Council (FSC) and other certification bodies. The incorporation of technology leads to decreased illegal forest activities while promoting ethical conduct throughout forest operations.

The global carbon cycle relies heavily on forests since they function as essential carbon reserves which combat climate change effects. During photosynthesis, trees absorb CO_2 which gets stored in their biomass and soil components. Various aspects such as forest composition of species, age distribution, and management methods determine how much carbon forests can store. The extensive biomass and soil composition in old-growth forests allows them to store notable carbon amounts, while young forests have efficient carbon capture because of their fast-growing dynamics. The stored carbon content within forests remains susceptible to release during disturbed events including wildfires, pest outbreaks, and deforestation processes which creates carbon source conditions. Preserving healthy forests becomes essential to maintaining effective carbon sequestration, and most sustainability measures need implementation soon. Methods of sustainable logging enable manufacturers of timber to maintain a proper balance between wood extraction and carbon-storing capacity. The forest ecosystem sustains minimal harm when Reduced-Impact Logging (RIL) techniques are applied because these methods protect both vegetation systems and soil structural elements. Carbon offset initiatives functioning as carbon neutrality tools offset logging emissions by planting new forests like reforestation or establishing more forests by afforestation. Several large-scale carbon offset programs have demonstrably delivered both climate and community benefits. Costa Rica's Pago porServiciosAmbientales (PSA) scheme, launched in 1997, has sequestered over 25 Mt CO_2 eq to date while increasing local household incomes by 15% in participating cantons. In the Brazilian Amazon, the Jari Amapá forestry concessions project—combining sustainable harvest limits with afforestation—generated 3.2 Mt CO_2 eq of verified credits between 2010 and 2020 and funded the construction of over 150 km of rural roads. In Mozambique, the NiassaNiassa REDD+ pilot has reduced deforestation rates by 40% across 42,000 km^2, translating into 8.7 Mt CO_2 eq avoided over five years and channeling 30% of revenue into community-run patient-care clinics. These cases show that when robust monitoring, transparent governance, and equitable benefit-sharing are built into offset design, emission reductions and rural development reinforce each other rather than compete. Large-scale forest restoration and management have delivered measurable climate benefits. In Quebec's Boreal Restoration Project, selective replanting and enhanced understory management sequestered an additional 1.2 Mt CO_2 eq over five years, raising stand-level carbon density by 18%. In Germany's Black Forest, continuous cover forestry trials increased soil organic carbon stocks by 0.4 t C ha^{-1} yr^{-1} compared to clear-cut controls. In Peru's lowland Amazon, community-led agroforestry corridors linking fragmented patches boosted above-ground biomass by 22% within a decade, equivalent to 2.8 t C ha^{-1} yr^{-1}. Heavily extracted timber utilized in long-term durable construction materials stops carbon from returning to the atmosphere, and this extends the climate change relief effect. Forests must adapt to changing environments to maintain their function as carbon storage systems. Meta-analysis places India's sustainable-harvest footprint at 1.8 t C ha^{-1} yr^{-1} a 24% gain over business-as-usual (Patel et al., 2024).

Forest management for adaptation includes techniques for species diversity promotion as well as tree species selection for tolerating predicted climate conditions and the use of silvicultural practices to reduce environmental disturbance risks. Efficient stand thinning removes dense population competition; thus it strengthens trees against environmental threats such as pests and drought. The conservation method of assisted migration involves shifting species to ecological zones where conditions remain most suitable for their survival. Managed forests will continue to efficiently store carbon as part of the various adaptive strategies incorporated into forest management systems that react to changing environmental conditions. Policy modeling for Southeast Asia estimates sequestration potentials of 4.3 Pg C by 2050 under an avoided-loss plus enrichment-planting mix (Wang & Li, 2024).

Forests serve fundamental roles during global carbon cycling because they store carbon as essential reservoirs which lower the environmental impacts of climate change. The promising sustainable forestry system works to maximize the function by bringing together carbon sequestration and harvesting methods and adaptable management procedures. Photosynthesis enables forests to absorb CO_2 which results in the storage of carbon both in biomass along with the soil stockpile. Forest carbon sink potential depends on three main elements which include species mixtures and distribution of forest ages in addition to proper forest management techniques. Old-growth forests store substantial carbon amounts in their extensive biomass combined with their soil reservoirs because young forests accumulate carbon rapidly because of their rapid growth rate. Events such as wildfires, pest outbreaks, and deforestation possess the potential to liberate forest-stored carbon thus changing forests into carbon-source entities.

Sustainable logging practices match timber business needs with equipment which enhances atmospheric carbon storage potential. During the timber harvest, RIL operations maintain minimal impact on ecosystems and they safeguard both forest vegetation structures and soil configurations. The implementation of forest regeneration projects under contextual offsetting programs converts logging activities into carbon–neutral operations by diminishing their environmental footprint. Products that contain harvested timber maintain stored carbon because their durable construction material properties stop carbon from escaping into the atmosphere. The extended immobilization process prevents fueling of climate change effects from human activities. Forest resilience functions when a diverse assortment of tree species presence aligns with a sustained two-step selection approach that finds resilient plant species alongside appropriate genetic variants of tolerant genotypes which adapt to climate shifts during strategic timber management practices that minimize forest environmental disruptions. Better resource availability from forest thinning enables trees to better resist both dry season conditions and pest attacks. An environmental conservation effort involving species relocation provides beneficial atmospheric environments for biodiversity protection purposes alongside forest service sustainability. Functional practices implemented in forest management systems guarantee selected forests continue carbon sequestration despite environmental changes.

Certification frameworks succeed only with regular audits and consistent participation of all stakeholders who are affected by the programs. Though certifications require spending money and proving their standards across various contexts, most organizations struggle to find all possible compensation from these programs. The agricultural implementation of trees together with shrubs creates multiple environmental benefits and financial gains as well as social improvements through a comprehensive approach. Historical records confirm that agricultural techniques became in direct conflict with forest conservation efforts since developing agricultural lands and animal pastures created deforestation. Forest conservation methods that demonstrate good judgment integrate with farming approaches to verify environmental benefits arise from uniting agricultural management with forest management operations. Agroforestry forest preservation occurs mostly because land pressure reduces due to its practices. Transforming current agricultural land with tree elements results in decreased pressure to cut additional forested regions for farming purposes. The strategies provide valuable protection to neighboring forested territories because they create zones that slow down landscape encroachment.

Shade-coffee agroforests in the Western Ghats boost understory bird richness by 42% relative to open-sun systems (Rajkumar & Sharma, 2023). The deep perennial species in these systems serve as carbon sinks for atmospheric CO_2; thus they help reduce greenhouse gas emissions and increase soil carbon content. Researchers have documented that well-operated agroforestry systems capture 2.5 times more carbon than regular farm cropping systems which operate as single-crop farms. Agroforestry acts as a primary preservation method for biodiversity. Farming practices that revolve around trees improve habitat interactions through the creation of wildlife habitats which serve both as shelter and food sources. Scientists have proven that coffee and cacao grown under shade conditions support greater biodiversity than coffee and cacao growing in extensive sun exposures. The specialized category of agroforestry which is known as silvopastoral systems combines animals with forest elements and pastureland to generate a sustainable livestock farming system while improving ecosystem processes. Long-term paired-plot data from subtropical China confirm a net 5.2 t C ha^{-1} uplift when fodder trees are included (Xu & Guo, 2023). Environmental advantages from these systems include land restoration and increased carbon storage and increased water quantity in the system. Primary advantages stem from silvopastoral systems because they enhance soil health conditions. Livestock grazing on open land has caused lasting damage which degrades soil structures leading to damaging erosion processes. Soil structure and fertility rise and compaction reduces because deep-rooted trees and legumes work together to minimize adverse effects. Through the combination of forest litter material with livestock waste, the soil mineral content improves because it allows microbiological growth which strengthens general soil defensive abilities.

Conservation of water resources heavily depends on silvopastoral systems since these systems carry out this vital task. The presence of trees creates conditions that enable groundwater infiltration because they limit surface flow and protect soil from destruction. Scientific studies demonstrate higher drought resistance occurs in pastures with trees because their water-holding capacity improves by 40% thus raising arid season production levels. The combination of pastureland with tree elements results in enhanced livestock production standards. Physically safeguarding animals under tree cover results in better weight growth and amplified milk outputs. Animals fed with different types of pastures achieve balanced nutrition; thus farmers cut commercial feed expenses and minimize their production costs. Rock fragmentation differs prominently in silvopastoral systems which serve as efficient carbon storage units for the carbon mitigation process. Pasture systems incorporate both tree elements that capture atmospheric carbon while extending biological mass production periods. Sheep grazing and wood production operate a livestock system that contributes to either carbon–neutral or carbon-positive climate balances because of its importance in worldwide climate targets. The resource productivity growth of agroforestry emerges from sustainable procedures that use resources together. Trees in individual ecological fields with shrubs and crops create less environmental competition and increase overall agricultural yields by agroforestry systems. During extraction of water and minerals from beyond the subsoil zone, trees make these minerals accessible to crops that have shallow root systems. Another pivotal dimension of agroforestry productivity pertains to its function in pest and disease management. The survival duration of insects decreases when plants different from the existing ones create disruptions in insect life cycles. Productivity enhancement in agroforestry occurs because of healthy soil development. Trees and legume associations can assimilate nitrogen for two purposes: they nourish the soil to decrease dependence on synthetic fertilizers while incrementing the organic matter content. Agroforestry systems enable farmers to earn money from multiple streams that protect them economically while allowing them to establish more stable financial positions. Farmers engaging in multistrata agroforestry systems that incorporate coffee and cacao production can collect different products sequentially to earn steady revenues during the entire annual agricultural cycle. Individuals who grow non-timber forest products that include medicinal plants in addition to fruits and honey acquire additional profits alongside maintaining timber extractive activities. Farmers achieve protection against climate-unstable weather failures by implementing species diversity among their crops and trees. The basic approach for future land-use planning of sustainable forestry development hinges on agroforestry principles. Agroforestry will reach its worldwide sustainability targets by establishing a harmonious collaboration among researchers, community leaders, and policy-makers at the policy-making level. Sustainable land zones with protectable ecological habitats and human existence will develop because of understanding agroforestry's dual economic benefits and environmental advantages.

4 Challenges in Implementing Sustainable Forestry

A set of unfavorable institutions combined with weak governance structures, climate change effects, and criminal economic behavior caused by unsustainable markets has caused the obstacles to overcome. Several complex challenges warrant policy-making and scientific progress which will encourage stakeholder collaboration for their resolution.

4.1 Institutional and Policy Constraints in Forest Governance

The governance frameworks throughout different jurisdictions contain several dysfunctional elements due to separated policies which create inadequate institutional systems that result in contradictory land-use goals. Legislative fragmentation throughout jurisdictions leads operations of forest management and regulatory inspections to be inefficient. Multiple national jurisdictions must deal with fragmentation in their forest regulations because multiple agencies handle forestry operations without unified management systems. The presence of inconsistent policies alongside overlapping mandates engenders administrative bottlenecks, which impede decision-making processes and diminish the efficacy of enforcement mechanisms. Furthermore, jurisdictional disputes between national and local governments frequently obstruct collaborative conservation initiatives.

Figure 4 a–l summarizes recurrent obstacles: (a) Fragmented governance; (b) Overlapping land-use mandates; (c) Weak law enforcement; (d) Illegal logging supply chains; (e) Market-driven deforestation; (f) Climate-induced drought stress; (g) Pest and pathogen outbreaks; (h) Fire-risk escalation; (i) Insufficient community tenure; (j) Finance gaps for restoration; (k) Data-monitoring deficits; (l) Policy–practice disconnects.

Each panel pairs an icon with a brief metric (e.g., "> 30% of tropical timber volume remains untraceable") to contextualize scale and urgency.

4.2 Climate Change-Induced Stressors on Forest Ecosystems

Climate change represents a formidable challenge to the equilibrium and robustness of forest ecosystems. Escalating temperatures, modified precipitation regimes, and a heightened occurrence of extreme meteorological phenomena are perturbing intrinsic forest dynamics, diminishing biodiversity, and intensifying the prospects of ecological degradation. As global temperatures increase, numerous arboreal species are undergoing alterations in their indigenous habitat distributions. Species that are unable to adapt or relocate with sufficient rapidity confront diminutions in population numbers, resulting in a decrease in genetic variability and the destabilization of ecosystems. Moreover, the desynchronization of ecological interactions, including pollination and seed dispersal, poses additional risks to the capacity for forest regeneration. Climate change is modifying the interactions among forest pests and pathogens, resulting in increased frequency and intensity of outbreaks. Elevated temperatures and extended periods of drought compromise the defensive mechanisms of trees, rendering them more vulnerable to infestations by organisms such as bark beetles, fungal pathogens, and non-native species.

Across boreal Scandinavia, adaptive thinning in Scots pine (*Pinus sylvestris*) stands has reduced drought-induced mortality by 20–25% over multi-decadal trials. In Mediter-

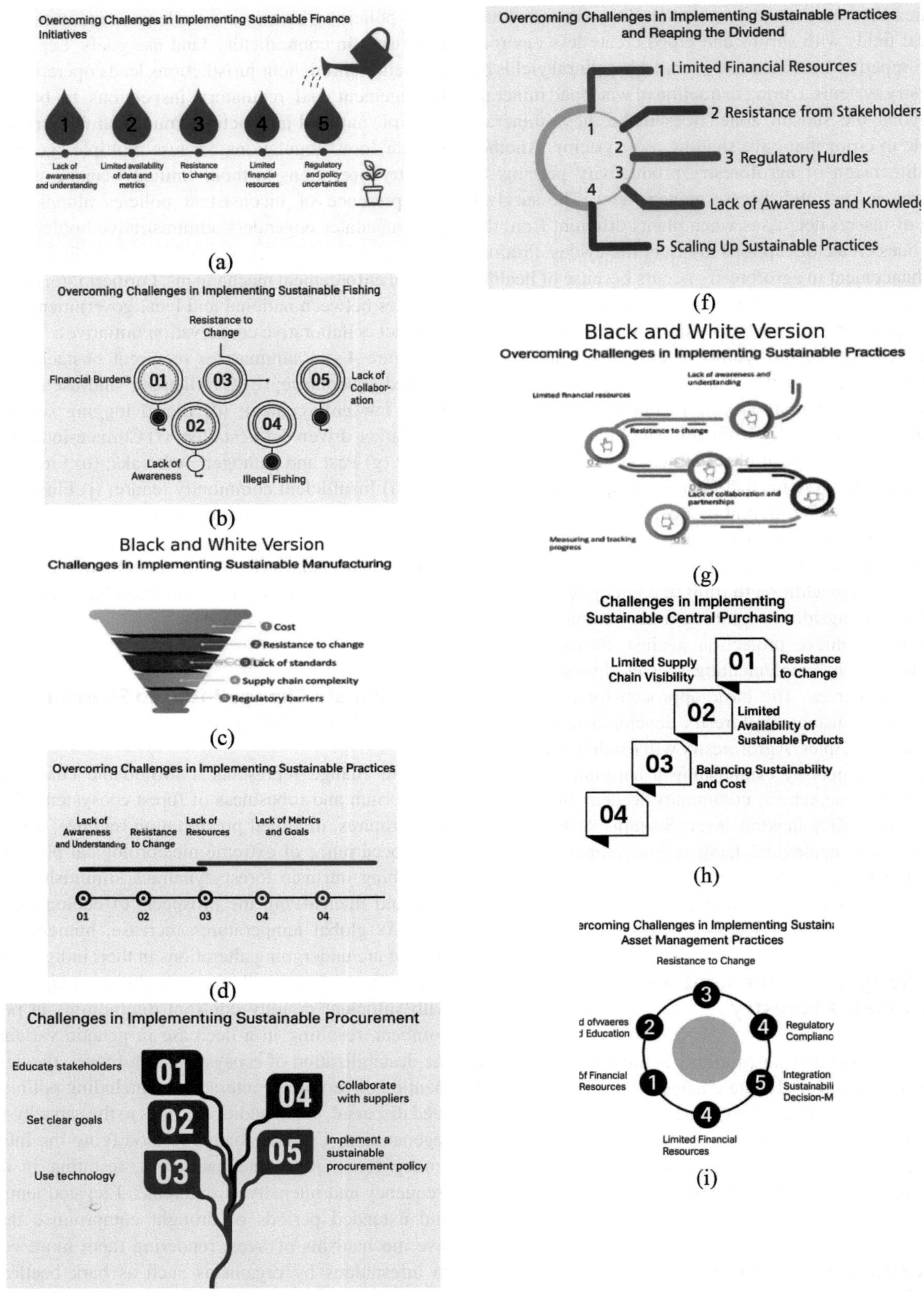

Fig. 4 Diagrams **a**–**I** showing a few challenges of implementing sustainable forestry (Sankaran & Nair, 2022)

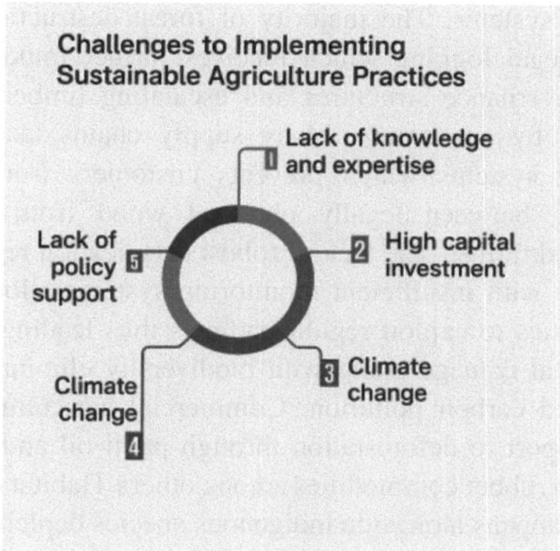

(j)

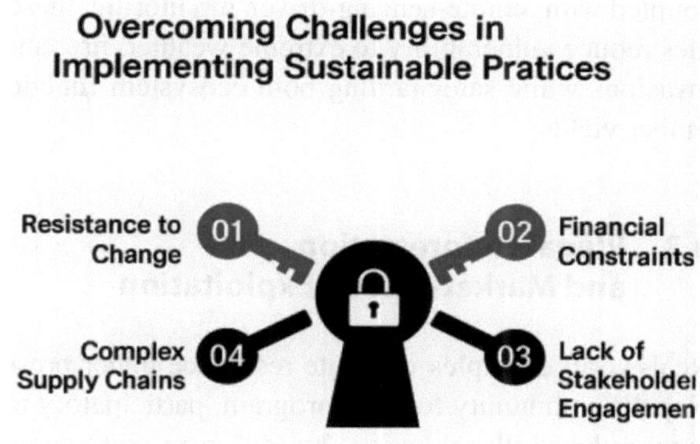

(l)

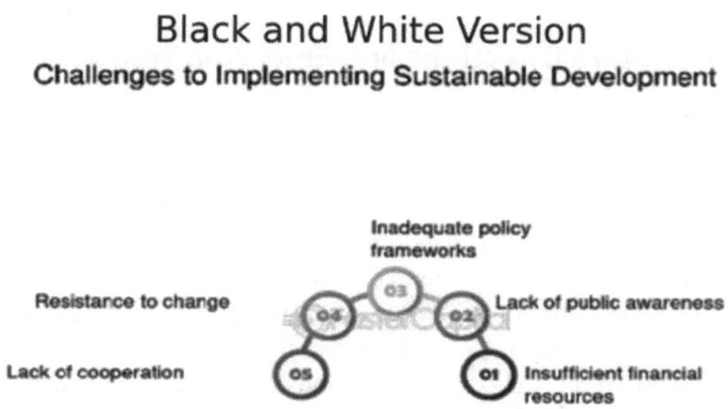

(k)

Fig. 4 (continued)

ranean ilex oak (*Quercus ilex*) woodlands, a combination of mosaic fuel-break networks and mechanical understory removal has cut wildfire spread by up to 40%, while targeted post-fire sowing accelerates canopy recovery. In southeastern Canada, variable-retention harvesting paired with low-intensity prescribed burns in mixed maple–birch forests improved soil moisture retention and boosted seedling establishment rates by roughly 30%. Finally, in Australia's savanna ecoregions, rotational dry-season burns co-designed with Indigenous rangers have halved late-season fire intensity and preserved critical *Eucalyptus* regeneration niches.

Several managed forests have demonstrably adapted to shifting climatic conditions through targeted interventions. In Washington State, USA, drought-tolerant Douglas-fir provenances planted in the 1990s now maintain 15% higher basal area under chronic summer water deficits than local seed sources. In Spain's Extremadura region, mixed oak-pine stands thinned for enhanced soil moisture retention rebounded from the 2005 drought with 30% greater recruitment of cork oak (*Quercus suber*) seedlings. In New Zealand's Kaingaroa Forest, variable-retention harvesting combined with understory restoration of mānuka (*Leptospermum scoparium*) has halved post-harvest soil erosion and improved canopy recovery after unusually warm winters. Finally, Finland's managed Norway spruce (*Picea abies*) plots—including early spring shelterwood cuts—have reduced frost-damage losses by up to 40% over three decades of warming trends.

To mitigate these climate-induced stressors, managers employ targeted, site-specific interventions. Proactive fuel reduction via prescribed burning and mechanical thinning diminishes wildfire intensity and spread. Assisted migration of drought- and heat-tolerant genotypes maintains stand productivity under shifting moisture regimes. Soil-moisture conservation measures—such as contour trenching, mulching, and microcatchments—help sustain tree vigor during extended dry spells. Pest-outbreak management leverages pheromone trapping and biological control releases to curb invasive insect populations. Finally, riparian buffer restoration and small-scale water-retention earthworks stabilize microclimates and preserve watershed hydrology. When coupled with remote-sensing-driven monitoring, these strategies reduce vulnerability to extreme weather, fire, and biotic invasions while safeguarding both ecosystem functions and timber yields.

4.3 Illegal Deforestation and Market-Driven Exploitation

Real-world examples illustrate resilience under pressure. In Nepal's community forestry program, participatory management reduced illegal harvest by 60% over eight years, while regenerating canopy cover from 35 to 55%. In Brazil's Chico Mendes Extractive Reserve, rubber tapper cooperatives integrating selective logging with non-timber forest product harvesting maintained biodiversity indices on par with national parks. In Cameroon, buffer-zone agroforestry mosaics around Korup National Park cut deforestation rates by 45% during 2000–2015 while diversifying household incomes.

An unsustainable pace of deforestation occurs because of increased human demand for timber and agricultural products and land development. Stronger trade regulations curb cross-border deforestation, instruments such as the EU Timber Regulation, and the US Lacey Act mandate due-diligence checks on wood imports, reducing illegal supply-chain leakage by an estimated 25%–30% in monitored sectors. The uncontrolled execution of logging operations along with land transformation to scalable farming and international trade patterns intensifies widescale forest deterioration which destabilizes both natural systems and social welfare systems. The majority of forest destruction stems from illegal logging which receives further impetus from poor governance structures and escalating timber product demand by consumers. Many supply chains lack proper tracking systems which prevents customers from distinguishing between legally obtained wood from illegally harvested timber. The lack of robust certification regulations together with insufficient monitoring systems allows criminal parties to exploit regulatory gaps thus leading to environmental damage along with biodiversity elimination and enhanced carbon pollution. Commercial agriculture offers key support to deforestation through palm oil and soy and beef and rubber commodities among others. Habitat fragmentation happens alongside indigenous species depletion when extensive land transformations occur owing to greenhouse gas emission increases.

5 Future Directions and Emerging Trends

National forestry management developments come from artificial intelligence (AI) and big data analytics in addition to new bioengineering results and genomic research and international partnerships which were made possible by international forestry agreements. The new technology delivered two advantages for forestry operations which improve practice efficiency and provide vital support to climate protection and ecosystem revival as well as responsible economic growth.

Figure 5 shows a left-to-right technology timeline that layers remote-sensing inputs (cubesats, UAV-LiDAR, IoT micro-sensors) onto an AI/ML analytics core (computer

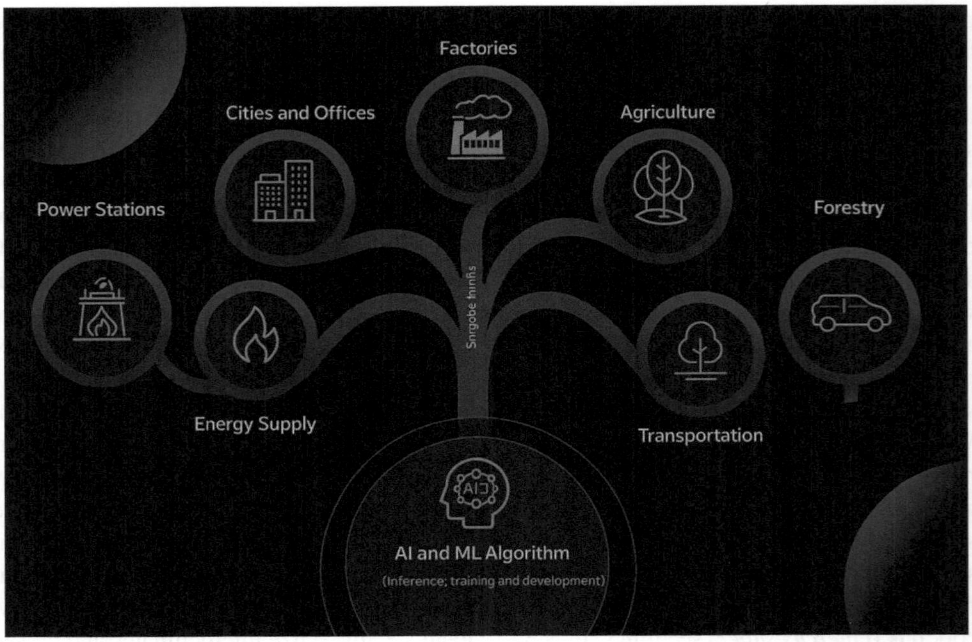

Fig. 5 Future trends using AI and ML in Sustainable Forestry (Sridhar & Basu, 2023)

vision for canopy health, predictive fire-risk modeling, blockchain-verified supply chains). Downstream arrows point to decision-support dashboards, certification smart contracts, and near-real-time community alerts—highlighting how feedback latency drops from months to < 48 h when hyperspectral cubesats and portable VOC sensors are deployed.

The implementation of sophisticated computational solutions helps forestry personnel increase the accuracy and operational effectiveness of their solutions for sustainability-focused adaptive-management programs. Artificial intelligence features of image processing software inspect large datasets by using automation to detect deforestation events as well as pest outbreaks and habitat fragmentation. Next-generation hyperspectral cubesats and portable VOC sensors are already shortening the feedback loop to < 48 h (Torresan et al., 2021). Forest management experts obtain immediate ecological change assessments through joint operations between UAV-based analytical programs and earth-based sensor systems. By running machine learning models, command systems develop predictive tools which measure forest development together with carbon sequestration capacity and fire danger potential. Yet ethical considerations must guide bioengineering. Transgenes and edited alleles may cross into wild populations, raising concerns about non-target ecological effects and loss of local genotypes. Social license for GM tree deployment remains low in many regions, necessitating transparent stakeholder consultations and robust impact-assessment frameworks. Regulatory harmonization—balancing innovation with precaution—will be key to ethically deploying next-generation forestry biotechnologies.

AI predictive analytics supports forestry experts by providing the detection of potential issues to implement protective actions that build forest health alongside reduced environment hazards. The development of biotechnological research generates exceptional means for battling both human-induced pressure. Ordinary forest bioengineering uses scientific investigations to create trees that grow faster while gaining disease immunity and absorbing higher volumes of carbon dioxide. Modern genetic modification of trees is achievable through clustered regularly interspaced short palindromic repeats (CRISPR) technology which enables better environmental resistance in modified trees. Early genomic-selection trials in *Pinus taeda* already cut breeding-cycle time in half while preserving allelic diversity (Zhang & Lee, 2024). The development of drought-resistant trees with pest-resistance and speedy growth capabilities by experts using CRISPR technology would enhance both reforestation and carbon storage operations. Genomic selection as an approach helps identify beneficial genetic characteristics within tree gene pools that then can be multiplied to boost sustainability together with adaptation abilities. However, data quality and accessibility pose challenges. Many remote-sensing archives are fragmented across vendors and lack standardized metadata, complicating large-scale model training. Ground-truth networks remain sparse in the Global South—fewer than 1 permanent plot per 10 000 km^2—leading to biases in AI-predicted biomass estimates. Open-data initiatives like FAO's Open Foris and NASA's GEDI are helping, but further investment in interoperable databases and community-driven sensor deployments is critical. Scientists use genomic technologies to study various trees for identifying DNA markers associated with drought adaptive deep roots as well as fast growing capabilities and fungal resistance. Forest breeding uses discovered genetic information in a selective tree breeding approach that develops powerful commercial operations and sustainable forest management systems. About three national borders frame forests so they require joint governance with international agreements to sustain their continued existence. Multiple global multi-stakeholder initiatives and different international frameworks act synergistically to create vital support for forest conservation efforts and financing systems along with equality measures for distribution.

6 Conclusion

The operational framework of sustainable forestry operates where environmental protection shares equilibrium with economic needs because it must fulfill social objectives and financial expectations. A sustainable forestry system should produce results for ecological preservation and community ecosystem use since communities need access to these ecosystems. Sustainable forest management has evolved to prove that protecting nature can strengthen economic benefits as much as it does the environment. Three methods for managing forests as ecological protectors and economic resources include timber responsibilities together with forest product ventures and service value assessments. Repeated logging methods as well as carbon emission market programs and forest certification systems display the possibility of linking monetary gain to extended forest ecosystem sustainability. The business strategy of entering agroforestry and non-timber forest products brings expanded income streams while decreasing forest dependency while supporting nearby communities. To obtain equilibrium between forest conservation and economic interests, an equitable distribution system needs to manage the benefits accrued from forest use. Financial and social programs should expand their scope to provide benefits to indigenous people as well as small landowners and protected environment defenders apart from commercial operators. Community-managed forests alongside programs compensating for ecological service values allow the development of socially balanced forestry systems that include all members of the community. Projects that combine sustainable governance with biodiversity protection mechanisms make forests into resources for safeguarding nature alongside human development opportunities. A novel forestry model should implement a unified framework which unifies natural system protection alongside cultural value protection through modern technological implementation.

The implementation of integrated policy participation between environmental agencies and forest-based communities and both the government authorities and private sector enterprises enables co-managed knowledge-sharing procedures. Total conservation methods included in this comprehensive approach make decisions through community-driven systems that lead to problem-solving and adaptive management and policy development based on data. Strategic planning for sustainable forestry demands equal distribution of forest resources and their corresponding advantages for such systems to form. The wisdom of Indigenous and local communities should lead all forest management practices because they have long maintained harmonious resource utilization systems. Ecological resource distribution and environment flexibility along with ecosystem protection are important aspects which traditional forest interaction methods transmit through their beneficial knowledge base. Sustainable forest protection depends on both formal legal standards and control of ownership rights since these basic elements protect forests from damage while ensuring their preservation. The main focus of development policies today should be implementing community-led policies for protection of territorial rights. Every forest management system needs social safety measures to give immediate benefits of sustainability plans to marginalized populations. The practice of forestry functions as the bridge which links natural forest ecosystems to modifications made by human beings in those environments. Sustainable forestry develops better results than technical standards through interdependent networks which sustain operations that value mutual respect. Sustainable forestry requires total connectivity between natural forest ecosystems and human cultures since this concept involves simultaneously enhancing forest productivity and reducing environmental destruction rates alongside building strong interdependent links between them. Sustainable development of society and financial stability will result from sustained connections between forest practices and local community progress together with ecological restoration initiatives to regenerate natural landscapes. Sustainable development of our future requires constant learning based on innovative collective work for all social communities. Sustainable forest management operates as an ongoing initiative that transforms because both ecosystem variations and social understanding shift. Flexible operational policies enable fast response capabilities during both emerging situations and new prospects.

References

Bian, S., Xie, Y., & Zhang, F. (2023). Re-imagining the future of forest management: An age-dependent approach towards harvesting. arXiv preprint arXiv:2308.03198. https://arxiv.org/abs/2308.03198

Cooper, L., & MacFarlane, D. (2023). Climate-smart forestry: Promise and risks for forests, society, and climate. *PLOS Climate, 2*(6), Article e0000057. https://doi.org/10.1371/journal.pclm.0000057

Engberg Sundström, L., &Shavazipour, B. (2024). Decision support for sustainable forest harvest planning using multi-scenario multi-objective robust optimization. arXiv preprint arXiv:2405.16612. https://arxiv.org/abs/2405.16612

Gunasekar, D. R., Devi, K. B., Singh, A., Selvam, P., & Chandra, A. (2024). An integrative review for the role of forests in combating climate change and promoting sustainable development. *International Journal of Environment and Climate.*

Hawks, B. S., Aust, M. W., Bolding, C. M., Barrett, S. M., & Schilling, E. B. (2022). Increased levels of forestry best management practices reduce sediment delivery from clearcut harvests. *Forest Ecology and Management, 507*, Article 120015. https://doi.org/10.1016/j.foreco.2022.120015

Huang, L. F., & Cao, Y. K. (2022). Review of the integrated development of ecological and cultural forestry. *Sustainability, 14*(11), 6818. https://doi.org/10.3390/su14116818

Holzinger, A., et al. (2022). Digital transformation in smart farm and forest operations needs human-centered AI: Challenges and future directions. *Sensors, 22*(8), 3043. https://doi.org/10.3390/s22083043

Kobal Grum, D., & Bončina, A. (2024). Integrating basic human values with forest ecosystem services: Pathways to sustainable forest management. *Frontiers in Psychology, 15*, 1312251. https://doi.org/10.3389/fpsyg.2024.1312251

Kotlarz, J. (2024). Application of Nash equilibrium for developing an optimal forest harvesting strategy in Toruń Forest District. arXiv preprint arXiv:2403.03555. https://arxiv.org/abs/2403.03555

Müller, J. M., Veile, J. W., & Voigt, K. I. (2020). Prerequisites and incentives for digital information sharing in Industry 4.0. *Computers & Industrial Engineering, 148*, 106733. https://doi.org/10.1016/j.cie.2020.106733

Parasnis, R., & Amin, S. (2024). Optimal interventions in coupled-activity network games: Application to sustainable forestry. arXiv preprint arXiv:2404.13662. https://arxiv.org/abs/2404.13662

Patel, R. H., Mukherjee, S., Sahu, B., Dash, B. P., Jaison, M., Avinash, K., & Singh, P. (2024). Sustainable forest management for carbon footprint and climate change mitigation. *Springer Journal of Sustainable Development and Biodiversity.*

Rajkumar, V., & Sharma, R. (2023). Enhancing biodiversity through agroforestry in climate-vulnerable zones. *Journal of Ecological Science, 18*(3), 234–249.

Saikia, R., & Baruah, A. (2023). Monitoring forest health using AI-powered geospatial analytics. *International Journal of Forest Technology, 12*(1), 45–58.

Sankaran, M., & Nair, P. K. R. (2022). Agroforestry for climate resilience and biodiversity restoration. *Agroforestry Systems, 96*, 75–90. https://doi.org/10.1007/s10457-021-00679-0

Sharma, S., & Kumar, A. (2023). Indigenous knowledge in sustainable forestry: A case study from Northeast India. *Journal of Environmental Management, 324*, Article 116363. https://doi.org/10.1016/j.jenvman.2022.116363

Singh, V., & Tiwari, R. (2023). CRISPR and climate-adapted trees: Emerging strategies in forest bioengineering. *Biotechnology Advances, 61*, Article 108094. https://doi.org/10.1016/j.biotechadv.2023.108094

Sridhar, P., & Basu, R. (2023). Blockchain applications in forest certification and anti-logging initiatives. *Forest Policy and Economics, 145*, Article 102894. https://doi.org/10.1016/j.forpol.2022.102894

Thakur, A., & Mohapatra, D. (2022). Evaluating forest landscape restoration strategies under the Bonn Challenge. *Land Degradation & Development, 33*(14), 2435–2448. https://doi.org/10.1002/ldr.4415

Thompson, I. D., & Finegan, B. (2023). Biodiversity and sustainable forestry in a changing climate. *Global Ecology and Conservation, 43*, Article e02453. https://doi.org/10.1016/j.gecco.2023.e02453

Torresan, C., et al. (2021). A new generation of sensors and monitoring tools to support climate-smart forestry practices. *Canadian Journal of Forest Research, 51*(12), 1751–1765. https://doi.org/10.1139/cjfr-2020-0295

Venkatesh, R., & Adhikari, P. (2023). Payment for ecosystem services: Implications for sustainable forest governance. *Ecological Economics, 206*, Article 107680. https://doi.org/10.1016/j.ecolecon.2023.107680

Wang, H., & Li, Y. (2024). Forest carbon sequestration potential and policy mechanisms in Southeast Asia. *Forest Policy and Economics, 148*, Article 102942. https://doi.org/10.1016/j.forpol.2024.102942

Xu, T., & Guo, L. (2023). Evaluating silvopastoral systems for carbon mitigation and soil restoration. *Environmental Research Letters, 18*(11), Article 115006. https://doi.org/10.1088/1748-9326/ad01b4

Zhang, J., & Lee, Y. J. (2024). Genomic selection for climate-resilient forestry. *Tree Genetics & Genomes, 20*(2), 42. https://doi.org/10.1007/s11295-024-01632-1

Smart Forestry: Leveraging Technology for Sustainable Resource Management

Ramaraju Nagarjuna Kumar, Satti Sudha Mohan Reddy, K. Yogeswara Rao, Isukupalli Kranthi Kumar, and Suresh Kumar Kanaparthi

Abstract

Industrial organizations use advanced technologies to build sustainable forestry practices that operate efficiently and show resilience. The foundation of smart forestry consists of precision resource management that uses technological tools to carry out accurate forestry operation assessment and monitoring and implementation activities. The role of sustainable timber harvesting is a core focus, with strategies like selective logging and reduced-impact practices being enhanced through technological interventions. Smart forestry's socioeconomic implications are discussed in depth, emphasizing the importance of involving local communities and stakeholders in forestry projects. The integration of technology not only improves efficiency but also reduces costs, allowing for the equitable distribution of benefits among all parties. Examples include mobile applications that connect smallholder forest managers to markets, certification programs for sustainably sourced timber, and training programs that empower local populations with new skills in technology-driven forestry. The future scope of smart forestry is vast, with emerging technologies promising to revolutionize the sector further. Supportive policies need to be implemented to promote sustainable technology investment as well as cross-border environmental challenge coordination programs. The experts advocate using smart forestry principles in international agreements and sustainability standards as a means to fulfill United Nations Sustainable Development Goals (SDGs).

Keywords

Sustainable resource management · Precision forestry · Remote sensing · Geographic information systems (GIS) · Internet of things (IoT) · Drone-based monitoring · Biodiversity conservation

1 Introduction

The necessary services for ecological balance introduction converge in three aspects that forests provide: carbon storage, aquatic system protection, and water resource preservation. The natural ecosystem serves as society's foundation for economic advancement since it allows enterprises to develop their activities like timber extraction as well as paper manufacturing along with pharmaceutical and eco-tourism industries. Rapid deforestation and unsustainable logging and climate change effects are causing harmful effects to habitats that result in biodiversity collapse, habitat destruction, and rising greenhouse gas releases (Ahmad et al., 2023). The techniques used previously in forestry management cannot effectively solve current environmental as well as financial problems. The traditional forest governance system uses personnel evaluations and periodic surveying efforts, but their time-consuming manual measurement methods render correct assessment very challenging. The use of efficient

R. N. Kumar (✉)
Computer Applications in Agriculture, ICAR-Central Research Institute for Dryland Agriculture (ICAR-CRIDA), Hyderabad, India
e-mail: nagarjunakumar191@gmail.com

S. S. M. Reddy
Department of ECE, SRKR Engineering College, Bhimavaram, Andhra Pradesh, India
e-mail: mohanreddysatti@srkrec.ac.in

K. Y. Rao
Department of Computer Science and Engineering, GITAM School of Technology, GITAM (Deemed to Be University), Hyderabad, India
e-mail: ykalla@gitam.edu

I. K. Kumar
Department of EEE, SRKR Engineering College, Bhimavaram, India
e-mail: kranthi.isukapalli@gmail.com

S. K. Kanaparthi
Department of Computer Science and Engineering, SRM University, Amaravati, Andhra Pradesh, India
e-mail: sureshkonline@gmail.com

M. Ashwin et al. (eds.), *Smart Forestry 5.0 for Powered AI Sustainable Environment and Climate*, Sustainable Artificial Intelligence-Powered Applications, https://doi.org/10.1007/978-3-031-99086-1_3

forest technology helps managers drive evolutionary change to establish new implementation methods. Forests achieve operational effectiveness for sustainability through modern remote sensing technologies that operate with GIS systems along with IoT devices combined with AI systems processing big data. Three main monitoring systems that include high-resolution satellite imagery united with LiDAR technology and Unmanned Aerial Vehicles (UAVs) are currently used for forest canopy density observation besides illegal logging detection and biodiversity change recognition. Multispectral cameras mounted on drones allow operators to spot early indicators of pest problems and canopy tension through their equipment, and thermal sensors can lead to wildfire warning signs detection. Machine learning models analyze the gathered data through these sources for forest health prediction and the generation of intervention recommendations. The combination between multispectral and hyperspectral imaging techniques improves forest disease and pest detection on large scales according to recent research findings (Li & Zhang, 2023). The analysis of contemporary forest data leads professionals to produce better decisions by using predictive modeling approaches. Strategic method produces efficient outputs which boost both environmental planning performance and operations of environmental resources. The fundamental goal of smart forestry involves managing financial production alongside environmental security. Sustainable forest activities combined with the best possible forestry and workforce efficiency in tree planting operations alongside adaptable forest management protocols give protection to the environment. Forest drones deployed by Isaiah employ vegetation planning methods together with land reclamation program procedures and AI monitoring systems to identify forest diseases before major damage becomes widespread (Beniwal et al., 2023). This work evaluates smart forestry standards that encompass tool assessment methods and benefit evaluations and future implementation potential of these tools. This text demonstrates how modern technology safeguards forests through two key implementations: it halts resource distribution while it counters climate change problems. Fire regime management continues to produce new practical applications that researchers discover today making it a proven approach to maintain forests for future generations.

2 Overview of Smart Forestry

The forest now receives its machinery supply through modern smart forestry systems. All living organisms in the forest—trees, streams, and wildlife—work as instant-messaging social networks which deliver continuous feedback about their status and necessities. The system incorporates contemporary instruments to enhance forest information which drives forest expansion. Data analysis provides forest management teams with exact guidance about priority areas through drone and sensor systems because these tools reveal both dangerous fire zones and dry sections and invasive species territories. The technology creates detailed environmental reports for every portion of the forest which enables managers to establish effective forest care decisions (Chen et al., 2022). The implementation of smart forestry brings various interconnected features to advance forest observation and administration. These include the following.

2.1 Remote Sensing and Satellite Imaging

Organizations using smart forestry methods heavily rely on remote sensing alongside satellite imaging for obtaining exact real-time detailed data about extensive forest areas. Large-scale monitoring of forested areas, biodiversity analysis, identification of degradation levels, and climate change investigation becomes possible because of these technologies. This system serves as an instrument for high-resolution scanning of forest covers and identifies vegetation health alongside forest destruction patterns. Light Detection and Ranging (LiDAR) functions as a Remote Sensing technology tool which provides accurate 3D forest maps suitable for biomass assessment together with topographic applications and many additional objectives (Singh & Verma, 2023). The system applies temperature detection through thermal imaging to monitor forest thermal variations which helps in detecting both wildfires in their early stages and identifies plant stress indicators. The monitoring of forest cover together with deforestation assessment becomes possible through satellite-based sensors which detect illegal logging operations and track forest changes from different time points. The best technique to fight climate change worldwide involves tracking forests with satellites for carbon sequestration assessments. Remote sensing technology in plant change analysis enables experts to secure natural areas from destruction while supporting the safety of wildlife habitats. The system enables disaster teams to monitor natural disaster development and wildfires to enable emergency response plan activation (Patel & Sengupta, 2023).

2.2 Integration with Geographic Information Systems (GIS)

Through modern high-tech magnifier, GIS technology managers obtain an absolute understanding of expanded forest territories despite their extended dimensions by viewing aerial imagery. This technology maintains easy-to-read detailed mapping information by transmuting miles of remote data

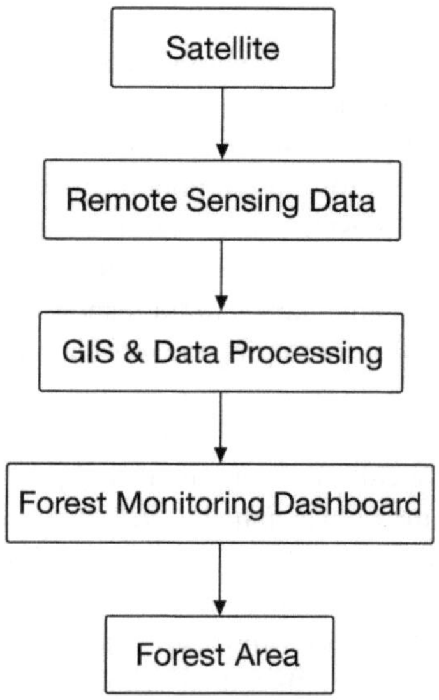

Fig. 1 Satellite-based forest monitoring system

broadcasts that originate from satellites along with high altitude drones (Chen et al., 2022).

The sequence of satellite-based forest monitoring operations appears in Fig. 1 which demonstrates the information progression starting from space observations until it reaches decisions for forest management applications. The system commences using satellite platforms to collect high-definition imagery together with environmental readings that cover extensive forested areas. Remote sensing data becomes usable by sending it to processing centers that perform Geographic Information System (GIS) functionality and execute advanced analytical methods. Data refinement processes at this stage are combined with spatial layer enforcement to detect forest vegetation variations as well as landscape transformation signals which lead to the identification of deforestation hotspots and forest health measurement elements. Data visualization occurs through Forest Monitoring Dashboard software which delivers viewing maps, graphical outputs, and predictive warnings to decision-makers. The dashboard allows real-time tracking while helping users conduct data-based conservation actions and planning and surveillance procedures. The developed insights guide crucial interventions for forest areas by offering instructions for reforestation planning alongside pest control, fire prevention, and ecosystem restoration. A sustained round of data sharing between satellites and analytics and direct implementation on forest areas supports a proactive approach to sustainable forest management. Through this technology, people can view comprehensive forest information in a manner similar to how floor plans are examined. Through this technology, managers make better planting choices for new trees, selection of tree planting locations, and protection decisions regarding endangered zones. The forest health planning requires the same level of precision as playing a detailed video game that requires exact calculations to drive forest development. The system reaches unparalleled precision through AI because its predictions concerning forest changes and needs prove highly accurate. The forests are now in the hands of forestry specialists who have all necessary tools to sustain and develop fragile ecosystems (Deng et al., 2022). The true strength of smart forestry lies not just in individual technologies but in their integration into cohesive systems. The combination of IoT sensors that measure environmental factors generates data which GIS platforms transform into geographic-tagged visual displays. Real-time geospatial information is processed by AI algorithms at the same time to search for signs of wildfires and diseases within the forest area. By placing drone images and satellite data atop GIS maps operators achieve two purposes: first they achieve confirmation of IoT warnings and second they create a ranking system for choosing critical areas to intervene. Forest managers obtain improved decision-making through the interconnectivity of satellite imagery, geospatial information systems, Internet of Things devices, and artificial intelligence analytic platforms to receive verified complete data instead of isolated data sources.

Table 1 sums up essential technologies utilized in contemporary forest monitoring systems through functional descriptions and application scenarios. High-resolution images captured through satellite technology let forestry authorities properly monitor forest clearing as well as track illegal hardwood removal activities. Performance monitoring of tree health relies on multispectral and hyperspectral imaging to detect physiological stresses as well as diseases in their early development stages. Through particular light wavelength analysis technology, we can identify pre-visual changes in vegetation health that traditional human observation cannot detect. The technological devices we mentioned work together to create smart forestry systems which permit immediate environmental surveillance together with ecological prediction and advanced forest leadership choices. Forest health monitoring systems gain a better capacity for quick threat response and sustainable planning through their adoption of these integrated technologies. Finnish boreal forests benefit from speedy forestry management that integrates LiDAR systems on drones and IoT sensors for the forest authority to assess forest conditions through data collection. The systemic operational information in GIS technology offers real-time management solutions which support forest planning and reforestation development and pest management operations. The intelligent forest management system enables increased forest yield by 20% and accelerates detection of possible diseases through fast monitoring.

Table 1 Forest monitoring technologies

Technology	Function	Application
Satellite imaging	High-resolution images of forests	Deforestation monitoring, illegal logging detection
Light detection and ranging (LiDAR)	3D mapping of forest structure	Canopy density assessment, biomass estimation
Multispectral and hyperspectral Imaging	Detection of tree stress and diseases	Early intervention in pest infestations

2.3 Geographic Information Systems (GIS)

Smart forestry implementation depends heavily on GIS as a core technology that helps forest managers deal with spatial data to build modern forest administration systems. The conservation of ecosystems and land-use planning and sustainable resource management benefit from different geospatial data sources that GIS systems utilize to simplify complex forest ecosystems. Through GIS technology, forest supervisors create extensive mapping systems that divide forests into areas with different vegetation densities and structures together with various land-use types. The zoning functions allow geographically precise representation of forest protection zones along with timber production zones and territories that protect forest ecosystems. The analysis of multiple satellite images through GIS allows the detection of environmental changes together with trends regarding deforestation and patterns of land degradation. The automated system provides live alerts through GIS to authorities who can use this information system to enforce policies promptly. GIS analysis of historical forest fire information along with vegetation moisture levels helps determine risk areas before potential wildfires. The system enables rapid response teams to watch active fire zones for developing the most appropriate containment strategies. GIS technology enables improvement of water availability assessment for sustainable forestry by identifying hydrologically relevant areas and performing hydrological modeling of watersheds. GIS functions enable the creation of soil erosion risk assessments that help prevent current threats against soil health by deploying suitable measures. Forest monitoring results achieve high accuracy levels by applying satellite data with ground observations within GIS technology. Smart forest monitoring operates in forestry domains with the help of sensors attached to IoT devices to collect ongoing environmental data for GIS system analysis. GIS data processing through machine learning and artificial intelligence enables predictions about deforestation patterns from assessing climate change damage thus leading to proposed forest plantation plans.

Figure 2 presents the GIS-based forest management system design which enables sustainable forestry operations by harmonizing spatial data assessments with decision support

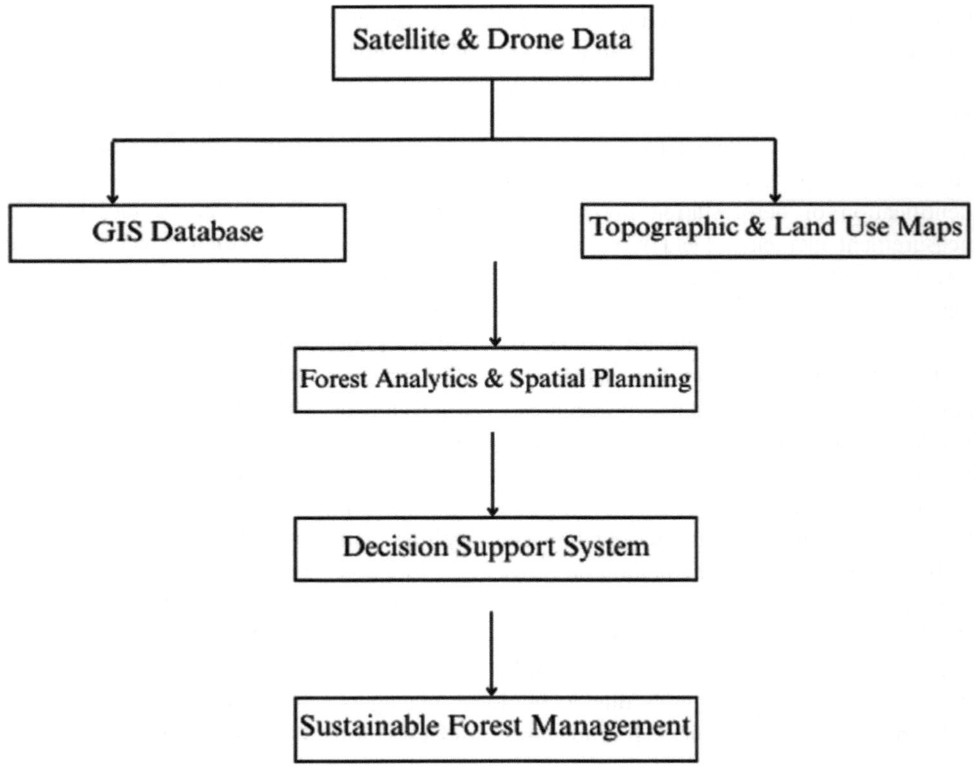

Fig. 2 GIS-based forest management system

tools. The system starts by collecting satellite along with drone data to visualize extensive forests areas inaccessible for field exploration. The system diverts data into two simultaneous components: a GIS database which maintains georeferenced data layers while topographic and land-use maps establish definitions for physical terrain, land classifications, and vegetation zones. All gathered inputs get processed at the Forest Analytics and Spatial Planning module through advanced spatial analysis for forest health indicator detection, time-based assessment, and land-use scenario modeling. The Decision Support System (DSS) depends on analytical results from processing geospatial data with ecological parameters combined with social parameters and economic parameters to make recommendations for targeted interventions such as conservation zoning and reforestation planning together with treatments for fire-risk mitigation and logging schedules (Jain & Srivastava, 2023).

2.4 Internet of Things (IoT) in Forestry

Internet of Things (IoT) uses real-time data collection automation for remote observation through decision systems (Deng et al., 2022). Forest managers deploy IoT sensors throughout ecosystems to better protect the environment while reaching higher operational outputs by improving natural resource management. The network of IoT environmental sensors tracks temperature measurements, soil moisture, environmental conditions, and humidity levels. The gathered data from sensors becomes available for future drought prediction by inspecting fire hazard threats and tree capacity to adapt to environmental changes. The IoT provides us with a system to connect multiple sensors and cameras which enable the tracking of wildlife movements and biodiversity population counts at certain locations. The system operates to monitor natural habitats automatically which tracks animal behaviors while recording species movements between biological terrains. IoT sensors allow for tied-up monitoring of wildlife through camera and sensor systems at various locations to automatically monitor animal behavior while keeping count of local populations. This system assists natural processes through unattended environmental animal observation across migration areas. The detection system instantly notifies staff members to initiate rushed environmental response procedures. The system includes automated resource controls which reduces operational costs while requiring fewer personnel members. The system allows managers to utilize data for better resource decision-making which leads to enhanced management results. Natural forests receive earlier protection through Emergency responders who can reach the forest faster because of early alerts about potential threats during fire and disaster situation. IoT relationships with environmental data produce copious sensor outputs that AI requires for processing environmental threats together with suitable solutions. Blockchain and IoT applications provide secure instrumentality to check IoT data entry into transactions, while these systems also enhance timber supply chain visibility together with sustainability verification (Smith & Green, 2023).

Figure 3 demonstrates an IoT-enabled forest data collection network which allows real-time forest monitoring through sensor technologies that link together. Multiple environmental sensors such as moisture systems and climate devices with wildlife cameras are put into operation across forest spaces. Environmental sensors operate at all times to record a wide range of ecological variables like soil hydration metrics as well as temperature readings, humidity data, rain totals, and animal conduct patterns. An edge computing device processes collected data at its location using basic filters and analysis methods in order to decrease latency and bandwidth needs (Zhao et al., 2023). The system exclusively transfers data that contains both relevance and high priority to reach the following processing phase. A centralized database accepts refined information for storage combined with management functions which enable users to access it for additional examination. Forest monitoring dashboard exists as the system's final component because it functions as a graphical interface which provides conservationists and forest managers real-time interpretation capabilities. Physical indicators, alerts, and spatial map features in this dashboard facilitate rapid decision-making about fire risks as well as poaching situations, biodiversity threats, and ecosystem variations. Forest health management systems benefit from IoT and edge computing integration through improved rapidness and are precise while sustaining their operational capability. The setup of IoT infrastructure with sensors and networks requires equal high initial financial investment. The available monitoring technologies face difficulties when working with networks requiring Internet connection because of huge distances combined with large continued forests. The security of IoT-generated data represents a vulnerability that needs well-designed protection measures to address it. Successful operation and interpretation of IoT systems demand specialized training that forestry personnel must obtain.

Table 2 describes the essential Internet of Things (IoT) devices employed for forest monitoring with their particular objectives and sustainable forest management advantages. Soil moisture sensors read the level of hydration in forest soils to create concrete water irrigation strategies and forestry reforestation strategies. These monitoring sensors give essential information which enables better water resource management especially in dry regions and ecological rehabilitation sites.

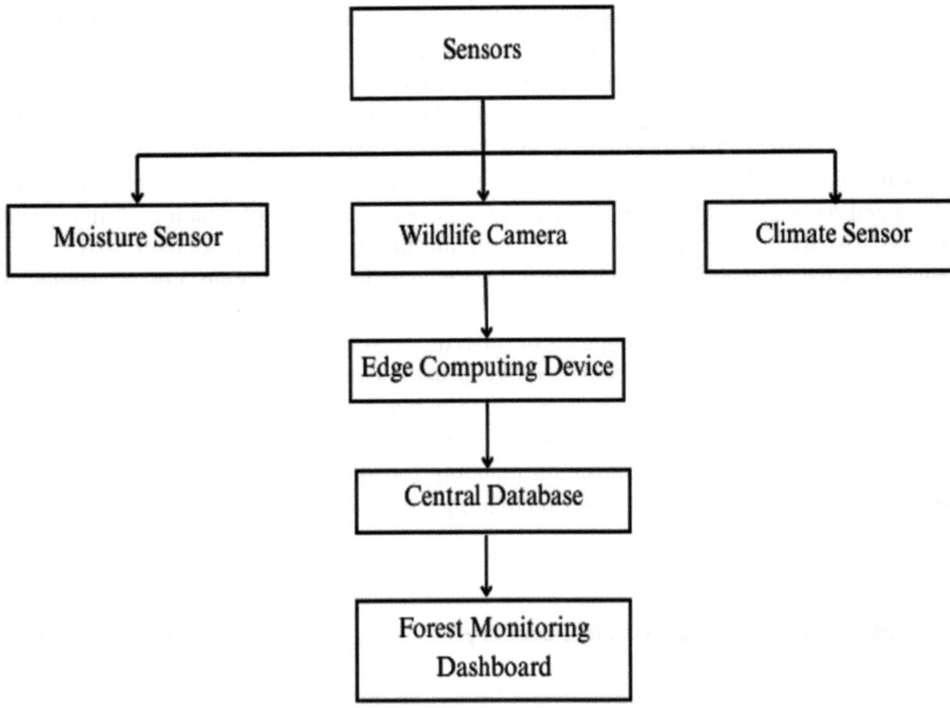

Fig. 3 IoT-enabled forest data collection network

Table 2 IoT devices for forest monitoring

IoT device	Purpose	Benefits
Soil moisture sensors	Monitor soil hydration levels	Optimize irrigation and reforestation strategies
Climate sensors	Track temperature, humidity, and rainfall	Assess climate change impact on forests
Acoustic sensors	Detect illegal logging sounds	Improve forest security measures

The data collected by Climate Sensors depends on temperature, humidity, and rainfall measurements in addition to other key environmental factors. Climate change assessment data from forests serves essential purposes for ecosystem management teams who can predict future dangers affecting biodiversity, tree health, and fire hazards.

The security functions of acoustic sensors rely on their ability to recognize non-standard sound signals that indicate forest intrusions such as chainsaws and moving vehicles which are linked to illegal logging operations. The devices help forest management teams conduct better surveillance through early detection of illegal forest operations which leads authorities to react quicker and secure forest resources more effectively. The Internet of Things devices function as a unified sensing system that improves real-time decisions, resource management, and forest ecological stability in modern forest management operations.

2.5 Artificial Intelligence and Machine Learning

The implementation of AI occurs throughout the stages of forest management activities. AI operates during planning through algorithmic analysis of historical multisource data to find optimal plantation spots as well as select species and predict ecosystem reaction patterns. Airborne technology along with satellite imagery is processed by AI systems during monitoring activities to spot tree health problems, unlawful timber harvesting, and forest diseases. During the intervention stage, AI decision systems offer precise suggestions for actions that include drone deployment at fire-prone areas and changes in irrigation methods. Parallel with data analysis, AI functions as an adaptive key resource that supports operations in forestry. Forestry management in contemporary times achieves its transformation by implementing artificial intelligence technologies which automate data systems while developing advanced operational frameworks and better procedures. Computing technology now utilizes extensive forestry databases with increased effectiveness for operations monitoring and risk assessment forecasting (Dubey et al., 2022). SDLC and operational efficiency grow together with AI and ML integration, which reduces management errors. The monitoring operation for forest health utilizes AI-based image analytics which uses drones and satellite imagery to detect unhealthy tree conditions together with pest infestations along with disease spread patterns throughout the forest. Briefing organizations about disease spread patterns through predictive machine learning models allows them to establish intervention plans that minimize widespread destruction. Artificial intelligence uses remote sensing technology to inspect past space imagery and current satellite images in order to detect illegal deforestation and unauthorized timber extraction. The automated notification system helps law enforcement agencies conduct swift responses to illegal deforestation events while they occur. Machine learning tools create forecasts about forest carbon storage potentials which aid government organizations to develop superior climate mitigation strategies. Machine learning system usage allows operators to analyze forest reaction to climate changes, and then suggests adjustments for forest management. AI operated through Remote Sensing provides enhanced accuracy when analyzing satellite and aerial imagery to reveal better vegetation alterations. The combination of smart sensors employing AI functionalities analyzes environmental data streams for detecting system issues and generates AI system-based risk predictions. Machine learning provides blockchain timber tracking systems with dual functionality for both data consistency verification and content authentication, which validates both data integrity and authenticity. The following is the upcoming challenge among several issues that emerge when AI solutions are implemented in agricultural and industrial sectors. Since experts lacked adequate training, they built numerous programs which exhibit limited understanding of new emerging technical tools leading professionals to remain incapable of running these programs.

2.6 Drones and Autonomous Vehicles

The entire forest management workflow depends heavily on drones for its operations. The preparation stage requires drones to employ LiDAR mapping for dimensional and biomass data collection. The operational phase of forest activities benefits from drone technology which distributes seeds while checking for pests and performing video monitoring of distant areas. The evaluation of damage caused by storms and fires together with burned area mapping serves to help plan reforestation activities. The combination of their quick movement and high data quality with widespread coverage minimizes human danger in forest care activities while improving decisions throughout the entire management process. Reliable forestry management results from drone and self-driving vehicle innovation to create improved forest management approaches with efficient methods and specific measured outcomes (FAO, 2023). Real-time monitoring together with fast data acquisition functions assist precision forestry methods by reducing human exposure to dangerous areas. Mobile seed distribution drones make reforestation projects faster to accomplish through automated operations despite needing less human workforce than manual tree planting activities do. AI mapping software through drones performs automated scans of environment factors that allow them to discover suitable planting locations based on soil quality data and water condition data and terrain observation data. Daily seed planting activities by reforestation drones reach thousands of seed amounts which decreases labor needs and expenses for more extensive operations. Drone elevation enables observers to perform forest pest inspections in combination with tree health assessments by surveying forest canopies. Three-dimensional mapping of forests occurs with LiDAR technology which drones operate. The crucial mapping systems facilitate biomass estimation and carbon storage research, which serves as a foundation for forest preservation used by environmental systems. Drone camera real-time videos flow through AI systems for fire spread prediction followed by automated fire protection method generation. Self-operating drone units function within high-risk fire zones to track environmental factors that send instant data to forest officials. The monitoring of undeveloped forest territories for illegal logging activities utilizes modern drone systems which integrate night vision and infrared cameras. The proper prediction of species movement patterns using drone-based inventory collections helps track wildlife populations and prevents human–wildlife conflicts. AI image recognition systems react immediately to pictures by performing instant image recognition.

A drone-based forest monitoring system operates as shown in Fig. 4 while using advanced aerial technologies for complete forest health assessment. A deployment of drones carries different sensors together with imaging systems including multispectral imaging, Light Detection and Ranging (LiDAR) scanning, and thermal imaging for the initial phase.

Advanced algorithms in the data processing and analysis module receive data streams from different sources to produce valuable insights through interpretation functions. Engineered information from the processing stage becomes critical to forest health evaluations that help track pests as well as monitor illegal logging events and observe drought effects and damaged areas. Results from processed information guide managers toward data-based decisions which lead to specific reforestation strategies along with immediate fire responses, biodiversity protection, and ecosystem benefit optimization. Drone technologies linked to intelligent analytics provide a fast-reacting and expandable forest surveillance method particularly for distant locations (Nguyen et al., 2023).

2.7 Blockchain for Supply Chain Transparency

The blockchain technology maintains a decentralized system of transaction records that provides unalterable verification

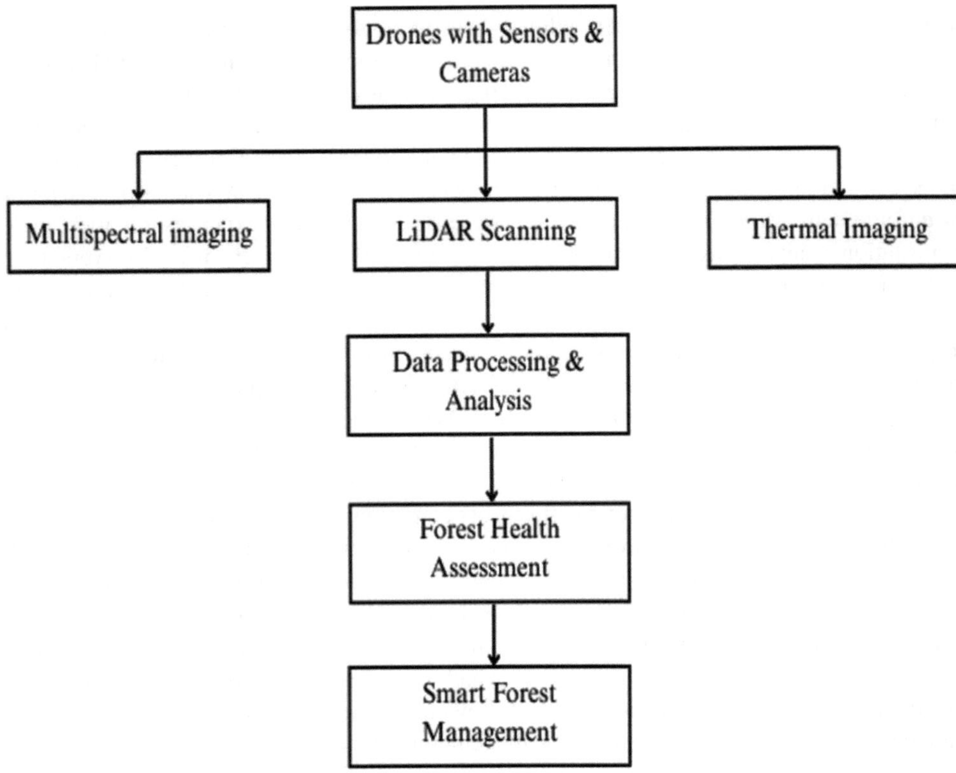

Fig. 4 Drone-based forestry monitoring system

thus transforming supply chain transparency for forestry businesses. The secure information is not easily accessible when burglary occurs or when other security threats affect timber supply from harvesting to product delivery (Ghosh & Roy, 2023). Stakeholders who use blockchain technology in smart forestry operations gain better accountability combined with protection against unlawful logging activities while creating trust with consumers for wood products from sustainable sources. Each action receives precise timing measurement starting from when products are logged and extending through their journey to retail stores. The blockchain majority status protects your confidential data from alterations which defends against data tampering and fraud attempts. The blockchain platform uses smart contracts to execute procedure validation during official document generation. Blockchain systems maintain unalterable records about timber sources and transfers, which decreases unregulated timber harvesting incidents (Mugo & Ndiritu, 2023). A unique digital code present on each wood batch allows customers and business clients to view information about the timber origin and processing details. Organizations allow customers and business clients to verify wood supply origins and management practices through device scans of product labels. The observance of forest protection regulations at active timber sites is confirmed by monitoring organizations which include state bodies. Illegal timber cannot infiltrate supply chains because the blockchain system uses track-and-trace procedures as its protection measure. The blockchain transaction system will authorize transactions of timber items which conform to FSC or PEFC certification mandates. The blockchain system performs live audit functions and sends automated warning notifications to stop unauthorized logging activities which helps stop ecological crimes while enhancing regulatory compliance. Paper-based verification systems together with certification procedures both require too much time and frequently produce mistakes (Gupta & Khatri, 2023). Through blockchain automation, companies minimize paperwork along with administrative overheads and decrease the time needed for processing. Smart contracts immediately verify timber supplier-transactions with buyers thereby eliminating intermediary costs and cutting down on expenses. The system shows precise logging records through automatic data transmission from IoT sensors to the blockchain platform. The system uses AI to analyze blockchain data in order to stop fraud while improving supply chain efficiency through business operations. The system enables blockchain monitoring of timber removal procedures through real-time drone GPS assistance whenever events occur. The implementation of blockchain infrastructure demands sizeable technological investments as well as necessary staff training to build secure data systems. Remote forest areas with poor Internet connectivity create several problems for both blockchain system data transmission and information maintenance efforts (Jain & Srivastava, 2023). The gap exists because traditional forestry organizations maintain their opposition to blockchain adoption primarily due to their unfamiliarity with digital transactions and information openness fears. Implementation of blockchain technology for forestry supply chain transparency faces various real-world problems that prevent its successful deployment. The integration of real-time blockchain data faces limited possibilities because remote areas experience both slow Internet connections and expensive setup expenses. Small-scale timber industry operators encounter technical restrictions which stop them from using these systems because their limited financial resources also get in the way. The international timber industry faces operational challenges from non-standardized blockchain systems that different nations maintain. The blockchain system proves its practical value because Cambodia's Ministry of Agriculture partners with international organizations for tracking community-managed forestry through its blockchain systems.

3 Advantages of Smart Forestry

Members of the forestry sector use modern management trends to improve processing methods and generate dual outcomes which conserve forests alongside expanded resource yields (Jain & Srivastava, 2023). Modern digital technology merged with traditional forestry methods through smart forestry practices enables forests to be sustainable while the product output increases. Traceability assessments occur immediately in forest monitoring through the combination of remote sensing systems and IoT sensors together with artificial intelligence tracking protocols. The systems help environmental authorities find and stop crimes while they occur because of their surveillance capabilities. The exact implementation of precise forestry techniques produces ideal tree planting practices that elevate the results of carbon sequestration systems. AI-based climate models assist predictions of woodland responses over changing environmental conditions to help prevent emission build-up. The combination between IoT tracking technology and AI functions enables timber harvesting efficiency improvement by reducing resource losses and maximizing resource utilization. The supply chain management system using blockchain technology enables organizations to track sustainably obtained timber through clear market pathways (Haque & Ali, 2023). Sustainability standard maintenance together with regulatory compliance helps organizations gain worldwide market success through trusted acceptance across global markets. Mobile applications make forestry data accessible to small-scale farmers which enables their sustainable forest management activities. Locals can participate in community-driven conservation programs through IoT-based monitoring systems because these systems let them report illegal logging and forest encroachments. More effective monitoring of sustainable forest management policies happens with the help of artificial intelligence analytics which provides data-driven insights. Several implementation barriers need solutions to ensure the success of smart forestry in obtaining community backing. Smart forestry encounters obstacles for successful deployment due to technical limitations as well as monetary restrictions, regulatory hurdles, and environmental problems (Haque & Ali, 2023). Multiple obstacles stand in the way of smart forestry implementation which serves as the most essential obstacle in order to execute this concept. Aforementioned technological solutions require substantial funding to acquire remote sensing systems with IoT devices, AI analytical tools, and blockchain systems. Developing country forestry businesses comprising small and medium concerns faces difficulties paying the expenditures needed to acquire, operate, and sustain advanced technology systems. Operations expense such as software licensing and system maintenance fees together with skilled personnel costs prove financially burdensome to forestry organizations. The three essential components for implementing smart forestry include advanced data analytics supported by AI modeling along with GIS-based mapping that requires specialized staff to operate (Hasan & Chowdhury, 2022). Professionals and stakeholders involved in forestry operations fail to utilize smart forestry potentials because they lack enough operational comprehension of these systems. The enhancement of digital competencies needs joint programs between capacity development initiatives and educational institutions to address current gaps in manager-level forest management capabilities. Data collection operations remain limited in underdeveloped forest areas because these regions lack essential power supply networks alongside Internet connectivity. Network failures restrict both IoT device installation with automated systems due to the unreliable nature of information transfer. The solution for connecting IoT devices depends on combining solar power technology with satellite communications infrastructure.

While smart forestry presents transformative opportunities, several challenges must be addressed as in the following table.

Table 3 presents significant obstacles together with proposed solutions for implementing advanced technologies within sustainable forestry practices. The table demonstrates three main problems: high expenses, inadequate specialized knowledge availability, and data protection risks and weak Internet connectivity in distant regions. The adoption of advanced technologies in sustainable forestry requires public–private alliances with addition of encryption features and satellite-based IoT systems as well as skill development initiatives. These operational strategies seek to improve both the deployment of smart forestry systems and resolve the system implementation challenges.

Good execution of smart forestry needs strong policy structures together with well-organized stakeholder involvement

Table 3 Transformative opportunities: several challenges

Challenge	Description	Potential solution
High cost	Advanced technology requires substantial investment	Public–private partnerships, government subsidies
Technical expertise	Use of AI, GIS, and IoT requires specialized training	Capacity-building programs, education initiatives
Data security	Cyber risks associated with digital forest monitoring	Secure data protocols, encryption methods
Connectivity Issues	Remote forests may lack Internet access	Satellite-based IoT networks

to succeed. Authoritative entities need to create supporting policies that facilitate digital platforms and establish standards for data sharing and ethical AI employments within forestry management. The development of solutions specific to individual regions requires platforms that join public agencies with private companies and local communities and research institutions and NGOs. Leadership through public–private coalitions facilitates innovative approaches along with diverse representations of society as an outcome. Community programs that monitor the environment create local ecological guardianship and trust. Scalable and practical solutions emerge in various nations to manage these problems. The Kenyan government launched a solar energy-powered IoT sensor test in forest reserves through partnerships with private technology organizations to handle network and power supply issues. The Forest Service of Brazil initiated mobile-learning GIS courses which trained rural forest workers to improve their technical expertise. The joint Indonesian program between NGOs and timber companies deployed affordable drone systems that received funding from green finance models to support community forest management and solve budget limitations. The implementation of smart forestry methods should receive international organization backing which would create standardized principles for borderless management of forest conservation initiatives.

The integration of modern digital technology with automated systems will transform forest management practices as well as conservation methods and utilization strategies when climate-sensitive methods are used for sustainable forestry operations in the future. Smart solutions hold a critical position for forest protection because sustainable forestry practices show increasing international interest worldwide. The field of smart forestry research generates multiple directions for its progressive development to become more effective. Tree growth, disease outbreaks, and environmental change assessment will be carried out by Artificial Intelligence (AI) systems that analyze large datasets for predictive decision-making in forestry. The integration of AI-based systems for timber harvesting optimization enables researchers to conduct sustainable forest management through real-time forest evaluation. Robotic machines utilizing the algorithms will undertake exact tree planting and pruning techniques and pest management operations automatically to deliver increased output while minimizing exposure risks for human workers in hazardous settings. The implementation of IoT sensors will increase while providing immediate environmental assessments including temperature measurement soil moisture levels, air quality parameters, and tree condition check-ups. Networks using IoT technology enhance disaster preparedness by collecting accurate bushfire data along with pest infection data as well as illegal logging activities.

3.1 Benefits of Smart Forestry

The deployment of smart forestry technology yields numerous advantages when elements of the environment, economy, and society undergo evaluation (He & Lin, 2023). Modern technology applications in forest management result in sustainable resources through coordinated output increases while protecting environmental stability. This paper investigates all key advantages that define smart forestry. Real-time forest monitoring occurs because smart forestry applies IoT-sustained sensors to combine with data collected from satellites. The implemented technologies enable the detection systems to detect illegal activities and environmental changes which triggers immediate protective measures for initiation. AI-based analytics enable the prediction of forest fires, disease outbreaks, and pest infestations so that suitable proactive protection can be implemented. The combination of drone technology and AI systems chooses suitable plant species to start tree development using a technique that optimizes carbon sequestration together with ecological equilibrium maintenance. Protection-based modern forestry methods dismantle forest destruction by creating necessary offsets of climate change which countries must build globally. The IoT system achieves accurate watering and fertilization methods by using soil testing with moisture measure analysis that safeguard forest health. The implementation of mapping technology allows for sustainable forest timber management through erosion-prevention measures and watershed basin protection programs to maintain forest health. Quantitative environmental advantages now appear in reality from the implementation of smart forestry tools. The integration of drone-based monitoring with artificial intelligence planning in Madhya Pradesh, India, led to a 17% betterment of tree canopy growth throughout three years. Smart forensic IoT systems in Swedish soil examinations reduced irrigation needs by 22% thus defending groundwater and improving tree growth areas. The fire monitoring system in California which integrates AI technology allowed firefighters to contain more than thirty percent of expanding wildfires by early detection. The described instances show that utilizing data throughout forestry operations generates immediate benefits that support biodiversity conservation and ecological restoration activities.

4 Case Studies and Applications

Smart forestry applications succeed in genuine forestry operations in various real-world examples. Various nation-states and organizations employ advanced technological systems to optimize forest management and sustainability tracking operations while building sustainable practices (Nguyen et al., 2023). There is a detailed exploration of smart forestry technologies' real-life operational success in the upcoming section. Different forest monitoring and decision-making operations utilize AI and IoT systems together in order to deliver better outcomes across the described case studies. Complete systems based on IoT sensors measure current data related to soil moisture in combination with temperature measurements and atmospheric observations before sending these measurements to centralized data centers. The collected data receives processing by AI algorithms to spot anomalies, forecast risks, and maximize intervention methods. The combination of sensor data analysis by AI systems activates early warning triggers to speed up drone operations for assessment and containment procedures in wildfire hazard zones. Smart forestry practice develops through the combined strength of sensing combined with intelligent analysis systems.

4.1 Case Study: Amazon Rainforest Conservation (Brazil)

According to its common name "The Lungs of the Earth," the Amazon rainforest faces a survival crisis from deforestation and illegal logging activities. Live forest deforestation assessment is conducted through satellite monitoring which Brazilian authorities accomplish with support from international organizations: High-resolution satellite imagery from NASA and European Space Agency (ESA). The detection system utilizes AI-powered models which examine alterations in forest coverage. Through IoT-based ground sensors, the monitoring of soil moisture and carbon measurements is possible. The implementation of the systems has resulted in over 60% reduction in illegal deforestation activities, achieved through strengthened law enforcement and improved accountability through blockchain-based timber tracking (Patel & Sengupta, 2023). AI systems will optimize the process of choosing tree planting locations for increased reforestation activities. National forestry reports indicate that sensored forestry technologies have resulted in greater than 60% decrease in monitored forest loss with 40% faster response rates to illegal logging.

4.2 Case Study: Canada's Precision Forestry Initiative

The nation of Canada improves timber extraction from boreal forests through environment-friendly preservation methods using precise forestry methodologies. The Canadian forestry industry incorporates suitable smart technologies to optimize forest resource management and protect environmental preservation. By using the Light Detection and Ranging (LiDAR) system, operators can obtain highly detailed three-dimensional forest mapping capabilities: Drone-based surveillance for pest and wildfire detection. Public officials employ AI predictive models to establish sustainable timber extraction approaches on extraction sites. These solutions would bring 25% higher timber production efficiency together with reduced environmental impact. The system will monitor forest health to better prevent occurrences of pests like spruce budworm outbreak. The increased application of sustainability certifications results in better market competitiveness through higher levels of compliance. Through augmented predictive models equipped with drone observing technologies, the wood harvesting productivity soared by 25% with simultaneous reductions of 30% in wildfire response actions in essential boreal territories.

4.3 Case Study: European Smart Forestry Certification

As part of environmental policies, Europe depends heavily on sustainable forest management as a fundamental practice. The Forest Stewardship Council (FSC) and additional certification bodies leverage digital technologies for the purpose of guaranteeing responsible timber supply. The system operates through blockchain tracking that verifies both the starting points of wood products and their eco-friendly criteria. Multiple IoT smart sensors track logging events while they track environmental soil conditions. Geographic information system mapping tools function as enforcers to check preservation law compliance through GIS-based advanced digital mapping tools, and increased consumer trust in certified sustainable timber reduced illegal logging activities through transparent supply chain tracking, and strengthened partnerships between governments, industries, and environmental organizations. Due to blockchain technology implementation with IoT monitoring, the reporting of illegal logging decreased by 45% while FSC-certified wood market sales increased by 20% throughout EU territories.

4.4 Case Study: Smart Reforestation in Africa

Different African nations have introduced technology-based tree plantation programs which function for desertification management while rehabilitating land resources. Other large-scale forestry projects need smart forestry systems to operate as basic infrastructure for initiatives like the Great Green Wall. The process of seed planting across extensive fields can be efficiently conducted with drone assistance. The implementation of artificial intelligence systems allows for soil analysis which identifies suitable tree species for distinct geographical areas. Tree growth monitoring and deforestation reporting functions are included in locally accessible mobile applications. The administration should promote major tree plantation growth initiatives for desertification repair across strategic regions, job creation and economic empowerment of local communities through sustainable forestry, and improved carbon sequestration, mitigating climate change effects (NASA, 2023). Drones used for seed planting processes restore projects at rates that are faster than manual efforts by 15 times while AI-driven approaches boost arid region sapling survival rates by 35%.

4.5 Case Study: Wildfire Prediction and Prevention in the United States

AI-based risk assessment tools became more essential as wildfire intensification rates within the United States persisted to increase. Together with tech operators and the U.S. Forest Service, scientists established contemporary wildfire risk estimation methods alongside emergency response preparedness systems. AI systems analyze historical wildfire data by using climate dataset information, IoT-enabled weather stations for real-time monitoring of temperature and humidity, and drone surveillance for early wildfire detection and response coordination. Faster wildfire alerts along with instant response capabilities reduce the amount of damage suffered by forests and local communities. Resource distribution for firefighting operations becomes more efficient because of this program. The land can benefit from improved risk reduction strategies against future wildfires. Recorded environmental and financial losses decrease due to the combination of AI and IoT tools that deliver faster wildfire detection and better containment solution accuracy by 50% and 28% respectively.

5 Challenges and Limitations

The implementation of smart forestry faces various difficulties alongside its strong developments in forest management. Several barriers originating from technical systems and monetary expenses together with environmental impacts and official regulations inhibit the successful deployment of smart forest solutions (IPCC, 2023). The implementation of smart forestry faces various difficulties alongside its strong developments in forest management. Numerous barriers block the successful deployment of smart forestry techniques because they originate from technological restrictions, financial limitations, environmental factors, and regulatory elements (Joshi & Singh, 2023). Tasks in the development of smart forestry require experienced personnel in AI and both GIS and remote sensing systems as well as data analytics skills. Workers in forestry especially those based in rural areas with developing nations need better knowledge to successfully operate smart forestry systems. Difficulties with implementing training programs combined with educational resources limit the removal of technical knowledge gaps. Research institutions together with public bodies need to conduct teaching programs and training sessions that help workers develop new abilities. Digital technologies integrated in forestry produce concerns about managing both data privacy and security-related matters (Kalita & Das, 2023). The data gathered via IoT devices combined with satellite imaging remains exposed to possible cyber threats combined with unauthorized extraction. Public along with private foresters protecting territorial control face cybersecurity threats because of both data abduction and unauthorized access that enable system tracking. Blockchain technology provides secure systems data and protects against unauthorized modifications through its implementation (UNEP, 2023).

The illustration in Fig. 5 demonstrates the main cybersecurity risks that affect smart forestry systems through their digital components including IoT devices along with cloud platforms as well as remote sensors combined with data analytics. The growing data collection and system interconnectivity of these systems create multiple cyber threats that endanger both forest observation capabilities and data credibility in addition to stakeholder belief. External threats known as data breaches and hacking use systematic vulnerabilities to extract sensitive data and perform unauthorized changes to it. This risk type consists of both unintentional and intentional data-related incidents perpetrated by organizational member personnel. Criminal activities occur through unauthorized access because of defective authentication procedures together with vulnerable passwords and unsecured access management. The lack of proper security measures puts smart forestry systems at risk for both compliance violations, requirements, and global data protection standards including GDPR. IoT Device Vulnerabilities emerge because remote forest sensors and cameras lack proper protection which leaves them vulnerable to hackers. System functionality as well as data accuracy face risks because of weak encryption combined with data tampering and poor firmware update practices (Singh & Bansal, 2023).

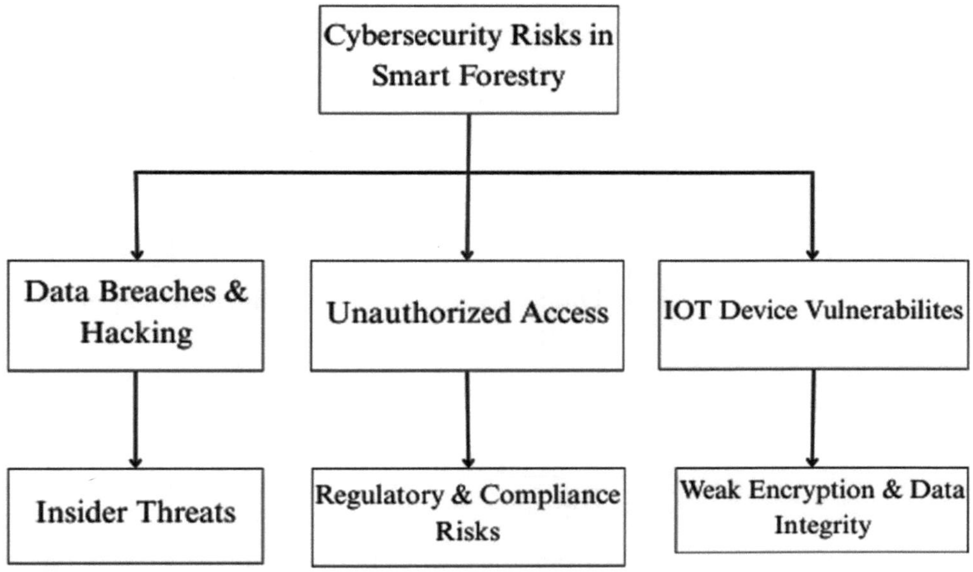

Fig. 5 Cybersecurity risks in smart forestry

6 Future Trends in Smart Forestry

Technological innovation will discover improved strategies for forest defense while guiding forestry management practices. Modern industrial patterns deliver improved methods to handle and defend forest areas alongside profitable resource extraction. Smart forestry science will advance toward these directions over the coming years according to our established list. The modern equipment allows forestry personnel to space tree plantations efficiently using minimal staff during plantation activities. This system uses its drones to scan for optimal planting sites after which seeds get distributed to each position. UGVs enabled with sensors will move through forests to obtain vital data about soil quality and moisture content and biodiversity information while preserving the habitat environment. AI models will achieve increasingly precise analysis of large databases and tree growth prediction as well as disease detection and deforestation risk assessment. Through machine learning algorithms, forest managers can optimize their decision-making because they improve timber yield management together with minimum ecological consequences. Forest managers will receive assistance from AI-driven virtual assistants for their planning activities along with risk assessment and regulatory compliance operations. Blockchain technology can explain supply chain forestry operations with systems that monitor starting locations while ensuring sustainability aspects and certification protocols. Blockchain-powered smart contracts enable real-time verification of forest origins for immediate wood source inspection purposes which reduces forestry crimes as well as fraudulent practices (Ramachandra et al., 2023). Through direct consumer access to records, consumers are able to verify wood product origins which builds stronger relationships of accountability between government institutions and businesses. The implementation of IoT devices will transform to become more advanced through interconnected sensor networks which deliver real-time measurements about temperature. humidity levels, tree development, and soil conditions. Self-adjusting AI-enabled IoT networks will optimize both water consumption and fire control programs while executing pest management protocols. The implementation of sensor-based systems allows for wildfire prevention since they identify irregular temperature changes and smoke patterns that trigger warnings before the fires grow out of control. Selection analytics allows timber harvesting to access forest resources for monetary gain by selecting trees from protected areas without hurting the environment. Two beneficial functions exist in remote-controlled harvesting technology: it supports both occupational risk reduction in logging operations and performance enhancement in work activities (Lin & Xu, 2023).

Figure 6 shows fundamental ethical aspects that need consideration during smart forestry system implementation and governance while noting the various difficulties which require management for achieving sustainable environmental results and responsible social practices. Three core ethical areas form the structure of this diagram which covers data privacy, security, indigenous rights, and concerns together with environmental impact. Data privacy and security administration centers on protecting significant environmental and geospatial information especially regarding regions with overlapping surveillance systems and community lands. Indigenous concerns and rights serve as a core ethical principle since numerous forested lands belong to indigenous territories. Smart forestry operations under the environmental impact dimension must avoid creating negative effects on

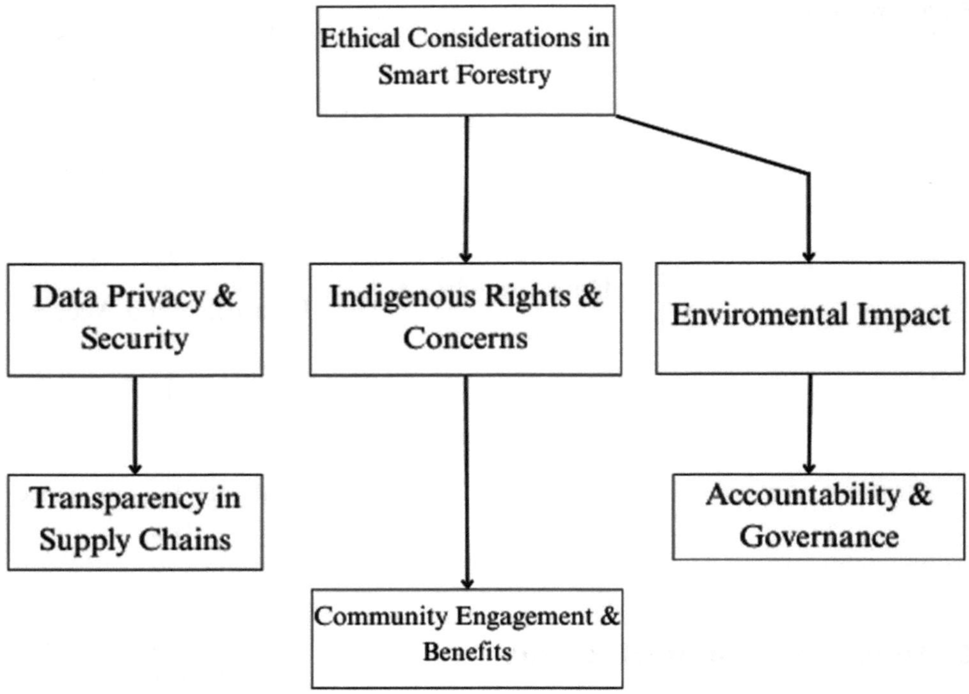

Fig. 6 Ethical considerations in smart forestry

the environment. The figure demonstrates the necessity to establish ethical foresight throughout the entire process of smart forestry technology development including operational deployment and implementation scale-up while prioritizing fair practices and embracing the principles of equity and environmental protection (Ramachandra et al., 2023). Advance research technologies create new ethical problems related to data protection and environmental damage that should be managed using adaptable policy systems (Zhao & Yang, 2023).

Several nations serve as proof that these upcoming trends already function in practice. The assessment of timber yield through AI-based forest growth models combined with optimized harvesting timeline prediction occurs in Norwegian forests. Plantation microclimate conditions in Andhra Pradesh receive monitoring through deployed IoT networks which help practitioners perform adaptive management. The Estonian government uses blockchain smart contracts to make timber deals more efficient and improve the process of obtaining certifications. Australia developed semi-autonomous bushfire management vehicles that use drone sensors as part of their controlled firefighting operations. These implementations showcase the real-world momentum toward technologically advanced and sustainable forestry systems. The successful utilization of smart forestry methods, along with the overcoming of associated challenges, requires a range of coordinated interventions. The following recommendations outline an approach which helps policymakers and industry stakeholders with research institutions achieve greater effectiveness and scalability in smart forestry implementation. The improvement of digital infrastructure within far-flung forested areas should receive funding from governments and businesses to enable real-time data collection together with analysis. Satellite and IoT networks need expansion for performing ongoing observations of forests together with their ecosystems (Liu & Wang, 2022). The forestry specialists should undergo training about specifications which incorporate GIS, AI, and IoT to enhance their current understanding and abilities. University teaching as well as research institutions need to implement smart forestry practices within their educational curriculum for developing future forest managers. The establishment of rules which integrate ethical and proper usage of AI and digital tools in forestry must be executed by governmental authorities (Mehta & Sharma, 2023). The formation of international organizations needs support and facilitation for developing practices in sustainable forest management and advancing key technologies related to sustainable forestry practice. The necessary innovation together with pilot project funding will emerge from industry support which comes from governments and high-tech private companies alongside environmental non-governmental organizations. Businesses implementing smart technologies in forest operations should receive tax assistance and financial subsidies as part of available facilities. Providing competent security measures for forestry organizations is essential to protecting data from cybercriminal and hacking attempts. The implementation of technology policies should be clearly defined to prevent any exploitation of indigenous knowledge resources and materials (Mugo & Ndiritu, 2023; NASA, 2023). The development of smart forestry technology demands technologies to progress at the same rate as environmental protection standards do. Ethical implementation of such procedures begins through comprehensive stakeholder involvement including Indigenous communities to keep traditional knowledge and their rights protected from data collection until resource management ends. The design criteria of technological equipment need environmental thresholds so habitat disruption stays prevented and operation systems use efficient energy consumption methods. AI and IoT technologies help forest managers boost their interventions by using sustainability criteria to conduct impact assessments coupled with adaptive feedback evaluation. Digital sustainable resource management practices will receive protection from proper regulatory control in addition to ethical frameworks for forest ecosystem preservation. A successful execution of smart forestry depends on well-established roles for stakeholders. Public administrations need to initiate policy backing, infrastructure financing, and regulatory framework oversight as their primary duties. Private sector entities play fundamental roles by carrying out technological innovation and providing necessary investments along with collaboration between public and private sectors. The necessary building of capabilities requires research institutions to provide both training sessions and customized solutions development. Co-managership between local communities and Indigenous groups includes participation in ecological knowledge sharing while upholding technology ethics. Establishing a bridge between organizations helps NGOs improve outreach to communities and enhances environmental impact monitoring as well as public awareness efforts. The establishment of clear responsibilities between stakeholders will develop a coordinated platform for inclusive sustainable implementation of smart forestry systems.

7 Conclusion

The existing evidence shows that advanced forestry practices succeed in fixing financial difficulties and handling environmental obstacles. GIS and LiDAR professionals deploy tracking systems which prevent forest clearing and deforestation through their appreciation mapping systems. AI-based risk assessments helps to protect forests by identifying environmental threats such as natural calamities, pest infestations and climate-related disturbances. Precise forestry systems lead to better woodcutting procedures because they establish exact practices for waste reduction and ethical forest management protocols. Advanced forest technology develops because of autonomous forest equipment which works in conjunction with blockchain tracking technology and artificial intelligence decision platforms. Sustainable forestry in the future will require worldwide smart forestry partnerships who actively implement their methods by engaging a large number of community members while developing new methodologies. Smart forestry adoption enables operations to shift into sustainable business approaches and manage current forested region challenges. The achievement of smart forestry requires joint collaboration between state enterprises and local communities working together with government agencies. Sustainable systems based on advanced responsible practices will create favorable coexistence opportunities among forests, industrial developments, and communities. The continued existence of the Earth depends on modern forestry methods since environmental issues are growing at a global scale.

References

Ahmad, I., Hussain, M., & Majeed, A. (2023). Application of IoT and AI in smart forestry. *Journal of Environmental Management, 312*, 114892.

Beniwal, S., Rani, R., & Singh, A. (2023). Role of GIS and remote sensing in sustainable forestry management. *Environmental Monitoring and Assessment, 195*(3), 456.

Chen, Q., Yang, L., & Zhang, Y. (2022). Remote sensing for forest fire detection using AI models. *Forest Ecology and Management, 510*, 120207.

Deng, Y., Liu, J., & Wang, Z. (2022). Blockchain in forestry: A new path for transparent timber trade. *Journal of Cleaner Production, 364*, 132699.

Dubey, R., Gunasekaran, A., & Bryde, D. (2022). Smart forestry systems: Sustainability through technology integration. *Technological Forecasting and Social Change, 175*, 121354.

FAO. (2023). *The state of the world's forests 2023: Forests and sustainable production and consumption*. Food and Agriculture Organization of the United Nations.

Ghosh, S., & Roy, A. (2023). Satellite imaging and forest mapping in the era of smart forestry. *Geocarto International, 38*(1), 55–70.

Gupta, N., & Khatri, A. (2023). Enhancing forest health monitoring through IoT networks. *Sensors, 23*(9), 4410.

Haque, M. I., & Ali, M. (2023). Artificial intelligence and big data in climate-smart forestry. *Sustainable Computing: Informatics and Systems, 36*, 100799.

Hasan, M. M., & Chowdhury, S. A. (2022). Smart contracts and timber certification in forestry supply chains. *Forest Policy and Economics, 139*, 102715.

He, Y., & Lin, Z. (2023). Autonomous drones for precision reforestation. *Drones, 7*(1), 27.

IPCC. (2023). *Climate change 2023: Mitigation of climate change*. Intergovernmental Panel on Climate Change.

Jain, T. R., & Srivastava, P. K. (2023). Geospatial technologies for sustainable forest management: integration of remote sensing, GIS, and decision support systems. *Journal of Environmental Informatics, 41*(2), 225–240. https://doi.org/10.3808/jei.202300226

Joshi, P. K., & Singh, T. (2023). Ethics and indigenous rights in digital forestry. *Environmental Ethics, 45*(2), 151–168.

Kalita, D., & Das, S. (2023). AI-driven fire risk prediction in forest ecosystems. *Ecological Informatics, 75*, 101852.

Li, R., & Zhang, H. (2023). Blockchain-based monitoring in environmental governance. *Nature Sustainability, 6*(2), 129–136.

Lin, W., & Xu, H. (2023). Integration of AI and GIS for sustainable forest management. *Remote Sensing of Environment, 292*, 113355.

Liu, Y., & Wang, J. (2022). Big data-driven forestry: challenges and opportunities. *Computers and Electronics in Agriculture, 200*, 107084.
Mehta, R., & Sharma, V. (2023). IoT and sensor networks in biodiversity conservation. *Ecological Engineering, 187*, 106874.
Mugo, R., & Ndiritu, J. (2023). Smart forestry for Africa's reforestation programs. *Environmental Research Letters, 18*(4), 045001.
NASA. (2023). *Earth observation for sustainable forest monitoring.* NASA Earth Science Division.
Nguyen, H. T., Pham, D. T., & Kim, Y. S. (2023). Drone-integrated sensing technologies for forest health monitoring: Applications and future prospects. *Remote Sensing, 15*(4), 1123. https://doi.org/10.3390/rs15041123
Patel, K., & Sengupta, S. (2023). Deforestation analytics using satellite and LiDAR fusion. *ISPRS Journal of Photogrammetry and Remote Sensing, 198*, 234–245.
Ramachandra, T. V., Bharath, H. A., & Varghese, S. (2023). Ethics and equity in forest monitoring: Integrating indigenous rights and data privacy in smart forestry. *Environmental Ethics and Policy, 45*(1), 51–69. https://doi.org/10.1016/j.envpol.2023.122876
Singh, A., & Verma, P. (2023). Forest monitoring using acoustic and visual sensors. *Journal of Forestry Research, 34*(3), 421–434.
Singh, R., & Bansal, R. (2023). Cybersecurity in smart environmental systems: Challenges and solutions for IoT-enabled forest monitoring. *Journal of Cybersecurity and Privacy, 3*(2), 215–232. https://doi.org/10.3390/jcp3020014
Smith, T., & Green, D. (2023). Legal and policy frameworks supporting smart forestry. *Forest Policy and Economics, 140*, 102735.
UNEP. (2023). *Sustainable forest management and digital transformation.* United Nations Environment Programme.
Zhao, J., & Yang, C. (2023). Real-time AI analytics in forest disease management. *Computers in Biology and Medicine, 157*, 106743.
Zhao, X., Liu, M., & Chen, Y. (2023). Real-time forest monitoring using IoT and edge computing technologies: An integrated framework for smart forestry. *Sensors, 23*(14), 6652. https://doi.org/10.3390/s23146652

Empowering Sustainable Forest Management: The Transformative Role of Artificial Intelligence and Machine Learning

M. Swarna Sudha, S. Manjula, and K. Vijayalakshmi

Abstract

For decades, forest communities have relied on forests for their livelihood, culture, and ecological balance, which has led to traditional forest management. Although such systems are indigenous knowledge-based and sustainable, they are now being threatened by population increase, commercialization, and climate change. Forest ecosystems are under threat from overharvesting, deforestation, and loss of biodiversity; these issues are exacerbated by socioeconomic issues, outdated laws, and poor governance. The increasing impacts of pests, invasive species, and climate change highlight the shortcomings of conventional forest conservation practices. To this extent, machine learning (ML) and artificial intelligence (AI) offer new mechanisms for ensuring forests are sustainably managed. These technologies enable real-time satellite monitoring, predictive modeling for climate impact analysis, and data-driven decision-making for resource optimization. Moreover, by enabling early identification of diseases, forest fires, and illegal logging, AI-based technology enhances governance and conservation. A data-driven strategy for forest sustainability can be encouraged by adopting AI-based methods, which can help fill the gap between traditional methods and the current needs for conservation. To maximize the use of AI and ML towards safeguarding the forest ecosystem for future generations, governments, scientists, and local people need to cooperate.

Keywords

Artificial intelligence (AI) · Machine learning (ML) · Sustainable forest management · Climate change mitigation · Predictive modelling

M. Swarna Sudha (✉) · S. Manjula · K. Vijayalakshmi
Department of Computer Science and Engineering, Ramco Institute of Technology, Rajapalayam, Tamil Nadu, India
e-mail: swarna@ritrjpm.ac.in

1 Introduction

Our world's forest well-being is at the core of the well-being of all life on our world. They are not only critical reservoirs of biodiversity, oxygen, and pure water but also play a critical role in Earth's climate control through acting as vital carbon sinks. Nevertheless, even though they hold such prominence, forests around the world are facing unprecedented challenges. Climate change, forest degradation, deforestation, and unsustainable forest management have battered these important ecosystems. Under risk of these increasing pressures, traditional methods of forest management are in most instances not sufficient to handle the complexity of today's issues. This is where artificial intelligence (AI) and machine learning (ML) come into play. As human intelligence-duplicating technologies, AI and ML usher in the latest solutions in understanding, managing, and preserving forests beyond imagination previously. With the ability to process huge amounts of data, predict environmental dynamics, and automate decision-making, AI and ML change the way we approach forest conservation and management.

The need for sustainable forest management has never been higher, and AI and ML are imperative drivers to make sure that forests do not only remain conserved but are better managed and more efficiently utilized as well. AI and ML can help in tracking the health of the forests, optimizing the utilization of resources, predicting fire risks, and maximizing carbon sequestering activities, among innumerable others. With the use of AI and ML, forest managers have been able to make a level of analysis that has gone way beyond what has been achieved by traditional techniques. However, with the application of AI and ML in forestry, there is no lack of issues. Data availability, computing capacity, and trained personnel are some of the challenges that need to be overcome. Despite these challenges, the role that AI and ML can play in sustainable forest management can be immense. This chapter introduces the revolutionary contribution of AI and ML to forest

M. Ashwin et al. (eds.), *Smart Forestry 5.0 for Powered AI Sustainable Environment and Climate*, Sustainable Artificial Intelligence-Powered Applications, https://doi.org/10.1007/978-3-031-99086-1_4

management, explaining how these technologies are shaping the future of forests.

1.1 The Growing Need for Sustainable Forest Management

Forest ecosystems are vital for preserving ecosystem balance, sustaining biodiversity, controlling the climate, and supplying resources that are necessary for human survival. However, environmental degradation, deforestation, and climate change are putting more and more strain on these ecosystems. Even though they have some value, traditional forestry management techniques rarely keep up with the quickly changing environmental conditions and can manage large, interconnected forest landscapes at a reasonable cost. By enhancing data collection, analysis, prediction, and decision-making procedures, artificial intelligence (AI) and machine learning (ML) technologies have the potential to revolutionize forest management (Wang et al., 2024). Massive volumes of environmental data from various sources, such as satellite imagery and remote sensing via internet-of-things devices, can be processed by AI and ML technologies. AI and ML can provide real-time insights into forest health, biodiversity, and ecosystem dynamics through their use of these data. One of the biggest benefits of AI and ML in forest management is the possibility to use them for predicting and modelling complex environmental phenomenon. For instance, predictive algorithms can help forecast forest fires, pest outbreaks, and the effects of climate change on forest ecosystems. Such predictive potential opens the way for pre-emptive actions to minimize large-scale environmental destruction. In addition, those decision support systems, powered by AI, can help designers optimize the allocation of resources so that they achieve efficiency within conservation efforts (Kolasani, 2024). Additionally, AI and ML support forest management: Precision forestry is a more sustainable approach to forestry management because it maximizes productivity at the same time as minimizing environmental impact from its data-based insights into planting, harvesting, and resource allocation.

Figure 1 shows how AI supports natural resource management across areas like resource monitoring, wildlife conservation, precision agriculture, disaster mitigation, pollution detection, and renewable energy optimization through applications such as forest tracking, soil analysis, wildfire detection, and air quality assessment. Forests are irreplaceable for life on Earth, as they supply vital resources like timber, food, fresh air, and water. They contribute significantly to biodiversity, climate regulation, and hosting millions of species, including human beings. Rapid urbanization, deforestation, climate change, and destructive farming practices are all seriously threatening forest ecosystems. Forest ecosystems are known to be harmed by a number of unsustainable land-use practices, such as mining and forest fragmentation. In addition to disrupting vegetational diversity and structure, these practices have detrimental effects on soil quality and regeneration (Forest for Sustainable Development, 2023). The Food and Agriculture Organization (FAO) estimates that 10 million hectares of forest are lost annually worldwide. This raises carbon emissions, which contribute to climate change, and has a significant impact on biodiversity. Forests play a crucial role in supporting life on Earth by providing essential resources

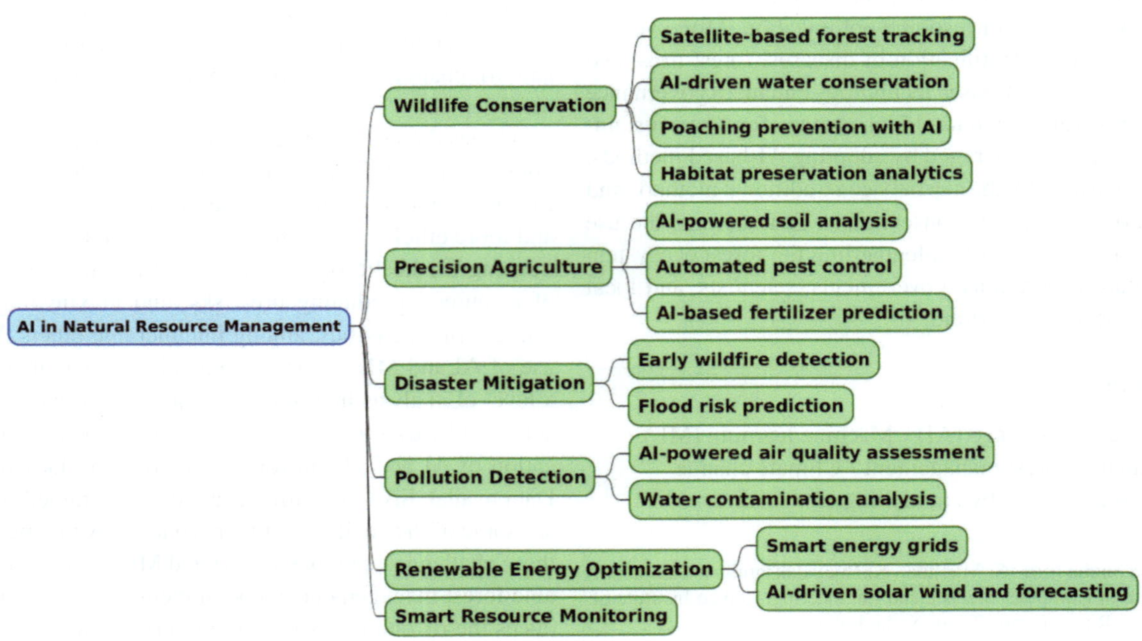

Fig. 1 Applications of AI in natural resource management

like food, fresh air, water, and lumber. They support millions of species, promoting biodiversity and preserving ecological stability and balance. Forests are crucial for regulating the climate and lessening the effects of global warming because they absorb carbon dioxide. They also protect soil by halting erosion and maintaining natural water cycles.

By preserving these ecosystems, forests promote environmental sustainability and the well-being of all living things. Apart from their ecological functions, forests are vital to the livelihoods of countless communities worldwide. They provide raw materials for industry, medicinal plants for healthcare, and food sources for both domestic and foreign markets. Furthermore, forests enhance human heritage and well-being by offering recreational opportunities and cultural value. They act as organic barriers against disasters like floods and landslides by stabilizing soils and regulating water flow. By maintaining the balance between environmental health and human needs, forest preservation contributes to the protection of the environment and helps to guarantee a sustainable future for future generations. Sustainable forest management techniques must be used in light of the growing environmental issues. In order to preserve and responsibly use forest resources, sustainable forest management aims to strike a balance between ecological, social, and economic needs. It comprises ongoing assessment and monitoring of ecosystem services, biodiversity, and forest health. Additionally, effective forest management helps local communities make a living, mitigates the effects of climate change, and ensures that natural habitats are preserved for future generations.

Accurate monitoring, prediction, and decision-making are made possible by the advanced data analysis capabilities offered by AI and ML. They provide real-time insights into forest dynamics, biodiversity, and ecosystem health by leveraging vast volumes of data gathered by drones, satellite photography, and sensor networks. Machine learning algorithms are able to identify trends, anticipate deforestation risks, and indicate changes in vegetation cover, hence improving the effectiveness of forest management plans. Additionally, AI systems enable multimodal information from a variety of sources to be combined and provide an overall perspective of forest ecosystems.

AI and ML integration into forest management for sustainability is not just a choice but a requirement. The technologies allow AI-powered and ML-enabled forest managers, ecologists, and policymakers to drive decision-making with facts, optimize the use of resources, and react to change in the environment quickly. In addition to this, AI-based solutions play a role in biodiversity protection, mitigation against climate change, and attainment of global sustainability targets, including the United Nations Sustainable Development Goals (SDGs) (Kumar et al., 2024a).

1.2 Challenges in Traditional Forest Management

For ages, communities have relied on forests for their subsistence, cultural values, and ecological balance, leading to the practice of traditional forest management. To ensure the preservation of forest resources, these techniques were frequently informed by sustainable use principles and indigenous knowledge. However, traditional approaches confront many difficulties due to the changing environment, economic pressures, and the increasing demands of modern society These issues pose a threat to biodiversity, the sustainability of forest ecosystems, and the welfare of communities that depend on forests. This chapter examines the major obstacles to conventional forest management and how they affect socioeconomic growth and environmental preservation.

Traditional forest management practices primarily favored maximizing resource extraction—timber, fuelwood, and non-timber forest products. Over time, rising population and commercial demand have emerged, compromising many of these practices to yield overharvesting, resulting in deforestation and soil degradation, and loss of biodiversity. In several areas, the absence of sustainable limits for harvesting has exhausted forest resources, further compromising the ecosystem's ability to regenerate (Sardjono et al., 2023). Forest policies in many countries are outdated and unable to address modern challenges. Weak enforcement and poor governance, increasingly aggravated by bureaucratic bottlenecks, deny participation from the local community in the decision-making process. Thus, at present, illegal logging, land encroachment, and unregulated resource extraction are widespread activities, all acts against sustainable forest management. Pests, diseases, and invasive species increasingly affect forests and are often aggravated by climate change (Yang, 2022). In traditional forest management, protection against diseases and pests often lacks scientific principles and contemporary approaches, leading to considerable ecological and economic losses.

Forest-dependent communities are often plagued by poverty, lack of education, and inadequate livelihood opportunities (Palaschuk et al., 2024). These socio-economic challenges create an environment that encourages illegal logging, overgrazing, and shifting cultivation, all of which devastate forests. Traditional values and community unity are steadily threatened by cultural assimilation and modernization. Ineffective governance structures, corruption, and poor institutional coordination cripple sustainable forest management (Modise, 2024). Lack of transparency and accountability in forest governance has contributed to policy failure and resource mismanagement. Furthermore, centralized governance tends to exclude local communities from decision-making processes. It is not betwixt and between any of the two features of ecological, socio-economical and governance policies; however, these challenges do address a challenge to paradigm shift from the idea of practice-oriented to develop sustainable forest management practices using scientific knowledge, modern technologies, and participation from the community. AI and ML hold a great promise. AI and ML can enhance forest monitoring through satellite images' analysis, prediction about the impact of climate change, and improved optimization of resource utilization by analyzing the complex environmental data. Early detection of pests and diseases, illegal logging activities, and biodiversity assessments can also then improve forest management efficiency and data basis.

1.3 Role of Artificial Intelligence and Machine Learning in Forest Management

The actual scenario of unprecedented environmental confrontation brings an ever-increasing need for effective and sustainable forest management. Traditional practices of managing forests are intricately linked to human errors, inherent inefficiencies, and challenges in accurately assessing and collecting scientific data. In the recent past, the advent of AI and ML has contributed to an ecosystemic transformation of the forestry profession, where new technologies are providing sophisticated tools for monitoring, conducting analysis, and making informed decisions. These technologies are now involved in ensuring sustainable resource utilization, aiding in biodiversity conservation, and strengthening measures to address climate change impacts. Forests hold ecological significance by sustaining biodiversity, sequestering carbon, and regulating the climatic conditions. Besides, they are a boon to economic growth, providing timber and non-timber forest products, thus sustaining the livelihoods of millions around the world. Deforestation, wildfires, and climate change are ever-increasing threats to the forest. This has called for advanced forest management systems that can provide accurate and timely information regarding decision-making. AI and ML have thus emerged to turn around these challenges, to analyses the huge amounts of data generated, and to generate meaningful insights for action (Sutriani et al., 2024).

Figure 2 highlights the role of AI in forest management and biodiversity conservation. It covers smart forest monitoring, wildlife protection, ecosystem health analysis, sustainable resource management, and climate adaptation. AI technologies like remote sensing, predictive analytics, and automated detection are transforming conservation efforts.

AI encompasses numerous technologies, from computer vision and natural language processing to expert systems and predictive analytics, and find application in forest management. Machine learning constitutes a sub-field of AI concerned with making systems learn from data and improve over time (Devi et al., 2024). Major machine learning techniques applied in forestry are supervised learning, unsupervised learning, and reinforcement learning. Algorithms such as decision trees, random forest, SVM, and neural networks assist in pattern recognition and prediction (Kumari et al., 2024). Deep learning, which constitutes an even higher order of machine learning, is able to carry out complicated pattern

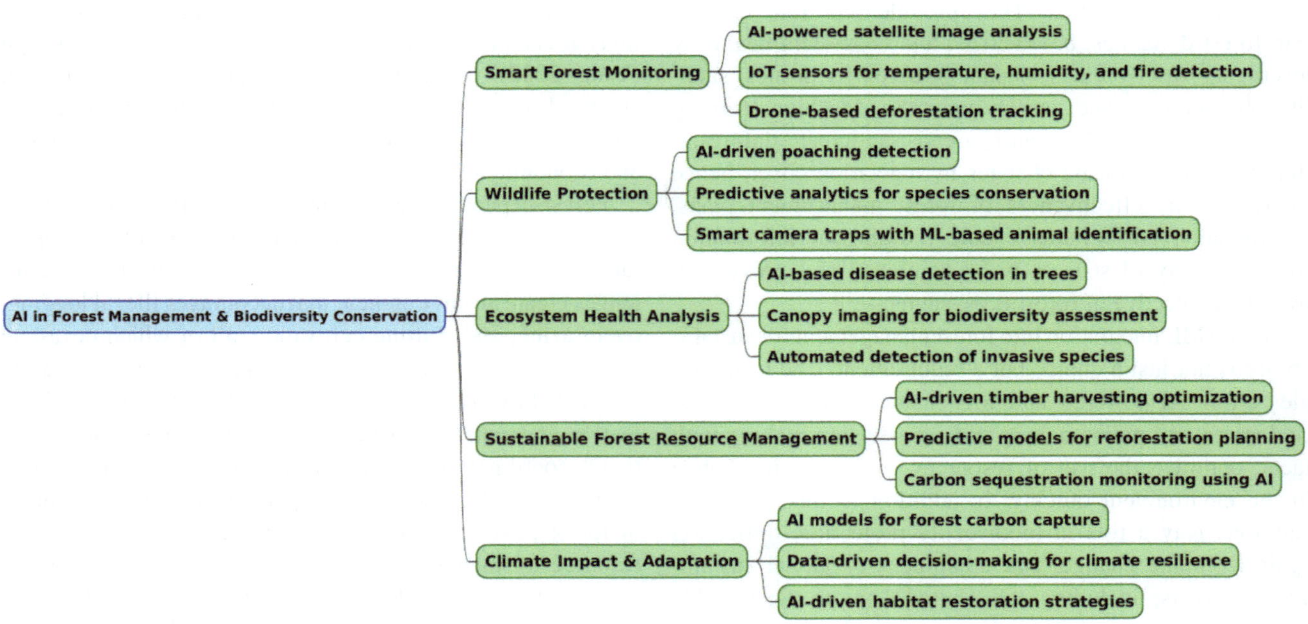

Fig. 2 AI applications in natural resource management

recognition tasks such as satellite-algorithm-based analysis and species identification. One of the larger applications of AI and ML in forestry includes remote sensing and forest monitoring. Satellites and unmanned aerial vehicles with advanced cameras and sensors were sending flooding images of forests in real time. This was analyzed through convolutional neural networks (CNN) to classify and segment images so that the accurate mapping of forest cover, vegetation health, and growth pattern could be established for better forest management. Species identification for richness and conservation of biodiversity is thus assisted by automated image recognition, which can enable researchers to observe the distribution and habitat suitability for such species. AI and ML also help to ensure forest health monitoring by detecting the early signs of diseases, pest infestation, and nutrient deficiencies. Through anomaly detection, unusual patterns are identified to protect forest health by intervening in time. Another important consideration is the prediction and management of wildfires, during which predictive models assess weather data in order to forecast the track of fire likely spread (Krishnamoorthy et al., 2023). Real-time fire detection using satellite imagery employing CNNs for alerts would provide timely notification, while Reinforcement Learning models can help optimally allocate resources and plan evacuations during wildfire emergencies. AI and ML help to ensure sustainability by optimizing timber harvesting schedules, thus reducing impacts on the environment and enhancing economic returns.

Predictive analytics-based decision support systems improve inventory management and resource allocation thereby contributing to sustainable development. These applications thus provide a glimpse into the enormous potential of further assisting in forest management through AI and ML. The real-world application of AI and ML in forestry has certain examples in practical implementation, such as AI systems monitoring deforestation in the Amazon Rainforest, whereas wildfire prediction models have been important for mitigation of disasters in California, USA. In Southeast Asia, deep learning models have been used in automated identification of species, which helps in conservation efforts for endangered species. The examples show some of the ways AI and ML can help in resolving some of the most critical environmental issues. Nevertheless, their implementation in forest management faces challenges as well. Data availability, quality, and bias affect model accuracy; computational burden and the need for advanced hardware present logistics hurdles; ethical issues of automation, such as loss of livelihoods; and overcoming barriers in coupling with established forest management systems also need consideration. Development in algorithms will bring about more accurate predictions and improved decision-making. Integration with IoT devices and edge computing will allow real-time monitoring of forest and anything related to ecosystems (Impana, 2024). AI for policy-making and decision-support systems would further promote sustainability and conservation. Collaborative platform for sharing data and models at country levels will strengthen global forest management regimes. AI and ML are set to redefine forest management through their innovative responses to complex environmental predicaments. They enhance sustainability, promote biodiversity conservation, and help better resource utilization. As these technologies evolve, more and more investments will be needed to prevent forests worldwide. To harness this potential to the fullest, inter-disciplinary research and collaboration should come into play.

1.4 Objectives and Scope of the Study

The objective is to provide a thorough overview of how ML algorithms and AI are used within the forestry industry and biodiversity conservation worldwide. Furthermore, this mainly examines the difficulties encountered while deploying AI technology in forestry and biodiversity. Improving the access to large amounts of data in the area of forests and biodiversity, and the use of cloud computing and digital and satellite technology, can make wider adoption and deployment of AI technology possible.

2 AI and ML in Ecosystem Management

The fine-tuned balance of ecosystems is the foundation of a healthy planet. From forests and wetlands to oceans and grasslands, ecosystems deliver vital services that support life—climate regulation, water purification, maintenance of biodiversity, and provision of resources for human well-being. But human activity is increasingly disturbing these systems, with habitat destruction, pollution, climate change, and a host of other environmental problems threatening their stability. Conventional methods to managing ecosystems, though necessary, have difficulty keeping up with the complexity and breadth of current environmental challenges. Managing ecosystems involves surveillance of large tracts, comprehending subtle relationships among species and their habitats, and anticipating long-term ecological patterns efforts that are increasingly becoming harder as the planet continues to evolve at an ever-increasing pace. AI and ML facilitate more accurate, effective, and adaptive management practices. From tracking biodiversity using computerized image identification to forecasting the effect of climate change on species distribution, these technologies open up new avenues for comprehending and managing ecosystems. Machine learning models can identify minute environmental changes, predict the effects of human activities, and optimize resource allocation, thereby making conservation efforts more focused and effective.

2.1 Basics of AI and ML

AI is a computer science discipline focused on creating machines with the ability to execute tasks normally involving human intelligence (Raj, 2024). Such tasks include problem-solving, decision-making, language understanding, and visual perception. AI seeps into various applications, ranging from health diagnostics, financial fraud detection, to autonomous cars (Murugesan et al., 2024). ML, one of the subfields of AI, enables systems to learn and improve their performance with experience. ML is mostly classified into three types: Supervised learning, in which models learn from labeled information to predict or classify; unsupervised learning, which involves finding patterns or groupings within data without predefined labels; and reinforcement learning, in which models learn through interacting with their environment to reap the most rewards. These techniques enable applications such as recommendation systems, image recognition, and predictive analytics. In machine learning, data are typically divided into training and testing sets (Bailey, 2023). The model identifies patterns from the training set and is evaluated on the test set to ensure accuracy and generalization. A significant challenge in ML is overfitting, where the model becomes overly complex and picks up noise in the data, and underfitting, where the model is too simple to recognize patterns. Techniques such as cross-validation enhance model performance by evaluating multiple data subsets. Common ML algorithms include linear regression, decision trees, neural networks, and SVM.

Figure 3 compares traditional and AI-based models in ecosystem management. The traditional approach relies on manual data collection and delayed reporting, while the AI-based model uses automated data collection, predictive analytics, and real-time monitoring for faster, adaptive conservation measures. Both approaches are evaluated for effectiveness and adjustments are made accordingly.

While traditional models rely on predefined rules and human intervention, AI-based models use data-driven learning, automation, and adaptability for improved accuracy and efficiency. AI and ML have transformed many industries by automating intricate tasks and facilitating intelligent decision-making. In healthcare, they aid in diagnos-

Fig. 3 Traditional model versus AI-based model

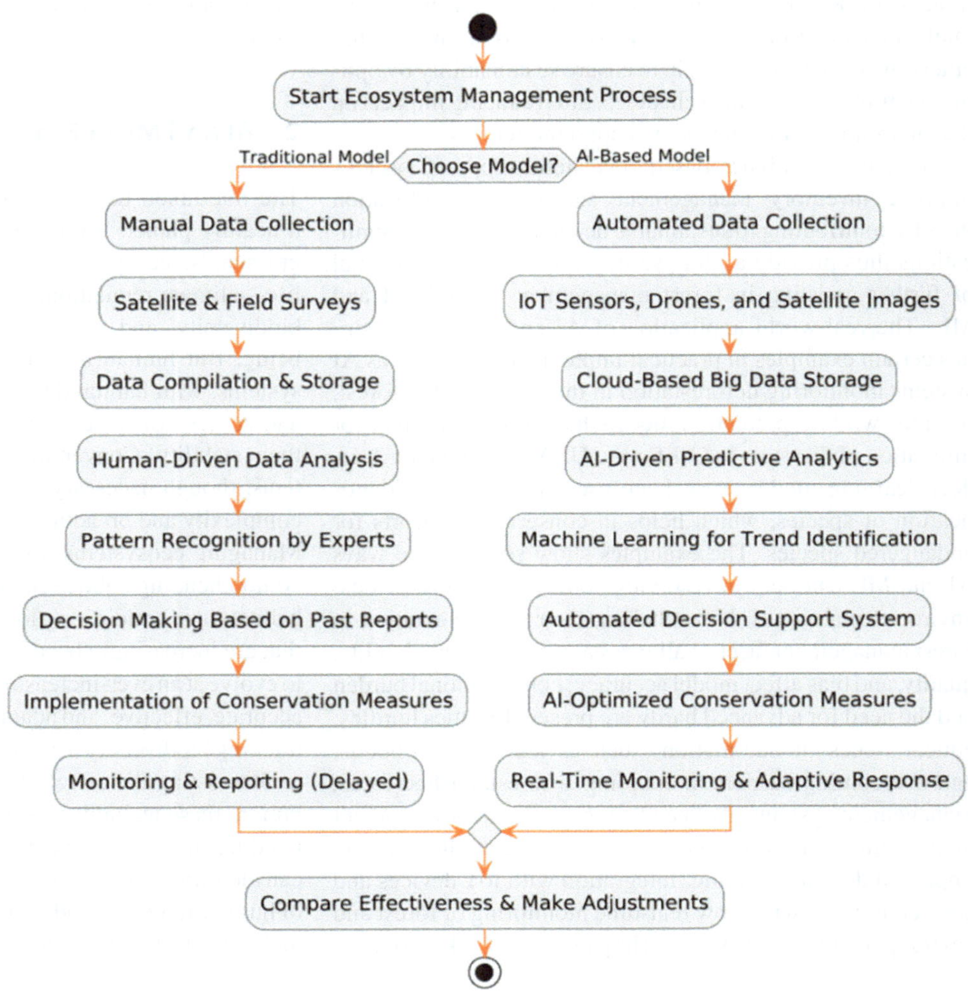

tics and personalized medicine. In finance, they are utilized for fraud detection and risk management. E-commerce platforms use ML algorithms for tailored recommendations. With progress in natural language processing (NLP) and computer vision, AI is becoming increasingly adept at understanding human language and visual data. As research advances, AI and ML are set to have an even greater impact on future technologies, including robotics, smart cities, and sophisticated autonomous systems (Lozinska et al., 2024).

2.2 Key AI and ML Technologies for Forest Management

AI and ML are playing a growing role in forest management, improving decision-making, monitoring, and conservation initiatives (Osco et al., 2021). These technologies facilitate the gathering and analysis of extensive datasets from sources like satellite imagery, drones, and ground sensors, offering precise and timely insights into forest health, tree species identification, and shifts in vegetation cover. Forest managers can better monitor deforestation, predict forest fires, and assess biodiversity by using AI and ML. This leads to more sustainable forest management techniques. ML is a data-driven approach where algorithms are trained to forecast or decide on the basis of historical data. It trains machines to improve with practice by using statistical techniques. ML is used in forestry to identify species, predict growth, and identify anomalous forest conditions.

2.2.1 Machine Learning Types in Forestry Applications

In sustainable forestry, machine learning is quickly becoming a valuable tool that offers more sophisticated ways to understand and preserve our forest ecosystems. Every type of machine learning, including supervised, unsupervised, and reinforcement learning, has unique advantages. Using pre-labeled data, supervised learning helps with tasks like identifying tree species from satellite imagery, measuring forest biomass, or spotting disease symptoms. When forest managers need reliable information to support conservation or take carbon storage into consideration, this is very helpful. On the other hand, unsupervised learning is useful when there are no explicit labels on the data. It can spot hidden patterns like classifying forest patches according to vegetation cover, spotting questionable alterations that might point to illegal logging, or establishing different ecological zones. Then comes reinforcement learning, which is even more advanced. These models improve their decision-making skills over time by experimenting with the world, much like trial and error. This is especially useful for tasks like figuring out the best way to deploy forest patrols, responding to wildfires, and scheduling harvesting schedules that balance sustainability and profitability. These three types of machine learning work together to create a strong, well-rounded approach to forest management that is more data-driven, responsive, and better equipped to handle the complex issues of environmental preservation and climate change.

2.2.2 Major ML Algorithms Applied to Forestry

Several advanced machine learning algorithms are leading the charge to transform how we observe and analyze forest ecosystems. Among the most well-liked are random forests and decision trees, both of which are excellent at classification tasks. Tree species are frequently classified using these algorithms based on a variety of characteristics, such as leaf color, shape, or spectral signatures from satellite imagery. In order for forest managers to prevent outbreaks, they are also used in disease detection by analyzing changes in vegetation health. CNN, a type of deep learning model that focuses on image identification, is another significant technological advancement. CNNs are frequently used to analyze drone or satellite imagery to map and identify forest cover, monitor changes over time, and even identify specific forest structures. CNNs are ideal for monitoring wide areas without the need for manual interpretation because of their ability to automatically extract image features. Recurrent neural networks (RNNs) are very helpful for forecasting changes in the environment over time. These networks are established to process sequential data and are therefore perfectly fit to predict trends like rainfall patterns, forest fire hazards, or plant seasonal changes. Through learning from past time-series data, RNNs can aid in anticipatory forest management policies and disaster planning. Finally, SVMs are commonly employed for both regression and classification problems in forestry. One of their applications is in predicting forest biomass, wherein SVMs are capable of processing a combination of spatial, spectral, and environmental features to predict the volume of vegetation mass within an area. With their versatility and good performance when dealing with high-dimensional data, they are a dependable option for modeling many tasks related to forestry.

2.2.3 AI and ML Tools Applied to Sustainable Forest Management

AI and ML are transforming the way we go about sustainable forest management, providing tools that improve both the extent and accuracy of decision-making. Perhaps the most significant application is in remote sensing and image analysis. AI-based systems can analyze huge volumes of satellite and drone imagery to track forest health, identify changes in canopy cover, and detect evidence of deforestation or forest degradation. These technologies enable forest managers to have nearly real-time visibility into extensive, generally inaccessible regions, allowing illegal logging, forest loss, and changes to ecosystems more easily to monitor with greater accuracy and rapidity.

Another key usage is in predictive analytics, whereby machine learning algorithms process past and environmental data to predict future occurrences. With these applications, one can anticipate growth rates in forests, the propagation of forest fires, and even pest infestations. By predicting such occurrences, forest managers and conservationists can activate early intervention measures—such as selective thinning, prescribed burns, or pest management thus mitigating damage and enhancing long-term forest resilience. Decision support systems (DSS) driven by AI are also being invaluable in sustainable forest management. Such smart platforms compile multiple sources of data, e.g., weather conditions, biodiversity indicators, and socioeconomic considerations and offer actionable intelligence to policymakers and forest planners. Through the analysis of different scenarios and the determination of likely consequences, DSS tools assist in making well-informed, fact-based decisions on land use, conservation priorities, and resource distribution. Essentially, such tools serve as a virtual compass, directing stakeholders towards decisions that reconcile ecological conservation with human necessity. Now that we have an understanding of the key AI and ML technologies used in forest management, it's important to see how these modern approaches stack up against traditional method.

2.3 AI Versus Traditional Methods: A Comparative Analysis

The comparative study of AI and conventional approaches in forest management indicates a substantial benefit of AI technologies in improving efficiency and precision. Conventional approaches like man-powered patrols and remote sensing tend to lack speed and accuracy, thus becoming less efficient in dealing with emergency situations like forest fires and deforestation. However, AI-based approaches utilize sophisticated data analytics and machine learning to offer timely and precise monitoring and management solutions. AI technology such as drones and satellite images supports real-time monitoring of forest cover, which lowers response times to such events as forest fires (Radzi et al., 2024). Machine learning models can process large databases and determine deforestation hotspots with over 85% accuracy compared to regression methods (Hasan et al., 2024). AI-driven decision-making models, including Random Forest models, handle complex data better than traditional methods, resulting in better management decisions for forestry. The inclusion of AI enables predictive modelling, which is able to predict environmental change and guide anticipatory management approaches (Dolle, 2024), whereas AI has revolutionary potential for managing forests, worries regarding dependence on data and human intervention remain. A reconciliation between AI's optimization and the tailored insights of conventional methods can prove to provide the best management measures.

2.4 Benefits of Integrating AI and ML into Forest Management

Integration of AI and ML into forest management presents many advantages, especially in maximizing predictive ability, streamlining resource management, and maximizing sustainability. The technologies enhance data-driven decision-making, an essential tool in the fight against forest fires and conservation of biodiversity. Forests are essential for ecological balance, biodiversity, and climate regulation. Deforestation, forest degradation, illegal logging, and climate change pose threats to these ecosystems. Through automation, predictive analytics, and data-driven insights, it has developed into a potent tool for bettering forest management. The benefits of applying AI and ML to forest management are covered here, along with how they affect sustainability, conservation, and policymaking. Through enhanced monitoring and surveillance, machine learning and artificial intelligence are transforming forest management. Satellite imagery and drone surveillance are technologies based on remote sensing that allow forest cover, logging, and land-use trends to be observed in real time. The ML models classify huge amounts of data derived from LiDAR, hyperspectral imaging, and thermal sensors and identify anomalies, deforestation rates, and encroachments accurately. Predictive analytics powered by AI aid forest conservation by analyzing historical and current data to predict wildfires, pest outbreaks, and disease epidemics. By combining climatic, soil, and vegetation information, these models enable forest managers to act preventively, reducing environmental and economic damage. In biodiversity and wildlife conservation, AI-powered image and audio recognition technologies aid in monitoring threatened species, monitoring animal migration, and studying habitat shifts. Technologies such as AI-powered camera traps, drones, and bioacoustics allow for efficient ecological data collection with minimal human intervention. AI also has a critical role to play in climate change mitigation by assessing levels of carbon sequestration and maximizing afforestation and reforestation. Combining AI with climate models enables policymakers to make wise decisions on sustainable forest management. Further, AI optimizes smart forest resource management through better efficiency in timber harvesting, supply chain management, and sustainable logging. ML algorithms facilitate reduced loss during harvesting while meeting environmental rules and regulations, and blockchain-integrated AI systems enhance transparency in timber deals and forest resource allocation. Disaster response and management are greatly aided by AI, as predictive modeling and automated alerts can enhance early warning of wildfires, landslides, and floods, allowing for interventions to minimize loss of life and biodiversity.

AI and ML also offer evidence-based insights for policy development and decision-making, enabling policymakers to examine trends and evaluate the effectiveness of conservation interventions, aligning strategies with sustainability objectives and global commitments like the UN Sustainable Development Goals (SDGs). Better fire management is another essential area where AI and ML models, such as convolutional neural networks (CNN) and ensemble models, improve early detection and prediction of forest fires, lowering response times considerably. More advanced models like XGBoost and gradient boosting machine show greater accuracy in classifying fire susceptibility, making targeted fire management strategies easier (Hang et al., 2024). AI is also enabled to support sustainable resource management as ML algorithms have the ability to forecast wild mushroom yields by considering climatic and forest structure data to ensure sustainable harvesting practices (Martínez-Rodrigo et al., 2024). AI technologies also optimize plantation mapping so that the precise crop health assessment and forecasting of yields become possible, thereby optimizing land utilization.

Improved ecosystem monitoring using AI ensures holistic control over ecosystem services, ensuring proper identification and management of non-wood forest products. In spite of these revolutionary improvements, challenges in the form of data quality, model interpretability, and necessitating specialized strategies specific to distinctive forest characteristics need to be overcome to best realize the potential of AI and ML in forest management (Udhaya Shree et al., 2024).

In areas with limited infrastructure, AI can start with low-cost mobile apps that sync when online, while developed regions can use AI for real-time monitoring. Blending AI with traditional practices ensures inclusive and adaptable forest management.

Case Study: In Brazil, AI-powered platforms are being used to monitor the forest in real time. Machine learning algorithms scan satellite images daily, identifying patterns that indicate illegal tree cutting. When suspicious activities are detected, alerts are sent immediately to enforcement teams. AI has broad applications across forest ecosystems, one of its most life-saving uses is in detecting and preventing wildfires. The next section dives into how AI is transforming wildfire management from the ground up.

2.5 Addressing AI Adoption Challenges in Developing Countries

The adoption of AI and ML in developing countries faces major challenges, including limited internet access and infrastructure in rural areas. In places like Kenya and India, mobile-based solutions that work offline have helped address connectivity issues. Data scarcity is another hurdle, being tackled through initiatives like Global Forest Watch and the development of cloud-based AI platforms. Capacity-building programs, supported by organizations like the World Bank, are training local experts to bridge the skills gap and make AI more accessible for forest management.

3 AI-Driven Forest Fire Detection and Prevention

Forest fires, sometimes referred to as wildfires, are uncontrolled flames that quickly spread across vegetation and frequently result in devastating economic and environmental harm. Conventional fire detection systems include human surveillance, satellite imaging, and ground-based sensors; however, the accuracy and response time of these approaches are limited. AI -driven solutions that use real-time data from many sources have become useful instruments for proactive fire prevention and early fire detection.

3.1 Fire Sense: The AI Revolution in Wildfire Detection

Over the past few decades, wildfires have become more frequent and powerful, driven by climate change, urbanization, and forest mismanagement. Conventional methods of detecting wildfires lookout towers, ranger patrols, and local reports usually arrive too late, when the fire has already grown out of control. Meet Fire Sense: A new generation of AI-based technologies to detect, forecast, and notify officials of imminent wildfires before they get too big to handle. It is not a system but an idea that embodies how artificial intelligence, when integrated with real-time data, can transform the way we watch out for forests. These smart systems look at enormous amounts of data from satellites, weather monitors, ground sensors, and remote observation drones to detect initial indications of fire whether it's a faint temperature increase, a quick surge in carbon monoxide, or the presence of smoke patterns in video feeds. Innovative AI-powered system that detects fires earlier than any current technologies by leveraging image analysis of high-definition panoramic cameras to identify smoke at the early stages which allows for faster responses to wildfires. The developed AI model scans images captured by the cameras every minute for emerging smoke indications which facilitates better communication with emergency services and more effective management of fire situations.

The AI model demonstrates high accuracy in detecting smoke patterns, ultimately aiming to reduce the damage caused by wildfires and protect human lives and property

(Remmelzwaal, 2023). Fire Sense's core lies in machine learning models that are trained to identify fire signatures. These algorithms are taught from past wildfire history, satellite imagery, and weather patterns, enabling them to forecast at-risk regions and identify suspicious patterns with remarkable precision. machine learning approach for real-time wildfire detection has demonstrated impressive accuracy, achieving a high detection precision of 98%. This system leverages the power of two specialized CNN models each tailored specifically for analyzing either daytime or nighttime imagery (Zhang & Zhang, 2022). By training the models on diverse datasets that reflect different lighting conditions, the system ensures accurate fire detection around the clock. This dual-model strategy significantly enhances reliability, enabling early alerts and faster emergency response, even in challenging visual environments. When coupled with computer vision, these programs can examine visual inputs in real-time, separating natural landscape and possible fire dangers, even in isolated or heavily forested locations. One of the biggest advantages of AI-powered wildfire detection is speed. While human spotters may take hours to detect and report a fire, AI can review thousands of data points in minutes and automatically send alerts to forest managers, emergency services, or even local community members. This enables quicker decision-making, more rapid containment, and, ultimately, less economic and environmental damage.

The AI-Based Forest Fire Early Detection and Response System demonstrated in Fig. 4 utilizes sophisticated technologies like satellites, drones, environmental sensors, and deep learning to provide early detection and prompt response to forest fires, especially in forest areas. Underpinning the Fire Sense AI Engine is processing real-time vision and sensor information to identify abnormalities like smoke, heat signatures, and gas release utilizing trained deep learning models. Satellite thermal scanning and infrared inspection give widespread monitoring, with close-range AI-driven monitoring provided by drones. Ground sensors monitor temperature, humidity, and gas concentration in real time to supply the system with essential information. All data are brought together in a centralized data repository where big data analysis and predictive modeling are used to produce risk maps and predict fire spread. The system initiates early warning notifications and initiates emergency response mechanisms, ultimately seeking to reduce fire damage, safeguard ecosystems, and save lives through proactive, data-driven action.

3.2 Eyes in the Sky: High-Tech Detection with Satellites, Drones, and Sensors

Forests are huge, frequently inaccessible, and difficult to track using conventional ground-based techniques alone. Luckily, the technological progress of today has given us quite literally "eyes in the sky," a potent combination of satellites, drones, and ground sensors all coordinated under the direction of AI to track forests in real time and identify wildfires before they get out of control. Satellites form the first layer of this high-tech surveillance network. They circle our planet, imaging persistent streams of information regarding land surface temper-

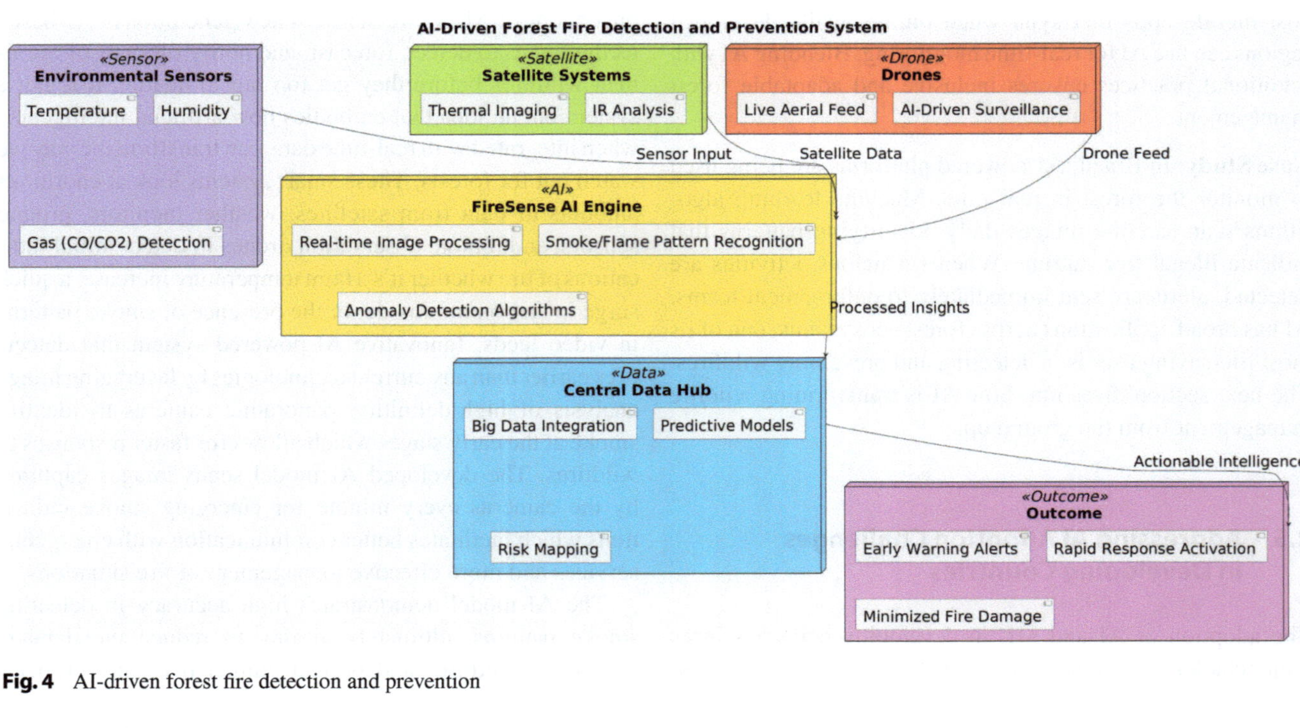

Fig. 4 AI-driven forest fire detection and prevention

ature, vegetation dryness, and detectable smoke plumes. It is AI here that plays its part by evaluating this huge bulk of satellite imaging and sensor records efficiently and efficiently. Machine learning algorithms fed with decades worth of wildfire history can analyze satellite imagery and pick up on fine details that hint at the beginning of a fire like a heat anomaly or a trail of smoke often even before it can be detected by the naked eye. Satellites, though, are not always sufficient, particularly when a fire is taking place under dense foliage or bad weather.

Drones equipped with AI-enhanced vision systems, thermal imaging sensors, and high-definition cameras can fly over delicate forest regions and gather extremely precise, real-time data. They can make burn maps, locate fire hotspots, and even track the paths taken by wildlife fleeing fire. Drones act as groundless fire scouts in emergency situations, reaching areas that are dangerous or impossible for people to access. The airborne layer is supported by a growing constellation of ground-level sensors, often disguised as weather stations, trees, or the ground. Environmental parameters like temperature, humidity, carbon monoxide levels, and wind speed are recorded by these sensors. Plugged into a central monitoring system powered by artificial intelligence, their data can activate automatic alarms instantly upon detecting irregular conditions enabling an enormously rapid, in-place response ability. Satellites provide the big picture, drones offer precision and flexibility, and ground sensors deliver constant, close-up monitoring. Together, they form a layered, intelligent surveillance ecosystem that continuously scans forests for the earliest signs of fire. And with AI coordinating the analysis and response, this setup can operate around the clock, covering areas that would otherwise go unmonitored.

The Flame and Smoke Detection Dataset (FASDD) comprises over 100,000 images, facilitating the training of deep learning models for fire detection Drones deployed on satellites, drones, and ground sensors, enabling collaborative fire observation and detection in a space-air-ground integrated network environment, enhancing wildfire detection capabilities across various platforms (Yu et al., 2023).

3.3 Building Early Warning Systems: From Alerts to Action

Where fighting forest fires is concerned, every minute counts. The earlier the fire is caught, the bigger the opportunity of confining the blaze before it can spread with full fury. This is where AI-based Early Warning Systems (EWS) fit in they act as a connector between data-based detection and on-ground response. These systems are engineered to detect danger at its earliest avatar much earlier than actual fire flames can be spotted. They operate by combining information from satellites, drones, ground sensors, and weather stations, inputting this data into machine learning models that have been trained to identify high-risk conditions and likely ignition points. AI doesn't merely detect that something is potentially amiss it looks at trends, cross-checks historical data, and foretells where a fire will most likely ignite, how it may spread, and what resources will be required for response. Among the strongest points about AI-based EWS is the fact that they can provide instantaneous alerts. Whenever preliminary indications of a fire such as an immediate temperature spike, a fall in humidity, or unusual concentrations of gases such as carbon monoxide are picked up, the system sends out alert messages automatically. The warnings are transmitted directly to forest departments, fire control authorities, local administrative offices, as well as onto the mobile phones of rangers and citizens operating in danger-prone areas. AI techniques, including machine learning and deep learning, are being utilized to analyze satellite images, weather parameters, and historical fire data, enabling accurate predictions of fire behavior and spread. These advancements have strengthened early warning systems, facilitating efficient firefighting and evacuation strategies (Zhang & Zhang 2022) In California, real-time sensor networks in forests have been deployed with AI-driven platforms. These systems not only respond to smoke and unexplained heat, but also alert firefighting crews automatically, saving precious minutes of response time. In Australia, machine learning models have already been applied to model fire spread in various environmental conditions, assisting in evacuation strategy and containment planning.

3.4 Fighting Fire with Data: Real-World Wins from AI-Powered Prevention

AI is no longer a concept of the future it's making a tangible impact today, especially in how we detect, manage, and prevent forest fires across the globe (Ramadan et al., 2024). From identifying high-risk zones to enhancing emergency response strategies, AI has emerged as a crucial tool in data-driven wildfire prevention. Among the strongest examples is Australia's deadly 2019–2020 bushfire season, notoriously called the "Black Summer." Amidst and after this record-setting disaster, ML and AI technologies proved instrumental in refining fire forecasting models, mapping evacuation routes, and coordinating emergency assets. Entities like the CSIRO (Commonwealth Scientific and Industrial Research Organization) worked with regional governments to infuse AI within their firefighting systems. These systems also drew on past fire data as well as live environmental conditions such as wind direction, humidity levels, and plant density to predict how fires could spread. This allowed quicker, wiser decisions about where to send resources and how best to evacuate fire-ravaged communities. Success for these tools speaks to the expanding role AI plays not only in responding to wildfires, but in actively mitigating their effect. In India, AI-powered systems are being piloted in states having high forest cover to track real-time environmental information from IoT-based ground sensors as well as drone monitoring. These systems enable detection of illegal operations such as burning and wood-cutting without permission, which are typical causes of wildfires. AI is also being utilized for sending notifications directly to regional language mobile apps-based local forest rangers, enabling rapid response even in remote tribal regions. At the same time, in Europe, initiatives such as advanced forest fire fighting (AF3) () are employing AI to manage aerial and ground firefighting operations. These systems process incoming data from satellites and manned aircraft to assist authorities in planning large-scale interventions with increased precision and efficiency. What is so successful about these AI tools is that they can learn and adapt. With each processed fire event, the system gets smarter—enhancing its predictions, minimizing false alarms, and better preparing for the future. UAV–IoT system enables rapid wildfire detection within 1–5 min of ignition, achieving 99.46% accuracy. This technology significantly enhances early fire detection, reduces response times, and minimizes potential damage, showcasing real-world benefits in forest fire prevention (Zhang & Zhang 2022). AI not only assists firefighting but actually assists in changing the whole approach from reaction to prevention. In wildfire-prone areas, AI-based fire detection and carbon modelling work together to protect forests. Early detection helps reduce emissions, while AI-driven recovery strategies guide reforestation, strengthening climate resilience and speeding up carbon restoration.

Case Study: In California, a region known for its frequent and destructive wildfires, AI-powered systems are now being used to predict fire risks and detect fires in their early stages. For instance, the use of satellite imagery combined with AI models helps analyze climate conditions, vegetation types, and historical fire data to predict which areas are most likely to catch fire. By integrating these predictions with real-time drone surveillance, authorities can deploy resources more efficiently, ensuring faster response times and reducing the overall damage caused by wildfires.

4 AI for Forest Carbon Sequestration and Climate Change Mitigation

Forest act play a very vital role in transferring carbon dioxide from the atmosphere and mitigating the climate change effect. However efficient tracking and management of forest carbon storage requires advanced technological capability. AI has emerged as a potent vehicle in this realm as AI provides inbuilt precise analysis, predictive modelling processes, and real-time monitoring of carbon dynamics. By utilizing machine learning algorithms, the AI can process large datasets from satellite imagery, drone surveillance and ground based sensors to estimate forest biomass, assess carbon sequestration rates and detect the deforestation trends. AI powered models can also predict the change in the carbon dynamics of forests following influences of climate change predictions enabling efficient data driven decisions by policymakers and conservationists for sustainable forest management (Ali et al., 2025).

4.1 Integrating Artificial Intelligence in Forest Carbon Sequestration Strategies

Forests are among the most significant natural carbon sinks, playing a crucial role in mitigating climate change by absorbing and storing atmospheric carbon dioxide (CO_2). Through photosynthesis, trees take in CO_2 and convert it into organic matter, storing carbon in their trunks, branches, leaves, and roots, as well as in the surrounding soil. This process, known as carbon sequestration, is essential for offsetting emissions generated from industrial activities, deforestation, and other land-use changes. However, the ability of forests to sequester carbon depends on several factors, including species composition, forest density, age, soil quality, and climatic conditions. Measuring carbon sequestration accurately is vital for effective climate policies, but traditional methods such as ground surveys and biomass allometric equations are often labor-intensive, costly, and time-consuming. AI is turning out to be a game changer. Through the use of AI technologies like machine learning, computer vision, and remote sensing analytics, scientists and environmental authorities can now track forests in real time, make more precise estimates of carbon storage, and identify changes in forest cover more quickly and accurately. AI models can also predict potential dangers, like fire breakouts or disease outbreaks, enabling proactive forest management. The integration of AI in forest carbon sequestration planning not only enhances data-driven decision-making but also maximizes the overall efficacy of climate action plans. AI-based technologies provide advanced solutions to improve the accuracy and efficiency of carbon stock assessment, making forest conservation efforts more data-driven and impactful AI models are able to track forest cover, estimate biomass, and quantify carbon stocks very precisely. AI also assists in the prediction and detection of forest fires, monitoring reforestation activity, and the identification of illegal logging activities, all of which are critical for maintaining and enhancing carbon sequestration. Implementation of nature-based carbon sequestration by introducing an AI framework that addresses key challenges in monitoring and verifying carbon capture processes (NetZeroCO_2., 2022).

Despite issues with data accessibility and the availability of technology, AI possesses significant tools to aid the global efforts in reducing carbon. Merging AI with forest management not only enhances conservation efforts but also speeds up the path to a sustainable and climate-resilient future.

4.2 AI Models for Estimating Forest Carbon Stocks

AI-driven models utilize machine learning (ML) and deep learning (DL) algorithms to analyze vast amounts of forest-related data, offering scalable and precise estimations of carbon stocks. These models integrate various data sources, including satellite imagery, light detection and ranging (LiDAR), and unmanned aerial vehicle (UAV) data, to monitor forest biomass and carbon storage capacity (Yan et al., 2024). Machine learning techniques such as regression models (e.g., random forest regression and gradient boosting) predict carbon stock levels by analyzing vegetation indices like the normalized difference vegetation index (NDVI) and other biophysical parameters. Deep learning approaches, such as CNNs extract spatial and structural patterns from high-resolution remote sensing data, further improving the accuracy of carbon stock estimations. SVMs classify forest types and predict carbon content based on spectral signatures derived from satellite images. Hybrid AI models, combining CNNs with recurrent neural networks (RNNs), enhance predictive accuracy by leveraging both spatial and temporal patterns in forest growth and carbon sequestration.

In Indonesia and Brazil, AI systems are applied to track shifts in carbon stock in tropical rain forests, whereas in Scandinavia and Canada, they are implemented to temperate and boreal forest landscapes. Such models facilitate near-real-time tracking of deforestation, forest degradation, and loss of carbon, making it easier for governments and conservation organizations to follow these activities. By making accurate, large-scale measurements of forest carbon, AI underpins carbon trading programs, AI isn't merely an upgrade in technology it's becoming an invaluable partner in the worldwide struggle against climate change. Additionally, AI-powered remote sensing technologies facilitate real-time carbon monitoring. LiDAR technology enables the estimation of tree height, canopy density, and biomass, providing accurate three-dimensional representations of forest structure. Hyperspectral imaging, powered by AI-driven spectral analysis, helps identify tree species and assess their photosynthetic activity, which correlates with carbon uptake. Synthetic aperture radar (SAR) data, processed through AI algorithms, is instrumental in estimating biomass in dense forests in thick and far-off forest areas, where optical imaging using traditional methods commonly fails under clouds or heavy canopies, (Colverd et al., 2024) SAR technology presents the breakthrough. Since SAR employs microwave signals that pass through rain, clouds, and vegetation to create high-resolution details of the Earth's surface, it proves to be particularly advantageous for tropical and evergreen forest tracking where other optical approaches are likely to fail. When integrated with AI algorithms, SAR data becomes even more effective. Machine learning algorithms can be trained on SAR-derived features like backscatter intensity and signal phase differences to estimate above-ground biomass and forest structure. These algorithms learn to identify intricate patterns in the radar data that are correlated with tree density, height, and moisture content—major determinants of biomass and, consequently, carbon stock. AI-augmented SAR analysis allows for large-scale biomass estimation with remarkable precision, even in regions inaccessible or dangerous for field survey teams. This method has already been implemented in places such as the Amazon, Congo Basin, and Southeast Asia, where dense forests contribute significantly to global carbon sequestration. Furthermore, the commissioning of missions such as ESA's BIOMASS satellite and NASA's NISAR (NASA-ISRO SAR) further emphasizes the increasing dependence on SA.

Case Study: Forests are one of our best tools for absorbing carbon dioxide and slowing down climate change. In Canada, researchers and forest managers are using AI to map and predict how much carbon forests can store over time. By analyzing satellite data, soil information, and tree growth rates, AI models can estimate the impact of different forestry practices on carbon storage. In British Columbia, for instance, these insights are helping design better forest management strategies that maximize carbon capture while maintaining biodiversity.

4.3 Impact of AI on Climate Change Mitigation Strategies

The integration of AI into forest carbon assessment has a profound impact on climate change mitigation strategies by enabling high-resolution, real-time carbon data analysis. These AI-driven insights enhance policy-making, conservation planning, and emission reduction initiatives. One significant application is AI-driven policy-making, where AI models assist in verifying carbon credits by accurately measuring and monitoring carbon sequestration. Carbon offset programs rely on precise forest data to ensure that emission reduction claims are valid. AI-powered simulations help policymakers design optimal conservation strategies by modeling potential deforestation scenarios and evaluating the effectiveness of different interventions. AI also supports the establishment of realistic emission reduction targets by continuously monitoring forest carbon contributions and updating climate action strategies accordingly. AI is also transforming deforestation monitoring by detecting illegal logging and land-use changes. Object detection models, utilizing CNN-based algorithms, can identify deforestation hotspots from satellite imagery, allowing authorities to take immediate action against unauthorized activities. Anomaly detection techniques, employing unsupervised learning algorithms, recognize unusual land cover changes indicative of deforestation or degradation. Predictive analytics powered by AI forecasts potential deforestation risks based on historical trends, climate patterns, and socio-economic factors, enabling proactive interventions.

The incorporation of AI into climate change mitigation measures represents a revolutionary chance to improve predictive capacity, optimize resource management, and inspire sustainability in different sectors. AI plays an important role in climate modeling through analyzing vast amounts of data, resulting in more precise predictions of weather catastrophes and environmental patterns, subsequently enabling improved planning and resource allocation for vulnerable populations (Pavapanunkul et al., 2024). In the energy industry, AI technology is being applied to streamline renewable energy systems, grid efficiency, and greenhouse gas emissions reduction, as presented in several case studies (Joshi, 2023). In supply chain management and agriculture, AI is supporting sustainability by enabling precision farming, reducing waste, and enhancing traceability, hence contributing to decreasing overall carbon footprints (Kumar et al., 2024b). Yet, the swift roll-out of AI also poses ethical challenges, such as possible data bias, biased access to technology, and even exacerbating current socio-economic disparities. These need to be rectified with robust ethical guidelines, open governance, and equitable approaches to guarantee that the advantages of AI are shared fairly. Finally, while there is tremendous potential with AI in combating climate change, its use needs to be bounded by fairness and social responsibility.

While AI and ML support forest health and carbon monitoring, challenges remain. Poor-quality data and sensor limitations can affect accuracy, and AI models must be continuously validated and adapted to different ecosystems to stay effective.

Beyond supporting climate strategies, AI also plays a growing role in evaluating carbon offset projects. Let's explore how technology is helping ensure these initiatives deliver on their promises.

4.4 AI for Assessing Carbon Offset Projects

AI specifically through ML application, is revolutionizing the evaluation, monitoring, and verification of carbon offset projects. Such projects—forest restoration and afforestation to regenerative agriculture and soil carbon sequestration—historically depend on human data collection and field visits, which are time consuming, expensive, and susceptible to errors. Using AI systems, huge amounts of data from satellite images, remote sensing, weather conditions, and land use history can be processed automatically to estimate levels of carbon sequestration more accurately and nearly in real time. Historical carbon and ecological data are used to train machine learning models, enabling them to forecast future carbon absorption capacity and identify inconsistencies or anomalies in reported results. AI is also crucial in detecting failing projects, detecting fraudulent claims, and efficient verification processes hence enhancing trust and transparency in carbon credit markets. By automating intricate analyses and delivering scalable technology, AI and ML are emerging as invaluable tools in guaranteeing the integrity and success of global carbon offset projects.

Carbon offset projects are crucial in balancing greenhouse gas (GHG) emissions through investments in forest conservation, afforestation, and reforestation initiatives. AI enhances the evaluation and monitoring of these projects by automating carbon accounting and ensuring transparency in carbon trading. Deep learning models analyze LiDAR and hyperspectral imaging data to estimate above-ground biomass, providing accurate assessments of carbon sequestration. AI-driven blockchain solutions facilitate secure and transparent carbon credit transactions, preventing fraud and ensuring that carbon offset claims are legitimate. Geospatial analytics, integrated with AI, assesses the long-term impact of offset projects by combining satellite data with environmental and socio-economic indicators. In sustainable forestry management, AI plays a pivotal role in optimizing conservation practices. Smart forestry drones, equipped with AI-powered image processing capabilities, assess tree health, detect pest infestations, and monitor growth rates, ensuring that forests remain healthy and resilient. IoT (Internet of Things) sensors, integrated with AI algorithms, measure real-time carbon flux levels, allowing for precise tracking of carbon sequestration efficiency. Climate-adaptive plantation models leverage AI to predict the most suitable tree species and planting strategies based on climate data, soil conditions, and biodiversity requirements, maximizing carbon sequestration potential. Figure 5 shows how AI can help estimate the amount of carbon stored in forests. It starts with different types of data like satellite photos, LiDAR scans, drone images, SAR data, and hyperspectral images. These data are processed using AI techniques to pick out important features such as vegetation health indicators (like NDVI), tree heights, biomass, spectral patterns, and how the trees are structured. These features are then used in various machine learning models—like random forests, gradient boosting, support vector machines, CNNs, and combined CNN + RNN models to accurately figure out how much carbon is in the forest.

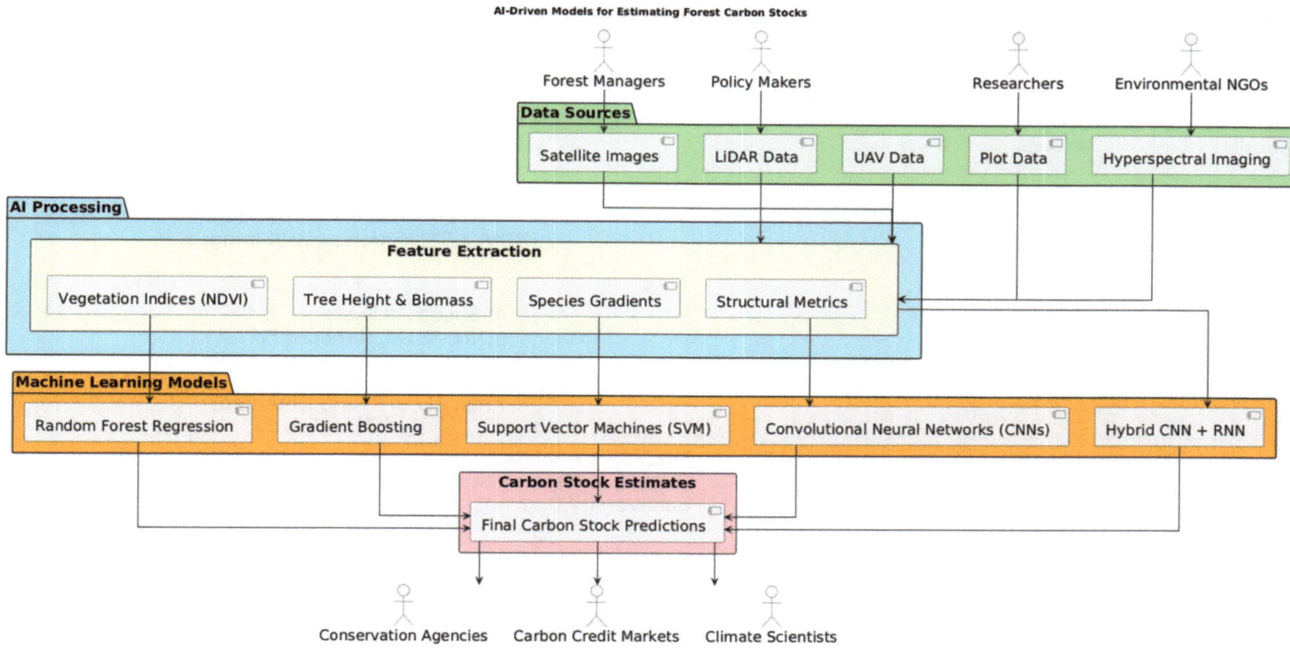

Fig. 5 AI models for forest carbon stock estimation

5 Bridging Tech and Trees—AI in Forest Management

AI and ML are transforming the way forest policies are formulated and implemented. Traditional forest policy development relied on historical data, expert opinions, and field surveys, which were often time-consuming and prone to biases. AI-driven policy frameworks, however, enable data-driven decision-making by integrating real-time ecological, climatic, and socio-economic datasets. Administrations and green agencies can leverage AI-driven models to evaluate deforestation threats, carbon sequestration capacity, and biological diversity conservation requirements, thus developing more efficient policies. Predictive modeling is one of the main uses of AI in policy-making. AI routines process immense volumes of remote sensing information, such as satellite images, LiDAR records, and aerial images collected by drones, to predict deforestation patterns and habitat loss. These models enable policymakers to foresee environmental issues and develop proactive conservation policies. Reinforcement learning (RL) algorithms maximize policy actions by modeling various conservation scenarios and learning the most efficient actions. For example, AI models can recommend optimal afforestation areas depending on soil conditions, rainfall intensity, and biodiversity levels, making reforestation policies effective and sustainable. AI also improves regulatory compliance and law enforcement within forest management. Through natural language processing (NLP) and computer vision, AI is able to analyze legal documents, determine conformity with global environmental legislation, and identify illegal logging from satellite imagery. Blockchain-based AI systems also promote policy implementation transparency through securely tracking timber trade, carbon credit exchanges, and conservation funding. Additionally, AI-powered platforms enable stakeholder participation through data-informed decision-making for policymakers, conservationists, and local communities. Artificial intelligence (AI)-based geographic information systems (GIS) facilitate interactive visualization of forest health indicators, enabling policymakers to make evidence-based decisions. These AI-based insights help governments frame sustainable forestry policies that reconcile economic progress with environmental conservation, ultimately promoting long-term environmental resilience. AI further optimizes forest inventory management by automating data capture and analysis. Deep Learning algorithms, including convolutional neural networks (CNNs), analyze high-resolution satellite imagery to estimate biomass levels, species distribution, and tree density. These results assist forest departments in planning sustainable timber yields, ensuring that economic utilization is balanced with ecological conservation. Furthermore, DSS tools are equipped with AI-driven spatial analysis models that determine land suitability for afforestation and reforestation purposes. Reinforcement Learning techniques simulate different reforestation strategies, optimizing species selection and planting patterns to maximize carbon sequestration. These AI-driven simulations help policymakers formulate scientifically sound reforestation policies that con-tribute to global climate change mitigation efforts.

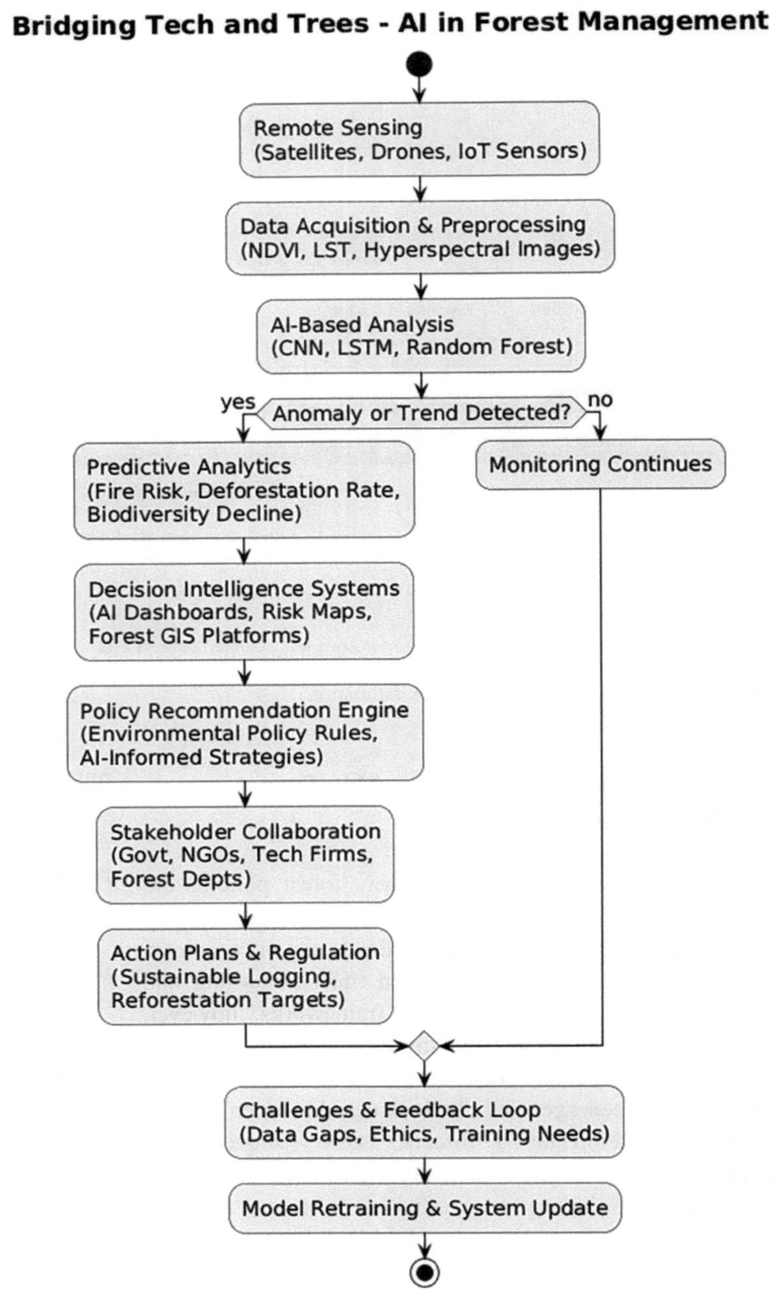

Fig. 6 Bridging tech and trees—AI in forest management

Figure 6 describes a technology pipeline for contemporary forest management with AI, remote sensing, and decision intelligence. It starts with remote sensing inputs satellite images, drone videos, and IoT sensors which record forest data in real-time. This raw data are processed with methods such as NDVI (normalized difference vegetation index), LST (land surface temperature) analysis, and hyperspectral imaging to derive significant environmental features. Then, machine learning algorithms like CNNs for image processing, long short-term memory (LSTM) networks for forecasting over a time horizon, and random forest classifiers for decision trees are applied to process this preprocessed data. If trends or anomalies such as rising temperatures, deforestation, or risk of fire are found, the system proceeds to predictive analytics predicting wildfire likelihood, loss of vegetation, or species extinction. These forecasts are mapped by decision intelligence systems, such as AI dashboards, heat maps, and geographic information system (GIS) platforms, that aid in depicting the scenario visually for forest officers and policymakers. These systems provide input to a Policy Recommendation Engine, which balances current environmental laws with strategic interventions like fire protection, reforestation zones, and sustainable logging limitations. To facilitate collaborative decision-making, the output is subsequently sent to relevant parties, including government organizations, non-governmental organizations, IT firms, and forest departments. Based on the insights, action plans are started for things like resource allocation, regulation enforcement, and forest restoration initiatives. At the same time, the system logs problems like missing sensor data, moral dilemmas (like surveillance), and incompetences among forest workers. These problems set off a feedback loop that ensures the system adjusts to forest conditions through improved AI performance, software updates, and model retraining.

5.1 Government Policies in AI-Based Forest Policy

By creating legislative frameworks, funding research, and facilitating AI adoption through national and international policies, governments can play a critical role in mainstreaming AI into forest management. By funding AI-conservation initiatives, providing public access to environmental data, and encouraging public–private partnerships, governments can increase the effectiveness of AI applications in sustainable forestry. A technologically driven approach to forest management includes AI-powered satellite monitoring, biodiversity conservation, and climate risk assessment regulations. With AI-based models, policymakers can assess the risks of deforestation, optimize afforestation efforts, and ensure adherence to international environmental regulations. In order for governments to implement scientifically supported reforestation policies that aid in reducing the effects of global warming, reinforcement learning algorithms also help with modeling conservation plans.

5.2 Legislative Frameworks and AI Integration in Sustainable Forestry

Legislative guidelines that facilitate AI-based forest management must be developed and implemented by governments. These regulations control the application of AI in conservation, guaranteeing appropriate implementation and preventing prejudice in judgment. By analyzing environmental regulations and detecting illicit logging through satellite imagery, AI-based technologies such as computer vision and natural language processing (NLP) help ensure legal compliance. Furthermore, by securely tracking carbon credit transactions, the timber trade, and conservation funding, blockchain-based AI systems promote transparency. The potential of integrating technology into forestry policies is demonstrated by government AI initiatives like NASA's Global Forest Watch and Microsoft's AI for Earth. In order to fully utilize AI's potential in sustainable forest management, guarantee environmental resilience, and advance the field of technology, In the future, improving cooperation among governments, NGOs, and the technology sector is crucial to achieving the full potential of AI and ML in sustainable forest management. Through coordinating technological innovation with policy structures and community participation, AI can be a successful instrument in safeguarding forest ecosystems, fighting climate change, and achieving environmental sustainability. Overall, AI-powered decision support systems enhance forest governance by improving resource allocation, risk assessment, and conservation planning. By leveraging AI-driven analytics, policymakers can implement sustainable forest management strategies that address ecological, economic, and social dimensions holistically.

5.3 Challenges in Integrating AI with Forest Governance

Challenges to integrating AI with forest management are data security issues, bandwidth constraints, interoperability problems, costliness, and the necessity of continuous research and dialogue among technologists, ecologists, policymakers, and local people to address these challenges in an effective manner (Ali et al., 2025). The most beneficial ailment that comes out of AI in forest governance is that it cannot be adopted in many cases for challenges. The dimension of the shortcomings is data provision and quality. Records of high-resolution satellite imagery, LiDAR data, and well-updated ecological databases rest on the lands of the majority of developing countries. Not having any streamlined methodologies for gathering data can cause historical records to become different from one another and hence cause false predictions from the AI. What is very important is the standardization on the data interoperability work in data as well as its open-access policy. Another awesome challenge in this field is that AI-induced decision-making is underpinned by various ethical and regulatory concerns. A machine-learning model always behaves as a black-box system due to which it is hard to understand the suggestions from policy makers and conse4rvationists. The absence of bioethics and transparency of this AI brings about the issues of accountability and bias in politics generated by AI. Explainable AI (XAI) technologies have come to the fore as a tool for more transparent AI governance and greater public trust in technologies. This is the last big obstacle digital divide comes in that forest management agencies within the developing world still lack the human resources and infrastructure to implement the needed applications. Capacity-building programs, technical training, and forming partnerships with institutions conducting research on AI are necessary to address this concern at its roots. Also, AI in the forest governance is an expensive exercise in machine learning, requiring a considerable amount of compute. While AI holds great promise in forest governance and policymaking, it also raises important ethical concerns. Issues such as algorithmic bias, lack of transparency, and the potential reduction of human oversight need careful attention. To ensure that AI-driven decisions are fair and accountable, it is crucial to maintain human involvement in reviewing and validating AI outputs. Balancing technological efficiency with ethical responsibility will be key to building trust and achieving sustainable outcomes.

Having explored various applications and impacts of AI and ML in forest management, we now turn our attention to summarizing the key findings and identifying future research directions.

6 Conclusion and Future Directions

The potential for AI and ML in forestry in the future is tremendous. AI technologies will increasingly transform conservation through enhanced real-time monitoring, predictive modeling, and decision-making. Future AI algorithms will probably include advanced deep learning algorithms to predict higher accuracy for forest health diagnostics, biodiversity monitoring, and climate change mitigation strategies. Combining AI with the IoT will make it possible to collect data automatically using smart sensors, drones, and satellite data, enabling more accurate deforestation identification and afforestation planning. AI-based GIS will also make land use planning easier by predicting the best locations for reforestation in terms of soil, water, and eco-factors. Governments and environmental agencies will also be required to utilize AI-powered digital twins' virtual representations of forest ecosystems to forecast the long-term effects of climate change and human activity on forest landscapes. With continued advancements in AI, technology-policy-community interface synergy will play a pivotal role in guaranteeing sustainable and resilient forest ecosystems.

Cloud-based AI platforms are helping low-income countries like Kenya and Indonesia monitor forests by analyzing satellite data for illegal logging and fire risks. Supported by initiatives like the World Bank and Global Environment Facility, these efforts are improving forest management and promoting sustainability in resource-limited areas.

6.1 Summary of Key Findings

The research emphasizes the revolutionary potential of AI and ML for transforming sustainable forest management. When conventional forest management is unable to deal with rising environmental and climate issues, the AI and ML technologies have evolved as effective devices to improve precision, efficiency, and decision-making. One key finding is how important AI and ML are when it comes to ecosystem monitoring and management. From pattern recognition and image recognition to predictive modeling, these are technologies that enable real-time monitoring of forest health, biodiversity, and environmental change. The research further indicates that supervised learning, unsupervised learning, and reinforcement learning are equally popular in forestry with random forest, support vector machines, and neural networks being the most successful algorithms in applications from species classification to disease identification. The study also delves into how AI has improved wildfire detection and prevention systems. Through the combination of satellite imagery, drone monitoring, and sensor data.

6.2 Challenges for Future Research

Despite its transformative potential, AI-driven forest management faces several limitations and challenges. One major challenge is data availability and quality. Many developing regions lack access to high-resolution satellite imagery and real-time environmental data, which can limit the effectiveness of AI models. Additionally, AI algorithms often require extensive computational resources, making their implementation costly and inaccessible for some forest conservation projects. There are also ethical implications for the role of AI in decision-making, including whether AI-based policies will exclude indigenous people or result in discriminatory conservation policies. A further constraint is the interpretability of the AI models most deep learning algorithms are "black boxes" that would confuse policymakers regarding the grounds for AI-suggested recommendations. Additionally, legal and regulatory hurdles need to be overcome so that AI technologies adhere to environmental legislation as well as data privacy regulations.

6.3 Last Words on the Future of Sustainable Forest Management

The convergence of AI and ML into sustainable forest management is a paradigm shift in forest monitoring, conservation, and restoration. Despite existing challenges, the future holds enormous promise for AI-based innovations to support conservation efforts, prevent deforestation, and reduce climate change. Governments, NGOs, and the tech industry need to join hands to bridge technology gaps, enhance data sharing, and guarantee ethical AI applications in forestry. AI research investment, encouragement of open data initiatives, and global cooperation will be critical to best leverage AI for forest conservation As AI advances, it will drive sustainable forest policies by supporting frameworks like the Paris Agreement and SDGs, enhancing carbon sequestration, forest resilience, and climate reporting to strengthen global climate action.

References

Ali, G., Mijwil, M. M., Adamopoulos, I., & Ayad, J. (2025). Leveraging the Internet of Things, remote sensing, and artificial intelligence for sustainable forest management. *Babylonian Journal of Internet of Things, 2025*, 1–65. https://doi.org/10.58496/bjiot/2025/001

Bailey, D. G. (2023). Machine learning (pp. 403–439). https://doi.org/10.1002/9781119819820.ch14

Colverd, G., Takami, J., Schade, L., Bot, K., & Gallego-Mejia, J. A. (2024). 3D-SAR tomography and machine learning for high-resolution tree height estimation. *arXiv preprint* arXiv:2409.05636. https://doi.org/10.48550/arxiv.2409.05636

Devi, R., Kumar, G., Narayana, P. A., Ramana, K. V., Kumar, K., & Rao, T. G. (2024). Forest fire prediction and management using AI (artificial intelligence), ML (machine learning) and deep learning techniques (pp. 324–327). https://doi.org/10.1109/icisc62624.2024.00062

Dolle, N. (2024). Comparison of classical and AI-based decision-making methodologies (pp. 179–188). Springer International Publishing. https://doi.org/10.1007/978-3-031-53598-7_16

Forest for Sustainable Development. (2023) (pp. 293–311). https://doi.org/10.1002/9781119910527.ch12

Hang, H. T., Mallick, J., AlQadhi, S., Bindajam, A. A., & Abdo, H. G. (2024). Exploring forest fire susceptibility and management strategies in western himalaya: integrating ensemble machine learning and explainable AI for accurate prediction and comprehensive analysis. https://doi.org/10.1016/j.eti.2024.103655

Hasan, R., Farabi, S. F., Kamruzzaman, M., Bhuyan, M., Nilima, S. I., & Shahana, A. (2024). AI-driven strategies for reducing deforestation. *The American Journal of Engineering and Technology, 6*(6), 6–20. https://doi.org/10.37547/tajet/volume06issue06-02

Impana, B. (2024). Integrating IOT and drone technology for real-time forest monitoring and fire detection: a sustainable approach for ecosystem preservation. *Indian Scientific Journal of Research In Engineering And Management*. https://doi.org/10.55041/ijsrem31933

Joshi, M. (2023). Harnessing the power of AI for sustainable climate strategies. *International Journal of Science and Research*. https://doi.org/10.21275/sr231214124609

Kolasani, S. (2024). Innovative applications of AI for sustainable supply chain management in environmental conservation (pp. 158–171). https://doi.org/10.1201/9781003497189-12

Krishnamoorthy, M., Asif, M. S. U., Nuvvula, R. S. S., Khan, B., & Colak, I. (2023). A design and development of the smart forest alert monitoring system using IoT. *Journal of Sensors, 2023*, 1–12. https://doi.org/10.1155/2023/8063524

Kumar, C. S., Mohan, C. R., Kumar, B. S., Singh, N., Lavanya, P., & Banerjee, D. (2024). Novel deep learning approaches to environmental management with sustainability. https://doi.org/10.1109/ic3se62002.2024.10593298

Kumar, C. S., Divya Vani, D. P., Damodar, K., Manoharan, G., & Veerapaga, N. (2024). Climate change mitigation through AI solutions (pp. 37–59). IGI Global. https://doi.org/10.4018/979-8-3693-4135-3.ch003

Kumari, P. L., Abid, Z., Abid, A., & Saxena, G. D. (2024). Robust forest fire detection using deep convolutional neural networks. *01*(01), 49–57. https://doi.org/10.58599/ijsmcse.2024.1111

Lozinska, T., Zadorozhnyy, A. I., & Масальський, В. П. (2024). Research of new technologies and innovations in the field of forestry. *Agrobìologìâ, 1(187)*, 268–276. https://doi.org/10.33245/2310-9270-2024-187-1-268-276

Martínez-Rodrigo, R., Fernández Águeda, B., Ágreda, T., Altelarrea, J. M., Fernández-Toirán, L. M., & Rodríguez, F. (2024). Assessment of mycological possibility using machine learning models for effective inclusion in sustainable forest management. *Sustainability, 16*(13), 5656. https://doi.org/10.3390/su16135656

Modise, J. M. (2024). Bad governance, corruption, nepostism, lawlessness, and goals for sustainable development. *International Journal of Innovative Science and Research Technology*, 2622–2633. https://doi.org/10.38124/ijisrt/ijisrt24jul973

Murugesan, S., Bharathiraja, N., Meenakshi, Ms. D., & Selvalakshmi, B. (2024). Artificial intelligence and applications. https://doi.org/10.59646/aip/221

NetZeroCO$_2$. (2022). NetZeroCO$_2$: An AI framework for accelerated nature-based carbon sequestration. In *Proceedings of the 2022 IEEE International Conference on Big Data (Big Data)* (vol. 55660, p. 10020359). https://doi.org/10.1109/BigData55660.2022.10020359

Osco, L. P., Junior, J. M., Ramos, A. P. M., de Castro Jorge, L. A., Fatholahi, S. N., de Andrade Silva, J., et al. (2021). A review on deep learning in UAV remote sensing. *The International Journal of Applied Earth Observation and Geoinformation, 102*, 102456. https://doi.org/10.1016/j.jag.2021.102456

Palaschuk, N., Gauthier, J., & Bullock, R. (2024). Developing community-based criteria for sustaining non-timber forest products: A case study with the Missanabie Cree First Nation. Forest Policy and Economics. https://doi.org/10.1016/j.forpol.2023.103104

Pavapanunkul, S., Raghav, A., & Raghav, R. (2024). AI in climate change with advancing mitigation tactics and adaptive strategies. *Practice, Progress, and Proficiency in Sustainability*, 1–16. https://doi.org/10.4018/979-8-3693-8684-2.ch001

Radzi, F. N. A., Zulkifley, M. A., & Kadim, Z. (2024). A concise review of artificial intelligence methods for forest-fire monitoring systems (pp. 1–5). https://doi.org/10.1109/icsgrc62081.2024.10691280

Raj, A. (2024). Artificial intelligence. *International Journal For Science Technology And Engineering, 12*(11), 646–655. https://doi.org/10.22214/ijraset.2024.64695

Ramadan, M. N. A., Basmaji, T., Gad, A., Hamdan, H., Akgün, B. T., Ali, M. A. H., Alkhedher, M., & Ghazal, M. (2024). Towards early forest fire detection and prevention using AI-powered drones and the IoT. *Internet of Things, 24*, 101248. https://doi.org/10.1016/j.iot.2024.101248

Remmelzwaal, L. A. (2023). An AI-based early fire detection system utilizing HD cameras and real-time image analysis. *Artificial Intelligence and Applications*, *3*(2), Article 975. https://doi.org/10.47852/bonviewaia3202975

Sardjono, M. A., Imang, N., & Saragih, B. (2023). The role of traditional local institutions in managing customary forest Hemaq Beniung in Juaq Asa Village, East Kalimantan, Indonesia. https://doi.org/10.24940/theijhss/2023/v11/i8/hs2308-0316

Sutriani, L., Impron, A., Saragih, V. B., Anggraini, S., & Suraji, S. (2024). Artificial intelligence in aspect of deforestation and plantation sustainability: bibliometric approach. *Literatus, 6*(2), 587–593. https://doi.org/10.37010/lit.v6i2.1583

Udhaya Shree, S., Usha, S., & Dharshana, S. (2024). A comprehensive machine learning approach for accurate forest fire forecasting (pp. 289–306). IGI Global. https://doi.org/10.4018/979-8-3693-2280-2.ch014

Wang, G. G., Lu, D., Gao, T., Zhang, J., Sun, Y., Teng, D., Yu, F., & Zhu, J. (2024). Climate-smart forestry: an AI-enabled sustainable forest management solution for climate change adaptation and mitigation.

Journal of Forestry Research, 36(1). https://doi.org/10.1007/s11676-024-01802-x

Yan, Y., Lei, J., & Huang, Y. (2024). Forest aboveground biomass estimation based on unmanned aerial vehicle–light detection and ranging and machine learning. *Sensors, 24*(21), 7071. https://doi.org/10.3390/s24217071

Yang, Z. (2022). Status, issues, and challenges of biodiversity: forest insects (pp. 325–362). https://doi.org/10.1007/978-981-16-9777-7_14

Yu, D., Foster, D. W., Fang, Y., & Flores, V. (2023). FASDD: an open-access 100,000-level flame and smoke detection dataset for deep learning in fire detection. *Earth System Science Data, 15*, Article 73. https://doi.org/10.5194/essd-2023-73

Zhang, A., & Zhang, A. (2022). Real-time wildfire detection and alerting with a novel machine learning approach. *International Journal of Advanced Computer Science and Applications, 13*(8), Article 801. https://doi.org/10.14569/ijacsa.2022.0130801

Climate Change Prediction Model on Smart Forestry 5.0 Using Novel Enhanced Streaming Gradient Boosted Trees-Based Optimisation Framework

S. Vidhya, Mishmala Sushith, Kandavalli Michael Angelo, and R. Geetha

Abstract

Smart Forestry 5.0 in conjunction with machine learning-driven optimisation is revolutionising forest management by enhancing resilience, sustainability, as well as efficiency. Predicting how climate variation will affect forests is crucial for proactive decision-making as the threat to global ecosystems grows. Advanced machine learning techniques improve carbon capture, forest development, as well as biodiversity preservation by making precise forecasts of environmental changes. However, it is challenging to predict how climate variation may affect intelligent forestry. Reliable projections are made more difficult by the complex nature related to climatic elements, the unpredictability associated with biological information, as well as the dynamic nature linked with forest ecosystems. Additionally, robust computational methodologies as well as adaptable methods are required for integrating remote sensing, Internet of Things (IoT), as well as big data analytics. Interdisciplinary approaches that combine Artificial Intelligence (AI), climate science, as well as sustainable forestry practices are needed to overcome these challenges. Hence, this chapter performs the Climate Change Prediction (CCP) model on Smart Forestry 5.0 with the help of innovative machine learning-based optimisation methodology. The data pertaining to the climate change on forests is first gathered from the sixth phase of the Coupled Model Intercomparison Project (CMIP6). The pre-processing of these gathered data is then accomplished using the Kalman filtering approach. From these pre-processed data, the extraction of the features is done by the Hilbert-Huang Transform (HHT) approach. Finally, the prediction of these extracted features is performed by the novel machine learning model called Enhanced Streaming Gradient Boosted Trees (ESGBT), where the parameter tuning of SGBT is accomplished by the optimisation algorithm referred as Wolverine Optimisation Algorithm (WoOA). Prediction accuracy maximisation and error minimisation is returned as the final fitness function. The findings demonstrate that the proposed ESGBT-WoOA for the CCP model on Smart Forestry 5.0 surpasses the other methods in terms of distinct error measures. The proposed ESGBT-WoOA for the CCP model on smart forestry is 13.06% and 61.65% better than the other state-of-the-art methods in terms of prediction accuracy and error, respectively.

Keywords

Climate change prediction model · Smart Forestry 5.0 · Novel enhanced streaming gradient boosted trees · Wolverine optimisation algorithm · Kalman filtering · Hilbert-Huang transform

S. Vidhya (✉)
Department of Computing Technologies, School of Computing, SRM Institute of Science and Technology, Kattankulathur, Chennai, India
e-mail: vidhyas2@srmist.edu.in

M. Sushith
Department of Information Technology, Adithya Institute of Technology, Coimbatore, India
e-mail: mishmalasushith_d@adithyatech.com

K. M. Angelo
Department of ECE, DVR & Dr. HS MIC College of Technology, Kanchikacherla, Andhra Pradesh, India
e-mail: kmichaelangelo@mictech.ac.in

R. Geetha
Department of Computing Technologies, School of Computing, SRM Institute of Science and Technology, Kattankullathur, Chennai, India
e-mail: geethar1@srmist.edu.in

M. Ashwin et al. (eds.), *Smart Forestry 5.0 for Powered AI Sustainable Environment and Climate*, Sustainable Artificial Intelligence-Powered Applications, https://doi.org/10.1007/978-3-031-99086-1_5

1 Introduction

A significant global concern, climate variation has a significant effect on forest ecosystems, affecting biodiversity, carbon storage, as well as the general health of forests. Forest sustainability is threatened by rising temperatures, shifting rainfall patterns, as well as extreme weather events, projecting the need for accurate climate prediction to support well-informed forestry management decision-making (Aun et al., 2021; Riutta et al., 2021). Through real-time monitoring as well as predictive analytics, Smart Forestry 5.0—a complex technical framework that integrates AI, the IoT, remote sensing, as well as big data analysis—is revolutionising forest conservation (Brodribb et al., 2020). However, the intricacy associated with climatic elements, changing environmental circumstances, as well as uncertainty in climate projections are challenging for traditional climate prediction methods to handle. Machine learning-driven optimisation is emerging as a viable way to address this problem by enhancing forecast accuracy, honing forest management techniques, as well as guaranteeing long-term sustainability. Climate modelling and machine learning work together to enable intelligent forest management, assisting environmentalists and policymakers in developing climate change-adaptive forestry solutions. Machine learning-driven optimisation can reveal hidden patterns as well as non-linear relationships within climate datasets, producing forecasts that are more robust as well as responsive to real-time information flows than existing statistical methods, which rely mostly on historical information as well as linear considerations. Furthermore, integrating large amounts of heterogeneous climatic information from many sources, including satellite imagery, sensor arrays, as well as climate modelling frameworks, is essential for Smart Forestry 5.0 (Chen et al., 2021; Clemmensen et al., 2021; Cosyns et al., 2020; Gao et al., 2021; Vries et al., 2021). Optimisation is crucial for efficiently handling high-dimensional datasets, reducing computing costs, as well as improving forecast reliability as the volume of climate information grows. As complex AI-oriented frameworks usually function as black-box methods, comprehending machine learning forecasts represents a major challenge in climate variation modelling for forestry. In order to address this, explainable AI (XAI) techniques are being researched to elucidate the ways in which climate influences forecast outcomes, improving the openness as well as usability of machine learning methods for forestry decision-makers. Additionally, the use of edge computing as well as IoT-based real-time information collection enhances model responsiveness, enabling forestry managers to take timely choices on the basis of continuously updated climatic information. Smart Forestry 5.0's CCP will be improved by the use of hybrid AI methods, which include deep learning, reinforcement learning, as well as evolutionary optimisation techniques. This will increase the prediction's dependability as well as flexibility in response to shifting environmental conditions. Machine learning optimisation is a groundbreaking approach to forestry CCP, offering unparalleled precision, effectiveness, as well as adaptability. As technology develops, reducing the adverse impacts of climate variation as well as ensuring the long-term health of global forest ecosystems will depend heavily on the integration of AI-powered optimisation as well as sustainable forestry practices.

1.1 The Evolution of Smart Forestry 5.0

Over time, forestry has made significant strides, moving from traditional methods to data-driven, technologically advanced solutions. The concept of Smart Forestry 5.0 represents the major enhanced stage associated with this evolution, combining machine learning optimisation, big data analysis, AI, remote sensing, as well as the IoT to create a more intelligent, effective, as well as sustainable forest management framework (Gong et al., 2021; Lull et al., 2020). The transition from manual labour as well as practical decision-making to data-centric, automated, as well as AI-supported approaches is seen in the development of Forestry 1.0 to Forestry 5.0. For forest management, Forestry 1.0 relied on traditional tools as well as human expertise, which frequently led to ineffective as well as unsustainable practices (Mayer et al., 2020). Mechanisation, as well as the first digital tools, was introduced in Forestry 2.0 and 3.0, increasing production but excluding real-time information integration. More precise forest information collection and predictive analysis were made possible by Forestry 4.0, which ushered in the era of automation, IoT-based monitoring, and Geographic Information Systems (GIS) (Soundararaj, 2024). In order to enhance biodiversity preservation, forest sustainability, as well as carbon capture efforts, Smart Forestry 5.0 now builds on these advancements by integrating AI-enabled automation, machine learning-oriented optimisation, as well as CCP methods. It makes it possible to follow forests instantly using IoT sensors, drones, as well as satellite photos, providing timely data regarding environmental variations, growth patterns, as well as forest health. By utilising cutting-edge technology to address climate variation concerns, improve resource effectiveness, as well as ensure the long-term health of forest ecosystems, Smart Forestry 5.0 represents the future of creative as well as sustainable forest management.

1.2 Understanding CCP Model on Smart Forestry 5.0

Forest ecosystems are significantly impacted by climate variation, which changes biodiversity, carbon storage, as well as the general health related to forests. As temperature fluctuations, precipitation cycle variations, as well as extreme weather patterns become more frequent, accurate prediction of these variations is essential for sustainable forest management. Real-time tracking as well as predictive analysis are made easier by Smart Forestry 5.0, an advanced technique that reduces climate-associated threats by combining AI, the IoT, as well as optimisation using machine learning. The CCP method in Smart Forestry 5.0 ensures improved prediction accuracy, proactive decision-making, as well as long-term sustainable forest management by combining predictive analytics, real-time information, as well as optimisation techniques. These methods will be essential in combating climate variation as well as enhancing global forest resilience as AI and machine learning continue to advance. The CCP model on Smart Forestry 5.0 can be diagrammatically shown as in Fig. 1.

Despite the promising capabilities of the CCP model in Smart Forestry 5.0, integrating machine learning into existing forestry management systems presents several real-world challenges. One key obstacle is the lack of standardised digital infrastructure in traditional forestry operations, making it difficult to collect and transmit real-time data at scale. Additionally, forestry stakeholders often face a steep learning curve in adopting AI-driven tools, which may require specialised training and technical support. Data heterogeneity—stemming from diverse sensor formats, inconsistent metadata, or incomplete records—can hinder model training and accuracy. Furthermore, the integration of high-frequency streaming data may strain existing computational resources and require system-wide upgrades. There are also concerns around model interpretability and trust, especially when decisions affect ecological conservation and local communities. Overcoming these challenges will require coordinated efforts in policy development, infrastructure modernisation, and cross-disciplinary collaboration between data scientists, ecologists, and forestry managers.

1.3 Traditional Smart Forestry Challenges and the Need for CCP Model on Smart Forestry 5.0

Many obstacles still stand in the way of Smart Forestry's ability to accurately predict the consequences of climate variation as well as ensure sustainable forest management, even with the advancements it has achieved in integrating remote sensing, IoT, as well as AI-oriented analytics. Various significant obstacles include:

1.3.1 Inaccurate Climate Forecasts

Traditional methods rely on historical climate data as well as statistical techniques that fail to take non-linear environmental changes into consideration. Forest managers are unable to take preventive measures to stop wildfires, deforestation, as well as biodiversity loss because of inaccurate climate projections.

1.3.2 Limited Real-Time Data Integration

Many traditional methods are based on static information and do not incorporate real-time environmental observations.

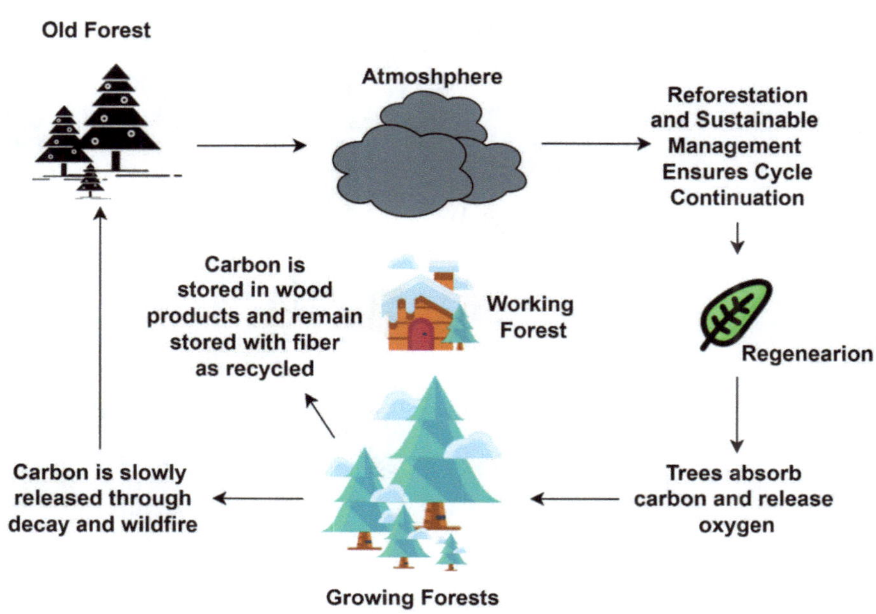

Fig. 1 CCP model on Smart Forestry 5.0

Predictions are out-of-date as well as unreliable without continuous information flows from IoT devices, satellite imagery, as well as climate archives (Arivumani & Nagarajan, 2024).

1.3.3 High Computational Needs

Complex climate methods require a lot of processing power, which makes more conventional methods slow as well as inefficient. Decision-making is delayed as a consequence of inefficient management of massive datasets.

1.3.4 Lack of Optimisation Techniques

Machine learning-oriented optimisation is not incorporated into numerous traditional forestry methods, which results in higher prediction errors as well as less flexibility.

In order to overcome these obstacles, Smart Forestry 5.0's CCP method integrates machine learning as well as optimisation methods to increase forecast accuracy as well as computational effectiveness (Janani et al., 2017).

1.3.5 Increased Predictive Accuracy

By significantly reducing disparities such as RMSE, MSE, as well as MAE, the WoOA-tuned ESGBT architecture makes climate predictions much reliable.

1.3.6 Real-Time Data Management

For continuous monitoring as well as adaptable decision-making, the framework makes use of IoT sensors, satellite information, as well as AI-oriented analytics.

1.3.7 Effectiveness Optimisation

Advanced metaheuristic algorithms improve the speed as well as scalability of information processing, making climate predictions much effective as well as feasible.

The CCP method in Smart Forestry 5.0 overcomes the shortcomings of traditional approaches as well as increases forestry's resistance to climate variation by ensuring accurate climatic predictions, proactive forest management, as well as long-term sustainability (Manoharan et al., 2023). Figure 2 depicts the traditional smart forestry challenges.

Fig. 2 Traditional smart forestry challenges

1.4 The Role of Machine Learning in CCP Prediction on Smart Forestry 5.0

CCP in Smart Forestry 5.0 relies heavily on machine learning, which enables accurate, data-driven prediction to enhance forest sustainability. Complex environmental interactions provide difficulties for traditional climate methods, but machine learning algorithms use pattern recognition, real-time information, as well as optimisation techniques to improve prediction accuracy.

1.4.1 Data Analysis and Feature Selection

Sizable datasets are collected for analysis from climate methods (such as CMIP6), IoT devices, as well as remote sensing. To remove noise as well as inconsistencies from climatic records, Kalman filtering is applied. Much accurate forecasts are produced by identifying important climatic features using the HHT.

1.4.2 Machine Learning for Climate Prediction

In order to identify trends in climate variation, ML algorithms look at temperature patterns, variations in rainfall, CO_2 concentrations, as well as soil moisture. By continuously adjusting to novel information, the ESGBT method increases prediction accuracy. ML techniques enable nonlinear modelling, which captures complex environmental interactions better than traditional statistical techniques.

1.4.3 Refinement for Increased Precision

By modifying hyperparameters, the WoOA boosts the CCP method's efficacy. High accuracy is ensured by optimisation techniques that reduce MSE, RMSE, as well as MAE. ML-based optimisation enables instantaneous adjustments for changing climatic constraints as well as faster computations.

In Smart Forestry 5.0, machine learning revolutionises CCP forecasts by offering real-time flexibility, increased accuracy, as well as efficient information administration. Smart Forestry 5.0 encourages sustainable forest management, proactive decision-making, as well as resilience to the impacts of climate variation by fusing machine learning algorithms with optimisation techniques.

While machine learning offers substantial improvements in climate prediction accuracy, several limitations persist that must be acknowledged. First, many ML models are data-hungry and require vast amounts of high-quality, labelled climate and forestry data, which may not be uniformly available across all regions. Second, ML models often operate as 'black boxes', making it difficult for forestry managers and policymakers to interpret how decisions are derived—a critical concern for explainability in ecological applications. Additionally, ML-based systems may struggle with outlier events, such as rare wildfires or abrupt climatic shifts, due to their under-representation in training datasets. Models trained on one geographic region may also fail to generalise across different forest types or climates. Finally, high computational requirements and energy consumption associated with training complex models like gradient-boosted trees or deep learning networks can pose sustainability and resource barriers. These limitations underscore the importance of hybrid approaches that blend machine learning with domain expertise, adaptive feedback, and transparent validation frameworks.

2 Literature Review

The strategy for selecting indicators that assessed the 'climate-smartness' related to forest management was described in this publication, along with the step-by-step procedure that produced a comprehensive as well as standard definition related to CSF (Bowditch et al., 2020). The CSF concept placed a strong emphasis on the need to integrate as well as stop the growth of adaptation, mitigation, as well as social components independently. In order to identify sustainable forest management indicators that were essential to CSF, a network analysis was used to support the participatory iterative process that involved a variety of experts in forest-associated fields from the CLIMO project. The concept developed here was an important first stage towards promoting CSF that would assist forestry sector practices. It might be adapted for particular local circumstances and used as a method across Europe.

It was proposed that frequent use of low tree harvest intensity might help CSF maintain stand density at moderate levels (> 60%, e.g. ~ 80%) (Sterck et al., 2021). Furthermore, it was proposed that the actual optimal density would change depending on the kind of tree, site features, as well as management techniques. The next phase was carrying out research from both above as well as below ground and combining mechanistic methods of carbon, water, as well as nutrient dynamics in trees as well as forest soils with forest trials. This strategy sought to move existing research—which usually focussed on either soil or trees and was often context-specific—into a more general forest methodology.

Both protected as well as unprotected oak (Quercus spp.) forests along a latitudinal gradient (55.6 °N–60.8 °N) in Sweden were compared for productivity (tree growth measured utilising dendrochronology) as well as biodiversity (flora evaluated via vegetation plots) and arthropods collected with Malaise traps and analysed through metabarcoding (Forsman et al., 2024). The region's past, present, as well as projected future climates were also looked. Yet, neither was higher in protected areas than in non-protected ones, nor was it associated with present ground temperature.

In contrast to the known global latitudinal diversity gradients, the species richness of plants as well as arthropods increased as one moved northward, probably describing recent climate-driven community redistributions. There was no difference in the species composition of arthropods among protected as well as unprotected sites.

In light of projected future temperature constraints, this research developed a novel site-specific approach that might identify the ideal tree species for certain planting places (Wenhuan et al., 2024). Utilising climatic niche forecasts for 100 key tree species, the feasibility as well as effectiveness related to this strategy at 10 typical sites in different climate zones in China was evaluated. The findings demonstrated the significance as well as effectiveness of this approach, which could choose the right tree species tailored for every potential planting site across China beneath a range of climate change scenarios. These findings demonstrated how important it was to use this technique in order to avoid choosing the wrong species or overlooking ones that could be suitable.

This perspective encouraged quick fixes to improve resilience through effective disaster response as well as preventive programmes created for this crucial sector (Cesco et al., 2024). Three case studies from the AFA area were included in the study that followed, which included elements associated with ontology as well as the conceptual methodology of an emergency management framework that addressed both prevention as well as restoration. Each case study described the emergency as well as recovery procedures, evaluated the risk prevention techniques that were implemented, highlighted the shortcomings in the response efforts, and was characterised by its location, chronology, nature, as well as results. The analysis highlighted the importance of acknowledging lessons learnt as well as potential future applications, and it incorporated a historical comparison element in line with the proposed conceptual methodology.

The purpose of the document was to outline the development of Industry 5.0 and its anticipated transition to Forestry 5.0 (Holzinger et al., 2024). Through a suggested methodology, the goal was to shed light on the state of these developments as of right now, identify helpful technologies—particularly AI—and highlight the challenges that stand in the way of their successful use in forestry. In order to successfully transition to Forestry 5.0, the article concluded by highlighting the significance of Human-Centred AI progress, which aimed to support human workers rather than replace them. To solve these issues as well as pave the way for a sustainable as well as technologically advanced future in the forestry sector, a cooperative approach including legislators, ecologists, technologists, as well as forestry experts was essential.

This concept paper: (1) described the origins as well as development of CSF, which integrated Carbon Forestry, Ecological Forestry, as well as Smart Forestry; (2) assessed the goals of CSF in enhancing the resilience related to forest ecosystems, increasing their capacity to store carbon, as well as encouraging the usage and development of cutting-edge methodologies in CSF, including robotics, AI, Light Detection and Ranging, as well as digital twins of forests; and (3) presented the latest applications of CSF that were based on previous research on the structure as well as function of forests using next-generation data technologies at Qingyuan Forest, China (Geoff Wang et al., 2024). The course for CSF's growth was suggested, taking into account key research topics as well as advancements in technology, on the basis of these practices as well as thoughts.

It was examined how U.S. CSAF incentives changed over time, moving from offset markets to biofuel laws (O'Hara, 2024). There were no biofuel requirements in the US twenty years ago, and offset markets were supposed to provide the mechanism for CSAF incentives. This implied that issues with additionality, which have impeded the development of CSAF offset standards, were less pertinent to biofuel projects. The development of carbon procedures in animal digesters, forests, as well as agricultural soils was specifically examined as instances. It was concluded by discussing potential future developments for CSAF credits.

As part of the methodology for Climate-Smart Forestry (CSF), the main goal related to the UrbanFoS project was to create indicators of soil deterioration in urban forests (Kašanin et al., 2024). Three stages would be used to achieve this: (1) Degradation-impacted important sensitive soil qualities were identified; (2) Field as well as lab climatic simulations were performed; and (3) Urban forest ecosystem services were evaluated. Since the sustainability associated with urban forests depended on the elements that defined soil sensitivity to degradation, the knowledge gained would be applied to develop a soil degradation index that was pertinent to urban forest soils that were struggling with climate variation.

This research looked at how bioengineering as well as AI might change sustainable forestry practices, emphasising how they could improve timber output, increase forest vitality, as well as preserve biodiversity (Mahmood et al., 2024). Forest managers might monitor tree health, assessed the results related to various management techniques, as well as gain instant insights into forest constraints by utilising AI-powered technology such as machine learning algorithms, remote sensing, as well as predictive analytics. Concurrently, advancements in bioengineering, such as Genetically Modified (GM) trees, have shown promising results in enhancing growth rates, strengthening resistance to diseases as well as pests, and improving adaptability to changing climate constraints, all of which contribute to an increase in the overall productivity of forests. The potential risks as well as moral ramifications of bioengineering in forestry were further examined in this paper, which emphasised the need for stringent regulatory frameworks to ensure environmental sustainability.

Recent studies emphasise that Smart Forestry 5.0 significantly advances traditional forest management and climate adaptation practices by incorporating Artificial Intelligence, remote sensing, and real-time data analytics. (Holzinger et al., 2024) highlights the shift towards Human-Centred AI in forestry, promoting adaptive, technology-driven solutions that support, rather than replace, human expertise. (Wang et al., 2025) demonstrates how AI-integrated platforms can enhance sustainability by forecasting climate impacts and optimising forest responses. (Georgiev et al., 2020) further illustrates the practical edge of smart forestry through a UAV-based early fire detection system, which drastically reduces response time compared to traditional methods. Collectively, these findings underline that Smart Forestry 5.0 not only improves prediction accuracy and operational efficiency but also introduces scalable, intelligent frameworks for managing the increasing complexities of forest ecosystems under climate stress.

Problem Statement: Forest ecosystems are seriously threatened by climate variation, which has an impact on biodiversity, carbon storage, as well as overall forest health. Traditional forestry management techniques are no longer sufficient to ensure long-term sustainability due to the unpredictability associated with climate variation, which includes extreme weather events, rising temperatures, as well as changed rainfall patterns. The methods used for climate prediction today rely on historical information as well as set algorithms that cannot adapt to variations in the environment in real time. Furthermore, complex interactions between forest ecosystems as well as climatic variables need for advanced computational methods to improve forecast accuracy. A promising method for real-time climate observation as well as responsive management is Smart Forestry 5.0, which blends AI, the IoT, as well as big data analytics. Important challenges still exist, though, including disparities in the information, high processing demands, as well as the need for effective feature extraction and optimisation techniques. Successfully addressing climate threats will be challenging for forest management activities without a robust as well as efficient climate prediction method. This study offers a CCP method that uses ESGBT improved by the WoOA to address these problems. In order to provide proactive measures for climate adaption in long-term forest preservation, this method aims to increase forecast accuracy, maximise computing efficiency, as well as encourage sustainable decision-making in Smart Forestry 5.0.

3 Proposed Work

The proposed ESGBT-WoOA for the CCP model on Smart Forestry 5.0 is involved of four modules such as data collection, pre-processing, feature extraction, and prediction. The CMIP6 is where the data about how climate variation affects forests is first gathered. The gathered information is subsequently pre-processed using the Kalman filtering technique. This pre-processed information is subjected to feature extraction using the HHT approach. In the end, the prediction of these derived attributes is carried out by the novel machine learning method called ESGBT, and the optimisation method WoOA is used to tune the parameters of SGBT. The ideal objective function is defined as maximising prediction accuracy as well as minimising errors.

1 There are four major parts to the structure. **Data Collection:** The CMIP6 provides the climate variation information for Smart Forestry 5.0. These datasets assist effective machine learning-driven optimisation for sustainable forest management by including historical as well as projected climatic parameters that are essential for predicting environmental changes.
2 **Pre-processing:** To eliminate noise as well as inconsistencies, the collected CMIP6 climate variation information is pre-processed using the Kalman filtering technique. By lowering variances, this method increases data accuracy as well as reliability and offers better input for machine learning-oriented Smart Forestry 5.0 improvement for precise climate variation projections.
3 **Feature Extraction:** Key characteristics are extracted from the pre-processed CMIP6 climate variability information using the HHT approach. This method improves the accuracy related to climate variation projections in Smart Forestry 5.0 by decomposing complex climate signals into intrinsic mode functions that capture non-linear as well as non-stationary trends.
4 **Prediction:** The ESGBT method uses extracted information to predict how climate variation would affect Smart Forestry 5.0. By modifying the method's parameters, the WoOA ensures accurate outcomes with little error. The results confirm that ESGBT-WoOA accomplishes better than alternative methods with respect to prediction accuracy.

These modules operate in a sequential yet interdependent manner, where each stage significantly influences the performance of the next. The process begins with data collection, where historical and projected climate data are obtained from CMIP6. This raw data is then passed to the pre-

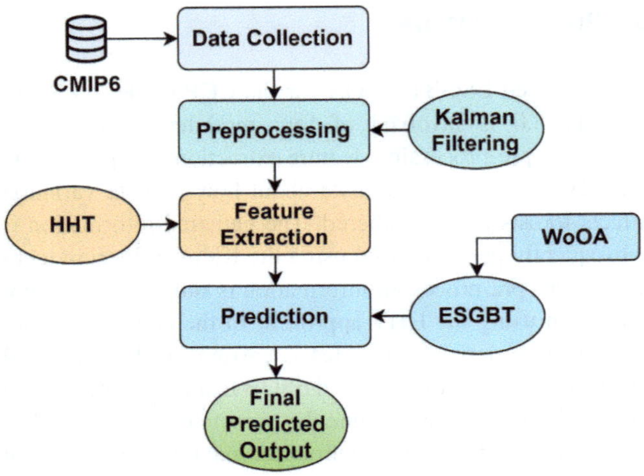

Fig. 3 Overall proposed ESGBT-WoOA for the CCP model on Smart Forestry 5.0

processing module, where the Kalman filtering technique is applied to eliminate inconsistencies and noise, ensuring that only clean, high-quality data progresses to the next stage. The feature extraction module, powered by Hilbert-Huang Transform (HHT), analyses this cleaned data to identify the most relevant temporal and frequency-domain features, thereby reducing dimensionality and focusing only on variables with predictive importance. These extracted features are then used by the **prediction module**, where the Enhanced Streaming Gradient Boosted Trees (ESGBT) model processes the input to generate climate forecasts. To further improve accuracy and robustness, the Wolverine Optimisation Algorithm (WoOA) tunes the hyperparameters of the ESGBT model in real time. This feedback loop between prediction and optimisation ensures that the model continuously adapts to new climatic data, thereby enabling accurate, scalable, and real-time decision-making in Smart Forestry 5.0.

The overall proposed ESGBT-WoOA for the CCP model on Smart Forestry 5.0 is diagrammatically depicted in Fig. 3.

3.1 Data Collection

The CMIP6 dataset offers both historical and projected climate data essential for climate variation research, supporting Smart Forestry 5.0. Generated by multiple Global Climate Models (GCMs), it provides valuable insights into the long-term impacts of climate change on forests. CMIP6 encompasses simulations from diverse climate methods, enabling comprehensive multi-scenario analysis of climate impacts. The dataset includes surface air temperature, rainfall patterns, daily temperature extremes, seasonal changes, and probabilities of extreme weather events. These factors are critical for assessing carbon storage, hydrological balance, forest growth, fire risks, and photosynthesis efficiency—key indicators of forest health and sustainability. CMIP6 incorporates various Shared Socioeconomic Pathways (SSPs), representing future climate scenarios influenced by human activities. These scenarios range from low emissions with moderate warming beneficial to forests, to current policy continuations with minimal climate effects, and worst-case scenarios marked by severe warming, droughts, and wildfires. Freely accessible via the Earth System Grid Federation (ESGF) and other repositories, this dataset supports machine learning-driven optimisation to predict climate impacts on forests. It aids decision-making related to afforestation, conservation, and sustainable forest management.

3.2 Pre-processing by Kalman Filtering

For CCP models in Smart Forestry 5.0 to remain reliable as well as accurate, pre-processing is necessary. Before applying machine learning-oriented optimisation techniques, gaps, noise, as well as inconsistencies in raw climate information—generally obtained from sources like the CMIP6—must be addressed. The climatic data sets are cleaned up of noise as well as fluctuations using the Kalman filtering technique. By predicting their expected values based on historical findings, this method actively evaluates missing or ambiguous data points. It increases the dataset's usefulness for accurate climate modelling by eliminating discrepancies. Strong pre-processing techniques enable Smart Forestry 5.0 to utilise machine learning-oriented optimisation to improve forest sustainability, improve climate variation projections, as well as hone decision-making for long-term forest management.

To make the climate data more accurate and usable, the Kalman filtering technique is applied as a pre-processing step. In simple terms, Kalman filtering acts like a smart filter that continuously improves noisy and missing data by predicting what the values should be based on previous data trends. If a sensor reading is missing or fluctuates too much, the filter uses mathematical models to estimate a more accurate value. This is especially helpful when dealing with real-world climate data, which can often be incomplete or inconsistent. By using Kalman filtering, the CCP model receives clean, reliable input data, which significantly improves the quality of subsequent feature extraction and prediction stages. The working scheme of Kalman Filter is shown in Fig. 4.

The Kalman filter (KF) is the least MSE estimator for estimating y_u from $z_{1:u}$,, which is the filtering job in linear Gaussian State-Space (SS) methods. To keep its complexity constant over time, the KF calculates y_u at every time step u using just the previous estimate $\hat{y}_{u-1}$ as a suitable statistic in addition to the novel observation z_u.

The first as well as second-order statistical moments related to the state, denoted at time u as $\hat{y}_{u|u}$ and $\Sigma_{u|u}$, accord-

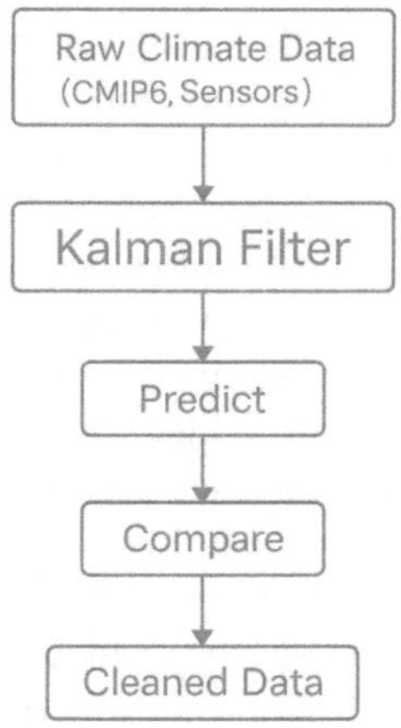

Fig. 4 Working scheme of Kalman filter in proposed work

ingly, are recalculated by the KF. These moments are placed to those of the starting state at time $u = 0$, namely $\hat{y}_0$ and $\hat{\Sigma}_0$. For time $u > 0$, however, the estimates are obtained in two steps:

Based on the previous a posteriori estimate, the initial stage predicts the first as well as second-order statistical moments related to the present a priori condition as well as observation. Specifically, the expected state moments at time u are computed using:

$$\hat{y}_{u|u-1} = G_{u-1} \cdot \hat{y}_{u-1|u-1} \tag{1}$$

$$\Sigma_{u|u-1} = G_{u-1} \cdot \Sigma_{u-1|u-1} \cdot G_{u-1}^{T} + R_{u-1} \tag{2}$$

The predicted observational moments are computed as:

$$\hat{z}_{u|u-1} = I_u \cdot \hat{y}_{u|u-1} \tag{3}$$

$$T_{u|u-1} = I_u \cdot \Sigma_{u|u-1} \cdot I_u^{T} + S_u \tag{4}$$

With the present finding z_u, the expected a priori moments are updated into the aposteriori state moments. The updated moments are computed as:

$$\hat{y}_{u|u} = \hat{y}_{u|u-1} + L_u \cdot \Delta z_u \tag{5}$$

$$\Sigma_{u|u} = \Sigma_{u|u-1} - L_u \cdot T_{u|u-1} \cdot L_u^{T} \tag{6}$$

In this case, L_u stands for the KG, which is described as:

$$L_u = \Sigma_{u|u-1} \cdot I_u^{T} \cdot T_{u|u-1}^{-1} \tag{7}$$

The innovation is described as the difference among the observed value and the expected observation, or $\Delta z_u \triangleq z_u - \hat{z}_{u|u-1}$. The forecasted state, represented as $\hat{y}_u \equiv \hat{y}_{u|u}$, is the KF output, and the error covariance is written as $\Sigma_u = \Sigma_{u|u}$.

In the context of forestry, Kalman filtering equations are particularly useful for managing sensor data collected from field-based IoT devices measuring temperature, humidity, soil moisture, and atmospheric CO_2 levels. For example, during wildfire season, temperature readings from a damaged sensor might suddenly spike or drop. The Kalman filter uses previous reliable readings (e.g. from the past 24 h) to predict what the current value should be, compares it with the faulty reading, and outputs a corrected value. This ensures that real-time climate predictions made by the model are not distorted by sudden noise or gaps in data, thereby enabling more reliable decision-making.

3.3 Feature Extraction by HHT

A key component of developing a successful CCP method for Smart Forestry 5.0 is feature extraction. It includes transforming raw climate information into meaningful representations that enhance machine learning-driven optimisation. The method can recognise patterns as well as trends that are crucial for predicting environmental changes in forests by recognising relevant characteristics from climate datasets, such as those from CMIP6. Climate information includes a number of variables, including precipitation, temperature, humidity, CO_2 levels, as well as soil moisture, many of which are non-linear as well as non-stationary. By finding underlying climatic trends, feature extraction helps to increase prediction accuracy, decrease dimensionality so the method can focus on the major pertinent data, as well as increase computing effectiveness by reducing processing time as well as model complexity. The pre-processed climatic data is used to extract characteristics using the HHT technique. This method works especially well for analysing time-varying as well as non-linear signals, which makes it ideal for climate information. In the face of environmental uncertainty, Smart Forestry 5.0 promotes more resilient as well as sustainable forest ecosystems by improving the accuracy associated with climate variation projections using feature extraction. One way to depict the phase $\theta_j(u)$ as well as amplitude $b_j(u)$ is as

$$b_j(u) = \sqrt{d_j^2(u) + I^2[d_j(u)]} \tag{8}$$

$$\theta_j(u) = \arctan\left(\frac{I[d_j(u)]}{d_j(u)}\right) \tag{9}$$

Thus, the instantaneous frequency $\omega_j(u)$ may be written as follows:

$$\omega_j(u) = \frac{d\theta_j(u)}{du} \tag{10}$$

The Hilbert-Huang spectrum $I(\omega, u)$ describes the name given to the time–frequency distribution:

$$I(\omega, u) = \text{Re}\sum_{j=1}^{o} b_j(u)\exp\left(k\int \omega_j(u)\text{d}u\right) \quad (11)$$

Here, $s_o(u)$, the rest, has been left out. Re$\{\cdot\}$ denotes a complex count's real component. By representing the amplitude as the height in the time–frequency space, Eq. (11) enables to show both the amplitude as well as the instantaneous frequency in a three-dimensional (3D) graph. Following the establishment of the Hilbert-Huang spectrum, the marginal spectrum, $i(\omega)$, may be described as:

$$i(\omega) = \int_0^U I(\omega, u)\text{d}u \quad (12)$$

Here, U stands for the information's total length. While the marginal spectrum offers a measure related to the total amplitude contribution for every frequency, the Hilbert spectrum offers a measure related to the amplitude contribution linked with every frequency as well as time. Consequently, each component's local marginal spectrum is written as:

$$i_j(\omega) = \int_0^U I_j(\omega, u)\text{d}u \quad (13)$$

The total amplitude contribution from the frequency is quantified by the local marginal $i_j(\omega)$ spectrum. The definition of the instantaneous energy IE(u) is:

$$\text{IE}(u) = \int_{\omega_1}^{\omega_2} I^2(\omega, u)d\omega \quad (14)$$

The instantaneous energy IE(u) provides information on the time evolution of energy.

The application of HHT in handling large-scale climate datasets, such as those sourced from CMIP6 or IoT-enabled sensor networks, is particularly advantageous due to its adaptability to non-linear and non-stationary signals—common characteristics in environmental data. Unlike Fourier-based methods, which assume stationary inputs and can become computationally inefficient with increasing data volume, HHT decomposes each time-series component into intrinsic mode functions (IMFs) locally and adaptively. This modular decomposition allows the system to process subsets of data in parallel, reducing memory overhead and making it scalable for high-resolution temporal datasets. For instance, when processing multi-year rainfall and temperature data across varied forest zones, HHT efficiently isolates meaningful trends and anomalies without the need for re-parameterisation or re-training. This decomposition capability makes HHT not only robust for pattern recognition but also computationally manageable when integrated with distributed or edge computing setups in Smart Forestry 5.0. Thus, it plays a critical role in transforming raw climate data into structured, predictive features suitable for real-time decision-making in diverse ecological contexts.

In forestry monitoring, many climate indicators such as precipitation levels and soil moisture show irregular patterns due to sudden weather events. The Hilbert-Huang Transform decomposes such complex, non-linear signals into components that reveal meaningful environmental patterns. For example, by applying HHT to rainfall data over dense canopy regions, the method can identify subtle shifts in seasonal rainfall intensity, which helps forest managers determine ideal planting times or predict drought stress. The instantaneous frequency and energy equations (Eqs. 10 and 14) thus provide dynamic insights that are vital for adaptive forest planning.

3.4 Prediction by Novel ESGBT

In Smart Forestry 5.0, prediction is crucial since it allows for proactive decisions in response to climate variation. Climate fluctuations such as CO_2 levels, rainfall patterns, temperature variations, as well as extreme weather events may be accurately anticipated by optimisation using machine learning. Forest managers can lower risks, distribute resources more effectively, as well as promote sustainability as a result. Analysing past as well as projected climate information from sources like CMIP6 is necessary to predict how climate variation will affect forestry. For this purpose, a novel machine learning method called ESGBT is used. By methodically reducing errors over a number of repetitions, gradient boosting improves forecast accuracy. Immediate learning as well as adaptation to new climatic data are made possible via streaming functionality. For large datasets, high effectiveness ensures rapid as well as extensible forecasts. Accurate climate predictions help with the following: Identifying temperature as well as humidity anomalies in a timely manner; predicting variations in ecosystems as well as tree development patterns; evaluating how climate variation affects carbon absorption; as well as supporting policymakers with adaptive forest management strategies. Smart Forestry 5.0 ensures accurate, real-time climate predictions with the combination of ESGBT-WoOA, promoting more resilient as well as sustainable forest ecosystems in the face of climate variation.

Smart Forestry 5.0's CCP is enhanced by SGBT, a sophisticated machine learning technique. It represents a more sophisticated kind of gradient boosting designed to handle large, continually updating information streams, making it very suitable for dynamic environmental forecasts. SGBT manages continuous information streams from climate sources including IoT sensors, CMIP6, as well as satellite images, as opposed to traditional gradient boosting methods that require fixed datasets. This ensures updated projections for forest management and allows for quick adaptation to changing climatic trends. Using sequential learning to improve weak methods over time, SGBT lowers prediction errors. This ensures better detection of long-term climatic patterns affecting forests as well as increased prediction accuracy for precipitation, temperature, as well as CO_2 levels. Large-scale datasets from remote sensing, IoT, as well as climate simulations are essential to Smart Forestry 5.0. SGBT has the following benefits: Quicker calculation by processing information in smaller segments, reducing memory usage; as well as the capacity to scale efficiently for handling large climatic datasets. Inaccuracies in measurements or malfunctioning sensors sometimes result in gaps or inconsistencies in climate databases. Even with insufficient data, SGBT can make accurate forecasts and is resilient to missing information. In the face of climatic uncertainty, this ensures sustainability as well as active forest care. One of the key strengths of the proposed ESGBT-WoOA framework is its ability to adapt to unforeseen climate anomalies through real-time learning and dynamic optimisation. Unlike static prediction models that rely on fixed historical datasets, the ESGBT model incorporates a streaming architecture capable of updating itself as new data becomes available from IoT sensors, remote sensing platforms, and climate databases like CMIP6. This continuous input enables the model to recalibrate its decision trees on-the-fly, maintaining high prediction accuracy even in the face of unexpected environmental fluctuations such as abrupt temperature spikes, sudden drought conditions, or unseasonal rainfall. Additionally, the Wolverine Optimisation Algorithm (WoOA) plays a critical role in tuning hyperparameters dynamically, ensuring that the model avoids overfitting to regular patterns and instead remains flexible enough to recognise outliers or rare climate events. This dual capability of adaptive learning and responsive parameter adjustment makes the CCP model resilient and suitable for handling the unpredictable nature of climate anomalies, which is vital for reliable forest management and disaster mitigation planning.

While SGBT is a valuable tool for predicting climate variation in Smart Forestry 5.0, it has limitations that can affect its effectiveness and efficiency. Its reliance on iterative modifications and sequential learning demands substantial computational power, making large climate datasets, like CMIP6 and IoT sensor data, resource-intensive to process. The complex method's high memory and processing requirements hinder real-time implementation. Effective hyperparameter tuning—such as learning rate, tree count, and tree depth—is essential for optimal performance. Improper tuning can result in overfitting or underfitting. Designed for streaming data, SGBT requires continuous climatic inputs for timely adaptation, but data irregularities or disturbances can reduce accuracy. Dependence on IoT devices and remote sensing makes forecasts vulnerable to disruptions caused by system failures. Although SGBT provides high accuracy, its decision-making process lacks transparency, limiting interpretability—a critical requirement for policymakers needing explainable AI models. Its black-box nature complicates understanding how climatic factors influence predictions. Additionally, the rarity of extreme weather events like wildfires and droughts results in imbalanced datasets, challenging SGBT's ability to predict significant but rare climatic anomalies. Addressing this issue requires weighted loss functions or resampling techniques. Despite its strengths, SGBT faces challenges related to computational complexity, data dependence, interpretability, and hyperparameter sensitivity. Incorporating hybrid AI frameworks, efficient data pipelines, and optimisation techniques can enhance its performance for sustainable forest management.

In Smart Forestry 5.0, the WoOA-enhanced ESGBT method is a sophisticated approach to CCP. Through the integration of real-time flexibility, precision, as well as computing economy, this hybrid method offers significant advantages over traditional approaches. ESGBT incorporates advanced learning techniques to enhance the fundamental SGBT paradigm. By modifying hyperparameters such as tree depth, learning rate, as well as estimator count, the WoOA optimisation technique improves prediction accuracy as well as lowers error rates. Continuous learning from real-time climate information collected from remote sensing, IoT devices, as well as CMIP6 datasets is made possible by the streaming capability. The method constantly improves prediction accuracy by modifying its forecasts in response to new climatic information. ESGBT-WoOA is appropriate for large-scale climate predictions because it reduces computing load by efficiently handling big datasets through incremental learning. Parallel processing speeds up inference as well as training, enabling timely decisions in sustainable forest management. WoOA ensures appropriate parameter modification, preventing issues like underfitting as well as overfitting. The dynamic adjustment of the optimisation process allows the method to function well under a range of climatic conditions. In Smart Forestry 5.0, the ESGBT-WoOA method provides an accurate, flexible, as well as computationally effective method for CCP. It describes a great choice for sustainable forest management because of its real-time learning capabilities, resilience to noisy information, as well as fine-tuning of hyperparameters.

Indicate y_j as the jth instance's characteristics as well as z_j as the associated target value related to a dataset with o examples. T sum functions may be used to represent a method ϕ in gradient boosting:

$$\hat{z}_j = \phi(y_j) = \sum_{t=1}^{T} g_t(y_j), g_t \in G \tag{15}$$

Here, the collection of regression trees is denoted by G. Every g_t in XGBoost corresponds to a unique tree configuration with leaf scores ω. For the jth instance, each regression tree has a continuous score ω_j at the leaf. To make predictions, the scores from every tree are combined at the leaves. Learning's objective is to lower the regularised objective:

$$M(\phi) = \sum_{j=1}^{o} m(z_j, \hat{z}_j) + \sum_{t=1}^{T} \Omega(g_t) \tag{16}$$

If the tree g's complexity is penalised by Ω:

$$\Omega(g) = \gamma U + \frac{1}{2}\beta\omega \tag{17}$$

In this instance, γ penalises the addition of a novel leaf as well as β forces minimum leaf forecasts. The number of leaves in the tree is denoted by U. The difference among the target z_j as well as the forecast $\hat{z}_j$ is measured by a differentiable convex loss function, denoted by m. Furthermore, the loss at the tth step is made up of the regularisation term, the loss from tree g_t, as well as the loss from the previous $(t-1)$ step:

$$M^{(t)} = \sum_{j=1}^{o} m\left(z_j, \hat{z}^{(t-1)} + g_t(y_j)\right) + \Omega(g_t) \tag{18}$$

The second-order Taylor approximation can be used to describe this loss as follows:

$$M^{(t)} \simeq \sum_{j=1}^{o}\left[m\left(z_j, \hat{z}^{(t-1)}\right) + h_j g_t(y_j) + \frac{1}{2} i_j g_t^2(y_j)\right] + \Omega(g_t) \tag{19}$$

Given the $t-1$th forecast, the first as well as second order (Hessian) gradient statistics associated with the loss are represented by $h_j = \partial_{\hat{z}^{(t-1)}} m\left(z_j, \hat{z}^{(t-1)}\right)$ and $i_j = \partial^2_{\hat{z}^{(t-1)}} m\left(z_j, \hat{z}^{(t-1)}\right)$. With weight i_j as well as goal h_j/i_j, this is referred to as a weighted squared loss:

$$\sum_{j=1}^{o} \frac{1}{2} i_j \left(g_u(y_j) - h_j/i_j\right)^2 + \Omega(g_t) + \text{Constant} \tag{20}$$

Instead of only using the streaming regression tree employed in XGBoost, Eq. (20) permits the usage of other streaming regression trees.

In a real-world forestry application, the ESGBT equations are used to predict forest vulnerability to climatic events like frost or drought. For instance, the loss function in Eq. 16 helps evaluate how far off the predicted risk of drought is from observed conditions. If the model predicts high vulnerability for a region and the actual drought index confirms it, the loss remains low. If not, the algorithm adjusts its prediction through sequential learning. This is especially useful for policymakers aiming to allocate water resources or initiate afforestation efforts in high-risk zones. Hence, the equations drive actionable decisions on the ground.

Furthermore, the regularisation term associated with the streaming regression tree can be different from that of XGBoost depending on how it is implemented. The prediction by novel ESGBT-WoOA for the proposed CCP model on Smart Forestry 5.0 can be diagrammatically illustrated as in Fig. 5.

3.5 WoOA Algorithm

For Smart Forestry 5.0's CP methods to be more reliable, accurate, as well as efficient, optimisation is crucial. In order to improve forecast accuracy, complex optimisation techniques are required since climate information is complex as well as includes a variety of environmental parameters, including rainfall, temperature, CO_2 concentrations, humidity, as well as soil moisture. Traditional machine learning methods may encounter difficulties when dealing with high-dimensional information, noisy inputs, as well as computational ineffectiveness. The method may adjust parameters flexibly, lower errors, as well as increase predicting accuracy by including optimisation methods. The method can analyse several solutions as well as identify the ideal parameters related to precise climate predictions by utilising evolutionary as well as metaheuristic optimisation techniques.

WoOA describes a biomimetic strategy that draws inspiration from wolverines' natural hunting behaviours. WoOA cleverly blends two essential techniques—hunting as well as scavenging—to mirror the wolverine's aptitude at locating carrion as well as pursuing living prey. The algorithm is unique because it accurately depicts these two approaches, which are theoretically constructed to efficiently optimise various issue types.

An evolutionary algorithm that depends on a wolverine population is called the WoOA technique. Through a repeating procedure, WoOA can develop effective answers for optimisation concerns by leveraging its members' search capabilities. As a WoOA participant, each wolverine sets up decision variables according on where it is in the problem-handling hierarchy. In essence, every wolverine represents a potential solution to the problem, which is represented as a mathematical vector. These wolverines together make up the WoOA population, and Eq. (21) provides a mathematical representation of their collective behaviour in the form of a matrix. At the beginning of the algorithm's execution, the wolverines' initial location inside the problem-handling region is chosen at random utilising Eq. (22).

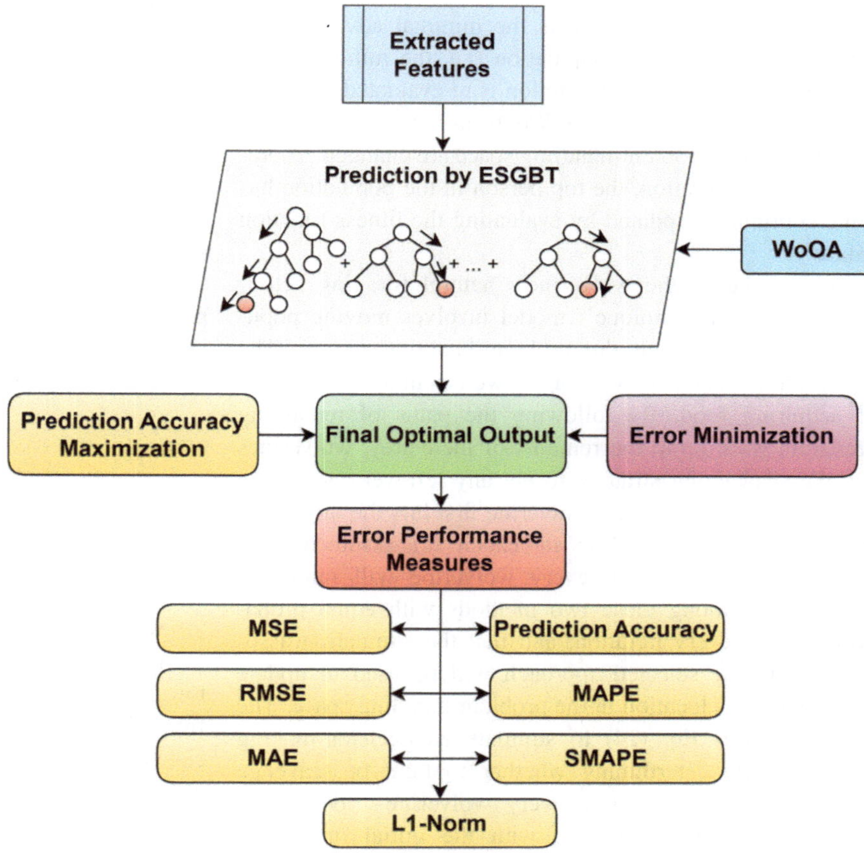

Fig. 5 Prediction by novel ESGBT-WoOA for the proposed CCP model on Smart Forestry 5.0

$$Y = \begin{bmatrix} Y_1 \\ \vdots \\ Y_j \\ \vdots \\ Y_O \end{bmatrix}_{O\times n} = \begin{bmatrix} y_{1,1} & \cdots & y_{1,e} & \cdots & y_{1,n} \\ \vdots & \ddots & \vdots & \cdot^{\cdot^{\cdot}} & \vdots \\ y_{j,1} & \cdots & y_{j,e} & \cdots & y_{j,n} \\ \vdots & \cdot^{\cdot^{\cdot}} & \vdots & \ddots & \vdots \\ y_{O,1} & \cdots & y_{O,e} & \cdots & y_{O,n} \end{bmatrix}_{O\times n} \tag{21}$$

$$y_{j,e} = LB_e + s \cdot (UB_e - LB_e) \tag{22}$$

Here, Y stands for the WoOA's population matrix, Y_j for the jth wolverine (basically, a potential solution), as well as $y_{j,e}$ for its value in the eth dimension associated with the search space (i.e. the decision variable). While n represents the total count of decision factors, O represents the count of wolverines. On the other hand, LB_e as well as UB_e denote the lower as well as upper bounds related to the eth decision variable, accordingly, while s represents a random value in the interval [0, 1].

According to this theory, placing each item in the area set aside for problem-handling serves as a workable solution to the current issue. Based on this understanding, the suggested values related to every element in relation to the choice factors aid in assessing the problem's fitness function. A vector format may thus be an effective way to display these evaluated values associated with the fitness function:

$$G = \begin{bmatrix} G_1 \\ \vdots \\ G_j \\ \vdots \\ G_O \end{bmatrix}_{O\times 1} = \begin{bmatrix} G(Y_1) \\ \vdots \\ G(Y_j) \\ \vdots \\ G(Y_O) \end{bmatrix}_{O\times 1} \tag{23}$$

Here, G_j is the assessed fitness function value that corresponds to the jth wolverine, and G is the vector related to the assessed fitness function values.

Significant details on the calibre associated with the potential solutions are provided by the fitness function's assessed values. The individual with the greatest fitness function rating represents the most advantageous member linked with the population (i.e. the best possible solution), and the person with the least rating represents the minimal advantageous member linked with the population (i.e. the minimal ideal possible option). The goal function is re-evaluated for every wolverine in every iteration of WoOA as the wolverine locations within the problem-handling space are changed. Therefore, in every iteration, the top person in the population has to be continually updated by evaluating the fitness function values.

By simulating the wolverine's natural foraging activities, the WoOA technique's model involves moving population members about the problem-handling region. The wolverine uses both scavenging as well as hunting as means of acquiring food. By following the paths of remaining predators who dump the remains of their prey, wolverines use the scavenging strategy to eat any leftover carrion. In the hunting technique, the wolverine first hits the target and then, following a battle, kills and eats it. It is assumed while constructing WoOA that every wolverine will randomly choose one among these two methods with equal probability during every iteration, and that the simulation associated with the selected approach will be used to update the wolverine's location in the problem-handling space. The wolverine uses Eq. (24) to simulate its decision-making process while determining whether to hunt or scavenge for food. This means that every wolverine's location is changed only in accordance with the initial or second scheme throughout every iteration.

$$\text{update procedure related to } j\text{th wolverine} Y_j : \begin{cases} \text{on the basis of scavenging scheme, } s_q \leq 0.5 \\ \text{on the basis of hunting scheme, otherwise} \end{cases} \quad (24)$$

Here, s_q stands for a random count selected from the interval [0, 1].

By simulating the wolverine's natural scavenging behaviour on corpses, the original WoOA technique involves shifting the population members' placements inside the problem-handling domain. By following the path taken by previous predators that have left their remains behind, the wolverine finds the carrion. An entirely different approach to exploring the solution space results from modelling the wolverine's behaviour when confronted with predators. This approach significantly alters population members' positions, which alters their trajectories in the problem-handling domain. The algorithm achieves a more thorough examination of potential solutions by employing this technique. In addition to broadening the search field, these significant locational adjustments enhance the algorithm's capacity to find a variety of ideal solutions. Consequently, the method improves the algorithm's ability to do a thorough worldwide search, which enables it to navigate via complex issue domains as well as find innovative answers. An exploring phase that mimics the wolverine's movements while looking for carrion is part of the scavenging strategy.

In the WoOA model, the sites of predators that want to provide the remaining kills for every wolverine are considered to be those of remaining members having higher fitness function values, as shown in Eq. (25).

$$\mathrm{DQ}_j = \{Y_l : G_l < G_j andl \neq j\}, \text{ in which } j = 1, 2, \cdots, O \text{ and } l \in \{1, 2, \ldots, O\} \quad (25)$$

If Y_l is the population individual with a higher fitness function value than the jth wolverine, G_l is its fitness function value, and DQ_j is the group of possible predator sites for the jth wolverine.

According to the WoOA concept, the wolverine travels towards a predator after choosing its location at random from the DQ_j group. A novel suggested location associated with the relevant individual has been determined utilising Eq. (26) when the wolverine moves towards the selected predator in an attempt to get the carrion. According to Eq. (27), this suggested location supersedes the related individual's previous location if the goal function's value increases.

$$y_{j,k}^{T1} = y_{j,k} + s_{j,k} \cdot \left(TQ_{j,k} - J_{j,k} \cdot y_{j,k}\right) \quad (26)$$

$$Y_j = \begin{cases} Y_j^{T1}, \; G_j^{T1} \leq G_j \\ Y_j, \text{ otherwise} \end{cases} \quad (27)$$

Here, TQ_j represents the predator that was chosen for the jth wolverine, and $TQ_{j,k}$ represents the predator's kth feature. The position related to the jth wolverine, as determined by the scavenging strategies employed in the proposed WoOA, is shown by Y_j^{T1}. Similarly, the kth component of this updated location is represented by $y_{j,k}^{T1}$. The fitness function value associated with the jth wolverine at its new position is denoted by the identifier G_j^{T1}. Additionally, $J_{j,k}$ describes binary random numbers that can have values of either 1 or 2, whereas $s_{j,k}$ are random values uniformly picked from the range [0, 1].

By imitating the wolverine's natural hunting behaviour, the second WoOA strategy involves moving population individuals about the problem-handling domain. This procedure mimics the wolverine's typical hunting style, in which it attacks live prey first, then engages in a battle and chase before catching as well as devouring its target. In accordance with this hunting technique, the population individuals are updated in two phases: (i) exploration, which involves modelling the wolverine's approach to its prey, and (ii) exploitation, which involves simulating the battle and chase among the prey as well as the wolverine.

By mimicking a wolverine's attack on its prey, the population's location in the problem-handling space is altered during this WoOA phase. The population's position is significantly altered by imitating the wolverine's hunting behaviour, which enhances WoOA's exploration skills when navigating the problem-handling area. The WoOA compares the position related to the prey with the optimal individual associated with the population. An innovative possible location is identified utilising Eq. (28) by imitating the wolverine's method of identifying the best individual (the 'prey') for every person in WoOA. According to Eq. (29), the newly proposed location will replace the previous one for the relevant individual if the goal function shows advancement.

$$y_{j,k}^{Q1} = y_{j,k} + s_{j,k} \cdot (\text{prey}_k - J_{j,k} \cdot y_{j,k}) \tag{28}$$

$$Y_j = \begin{cases} Y_j^{Q1}, & G_j^{Q1} \le G_j \\ Y_j, & \text{otherwise} \end{cases} \tag{29}$$

Here, Y_j^{Q1} represents the newly determined location associated with the jth wolverine, determined by the first stage related to the hunting scheme in the proposed WoOA, prey describes the ideal population individual as prey, and $prey_k$ describes its kth dimension. Similarly, the kth dimension of this revised location is indicated by $y_{j,k}^{Q1}$. The fitness function value related to the jth wolverine at this updated position is denoted by G_j^{Q1}. Additionally, $J_{j,k}$ represents binary digits that are randomly assigned as either 1 or 2, while $s_{j,k}$ represents random values selected from the interval [0, 1].

By simulating the hunting as well as pursuing interactions among the wolverine as well as its prey, the population individual's locations inside the problem-handling area are altered during this WoOA stage. The population individual's placements are somewhat altered by imitating the wolverine's movements while pursuing, which enhances the WoOA's capacity for local search in the context of problem-handling. These encounters are assumed to occur close to the hunting site in the WoOA concept. For every WoOA individual, a novel site is identified using Eq. (30), which mimics the wolverine's movements while hunting as well as pursuit. This novel job replaces the individual's previous role if it increases the value related to the fitness function as determined by Eq. (31).

$$y_{j,k}^{Q2} = y_{j,k} + (1 - 2s_{j,k}) \cdot \frac{\text{UB}_k - \text{LB}_k}{u} \tag{30}$$

$$Y_j = \begin{cases} Y_j^{Q2}, & G_j^{Q2} \le G_j \\ Y_j, & \text{otherwise} \end{cases} \tag{31}$$

Here, the second stage related to the hunting scheme used in the WoOA determines the newly described location associated with the jth wolverine, denoted by Y_j^{Q2}. The kth dimension of this updated location is denoted by $y_{j,k}^{Q2}$. The value related to the fitness function associated with the jth wolverine at its modified position is denoted by G_j^{Q2}. Additionally, u is the iteration counter, and $s_{j,k}$ are random values selected from the interval [0, 1].

The first iteration of WoOA ends when entire wolverine locations in the problem-handling space have been recalculated. The method next proceeds to the following iteration using the updated values, continuously changing the wolverines' positions until the algorithm's last iteration, as per Eqs. (24) to (31). In each iteration, the evaluated values related to the fitness function are used to revise the best candidate solution. The optimal solution discovered throughout the algorithm's iterations is disclosed as the WoOA solution for the specified problem once the algorithm has finished running. In Algorithm 1, the WoOA pseudo-code is illustrated.

Algorithm 1: *WoOA*

Begin WoOA

Input problem data such as fitness function, variables, and constraints. [extracted features of the proposed CCP model on smart forestry 5.0]

Set the count of iterations (U) *and WoOA population size* (O)

$$y_{j,e} = LB_e + s \cdot (UB_e - LB_e)$$

Evaluate the fitness function [prediction accuracy maximization and error minimization of the proposed CCP model on smart forestry 5.0]

Describe the optimal candidate solution

For $u = 1$ *to* U

For $j = 1$ *to* O

$$update\ procedure\ related\ to\ j^{th}\ wolverine\ Y_j: \begin{cases} on\ the\ basis\ of\ scavenging\ scheme, \\ on\ the\ basis\ of\ hunting\ scheme, \quad ot \end{cases}$$

If $s_q \le 0.5$

$$DQ_j = \{Y_l: G_l < G_j\ and\ l \neq j\}, in\ which\ j = 1,2,\cdots,O\ and\ l \in \{1,2,\cdots,O\}$$

$$y_{j,k}^{T1} = y_{j,k} + s_{j,k} \cdot (TQ_{j,k} - J_{j,k} \cdot y_{j,k})$$

$$Y_j = \begin{cases} Y_j^{T1}, & G_j^{T1} \le G_j \\ Y_j, & otherwise \end{cases}$$

Else

Perform exploration stage

$$y_{j,k}^{Q1} = y_{j,k} + s_{j,k} \cdot (prey_k - J_{j,k} \cdot y_{j,k})$$

$$Y_j = \begin{cases} Y_j^{Q1}, & G_j^{Q1} \le G_j \\ Y_j, & otherwise \end{cases}$$

Perform exploitation stage

$$y_{j,k}^{Q2} = y_{j,k} + (1 - 2s_{j,k}) \cdot \frac{UB_k - LB_k}{u}$$

$$Y_j = \begin{cases} Y_j^{Q2}, & G_j^{Q2} \le G_j \\ Y_j, & otherwise \end{cases}$$

End if

End

Save the optimal candidate solution attained so far

End

Output the best quasi-optimal solution attained by the WoOA [solution that returns maximized prediction accuracy and minimized error of the proposed CCP model on smart forestry 5.0]

Stop

4 Results and Discussion

The proposed ESGBT-WoOA for the CCP model on Smart Forestry 5.0 was implemented in MATLAB and the findings were discussed. The population size was placed to be 10 and the total number of rounds was fixed to be 200. The proposed ESGBT-WoOA was compared with distinct models like Site-based approach (Wenhuan et al., 2024), GM-AI (Mahmood et al., 2024), NSGA-II, and WSN-IoT with respect to analysis such as prediction accuracy, Mean Square Error (MSE), Mean Absolute Error (MAE), Mean Absolute Percentage Error (MAPE), Root Mean Square Error (RMSE), Symmetric Mean Absolute Percentage Error (SMAPE), and L1 Norm to demonstrate the betterment of the recommended CCP model on Smart Forestry 5.0.

This approach is compared with four contemporary state-of-the-art methods to confirm the effectiveness related to the proposed ESGBT-WoOA for the CCP model in Smart Forestry 5.0:

1. **Site-Based Approach** → In light of climate variation, this concept offers a location-specific approach to selecting climate-resilient tree species for reforestation. Unlike traditional species-oriented methods, this method uses climatic niche projections to identify the optimal tree species for specific planting areas. The work uses machine learning as well as climate modelling techniques to predict species compatibility as well as evaluates its feasibility across ten different climatic zones in China. Results show how climate variation affects the survival

of species, underscoring the need for adaptable forestry practices. The approach improves biodiversity, ecological restoration, as well as forest resilience while offering a data-driven resource for sustainable forest management.

2. **GM-AI →** This concept looks at how bioengineering as well as AI are transforming smart forestry for biodiversity protection as well as wood production. It looks at how AI-powered technologies, like predictive analytics, machine learning, as well as remote sensing, optimise wood output, enhance forest monitoring, as well as advance biodiversity conservation. It also looks at how GE trees help with faster growth, insect resistance, as well as climate adaptation. The concept highlights the potential benefits, challenges, as well as moral ramifications of integrating biotechnology as well as AI into sustainable forest management.
3. **NSGA-II →** With a focus on Human-Centred AI in forest management, this concept explores the transition from Industry 5.0 to Forestry 5.0. It highlights the ways in which AI-driven methodologies, like precision forestry, automation, as well as predictive analytics, may raise production as well as sustainability. The concept identifies key enabling methodologies, implementation challenges, as well as a Forestry 5.0 framework that balances economic effectiveness, environmental preservation, as well as worker safety.
4. **WSN-IoT →** In order to support CSF, this concept presents a novel line of sensors as well as monitoring tools. It looks into how the IoT as well as WSN might enhance real-time forest monitoring as well as provide early warnings for ecological variations. The concept demonstrates how in situ as well as remote sensing data may be integrated to improve decision-making as well as shows significant technology advancements in monitoring forest production, health, as well as biodiversity. It also explores public involvement in extensive information collection as well as emphasises data-driven strategies for flexible forest management.
5. **Proposed ESGBT-WoOA →** In Smart Forestry 5.0, the WoOA is used to refine the ESGBT method, which enhances CCP. By utilising adaptive learning, hyperparameter tweaking, as well as real-time data processing, ESGBT-WoOA improves forecast accuracy, lowers errors, as well as encourages sustainable forest management in the face of climatic variability.

The simulation settings for the ESGBT-WoOA framework, including parameters such as population size, number of rounds, and performance metrics, are summarised in Table 1. In order to evaluate the efficacy associated with the proposed ESGBT-WoOA for the CCP model in Smart Forestry 5.0, the following significant assessment indicators are considered:

Table 1 Simulation settings

Parameter	Value
Simulation tool	MATLAB
Population size	10
Compared models	Site-based approach, GM-AI, NSGA-II, WSN-IoT, Proposed ESGBT-WoOA
Total rounds	200
Performance metrics	Prediction Accuracy, MSE, MAE, MAPE, RMSE, SMAPE, L1 Norm

- **Prediction Accuracy**: The accuracy of the ESGBT-WoOA-oriented CCP method for Smart Forestry 5.0 in forecasting variations in the climate that affect forests is evaluated. It supports proactive decisions in the context of climate variation as well as sustainable forest management by ensuring high accuracy, minimal errors, as well as reliable predictions through the use of sophisticated machine learning techniques.
- **MSE:** The ESGBT-WoOA-oriented climate variation prediction method for Smart Forestry 5.0 uses MSE, which computes the average squared variances among observed as well as predicted data, to evaluate prediction precision. A lower MSE indicates improved method accuracy, reducing climate variation predicting errors as well as promoting informed decision-making as well as sustainable forest management.
- **MAE:** The average absolute difference among actual as well as predicted values is measured by the MAE in the ESGBT-WoOA CCP method for Smart Forestry 5.0. Higher forecast accuracy is shown by a lower MAE, which ensures precise climate projections that promote climate resilience as well as sustainable forestry management.
- **MAPE:** By calculating the average percentage difference among actual as well as predicted values, the MAPE in the CCP method on the basis of ESGBT-WoOA for Smart Forestry 5.0 evaluates prediction accuracy. Higher reliability is shown by a lower MAPE, which makes it easier to make appropriate climatic projections for conscientious forest management as well as proactive climate adaption strategies.
- **RMSE:** The square root associated with the mean squared deviations among observed as well as estimated values is determined by the RMSE in the ESGBT-WoOA-oriented CCP method for Smart Forestry 5.0. Higher method accuracy is shown by a lower RMSE, which ensures precise climatic projections for climate adaption as well as sustainable forestry management plans.
- **SMAPE:** In the ESGBT-WoOA-oriented CCP method for Smart Forestry 5.0, the SMAPE calculates the percentage difference among observed as well as

expected values, symmetrically adjusted, to assess prediction accuracy. Enhanced prediction accuracy from a lower SMAPE supports efforts at climate adaption as well as sustainable forestry.

- **L1 Norm:** In the ESGBT-WoOA-oriented CCP method for Smart Forestry 5.0, the L1 Norm provides resistance against outliers by quantifying the sum of absolute differences among observed as well as estimated values. Enhanced prediction accuracy is shown by a lower L1 Norm, which supports sustainable forest management choices as well as precise climate forecasts.

4.1 Prediction Accuracy Analysis

The accuracy analysis associated with the Smart Forestry 5.0 CCP method, which employs five different approaches across 200 iterations, is shown in Fig. 6. The accuracy percentage is used to assess every method's efficacy and provide information about how reliable they are in predicting the climate. The site-based approach's accuracy ranges from 81.892% to 83.457%. It shows constant but diminished effectiveness, indicating that complex climatic changes are hard to capture with traditional site-based approaches. For GM-AI, the accuracy falls between 85.432% and 86.435%. It uses AI approaches to exceed the site-based technique, although it still lacks the optimisation efficacy of more sophisticated methods. The accuracy achieved by NSGA-II ranges from 87.653% to 88.678%. It uses an evolutionary approach, which produces notable improvements but not quite as much as IoT-driven solutions. For WSN-IoT, the accuracy ranges from 89.214% to 90.456%. It provides more accurate predictions than traditional AI as well as optimisation techniques using real-time data collection. The proposed ESGBT-WoOA achieves an accuracy of 92.876% to 94.102%, outperforming entire other methods. It ensures robust as well as precise climate predictions by combining an optimisation technique with advanced gradient enhancement. The proposed ESGBT-WoOA method consistently achieves the highest accuracy, outperforming complete other methods. The entire methods continue to enhance, suggesting that more training leads to better effectiveness. Although they are more flexible than traditional AI-oriented methods, the WSN-IoT as well as NSGA-II techniques are still inferior to the suggested method. In Smart Forestry 5.0, the ESGBT-WoOA method represents the major effective way to forecast climate variation, offering improved precision as well as robustness in environmental monitoring as well as decision-making. The proposed ESGBT-WoOA for the CCP model on Smart Forestry 5.0 with respect to prediction accuracy is 13.06%, 9.09%, 6.30%, and 3.92% better than site-based approach, GM-AI, NSGA-II, and WSN-IoT, respectively.

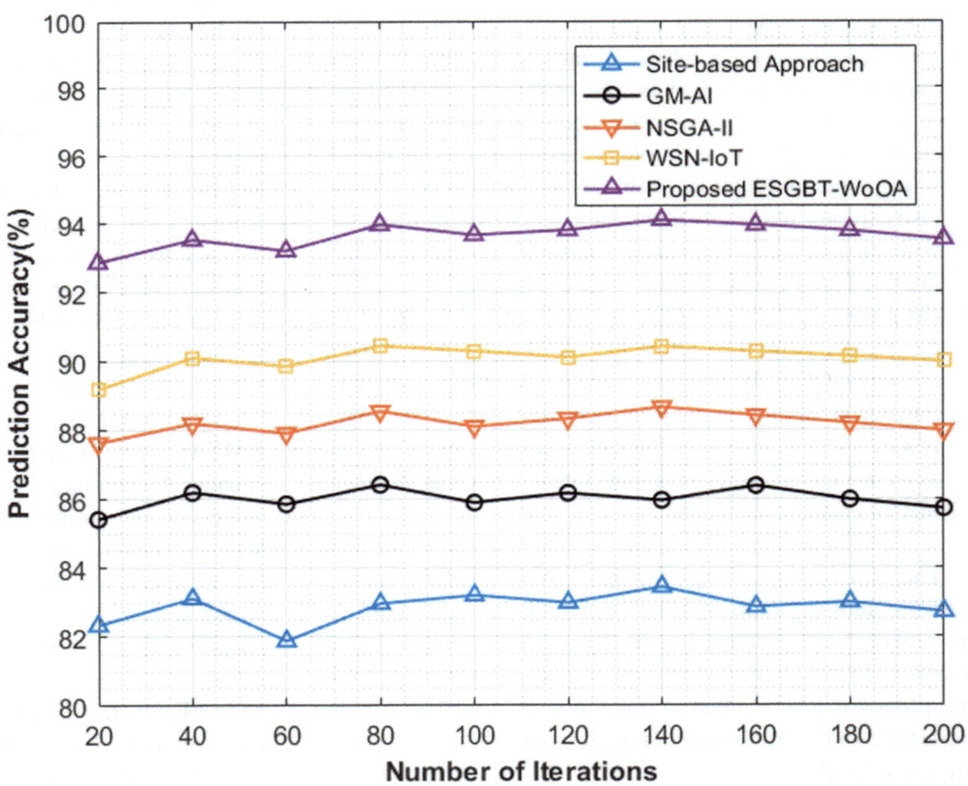

Fig. 6 Prediction accuracy analysis

4.2 MSE Analysis

The efficacy associated with five different CCP methods across 200 iterations in Smart Forestry 5.0 is evaluated by the MSE analysis in Fig. 7. The average squared deviation among expected as well as actual values is measured by the crucial MSE metric; lower values indicate better prediction accuracy. The site-based strategy had the highest MSE values (ranging from 0.127 to 0.139) associated with the five techniques examined, highlighting its limitations in accurately forecasting the consequences of climate variation. Although the GM-AI technology improves the site-based technique via AI, it ultimately falls short of peak accuracy, with MSE values ranging from 0.107 to 0.114. With MSE values ranging from 0.092 to 0.099, the NSGA-II approach demonstrates further error minimisation by utilising its evolutionary optimisation advantages. With an MSE range of 0.074 to 0.081, the WSN-IoT methodology, which combines WSNs with IoT technologies, significantly improves effectiveness as well as highlights the advantages of instantaneous environmental information collection. However, the proposed ESGBT-WoOA outperforms entire other methods and achieves the least MSE values, ranging from 0.048 to 0.054. This exceptional effectiveness shows how effective gradient boosting as well as optimisation techniques are in lowering prediction errors as well as improving method accuracy. The proposed ESGBT-WoOA method represents the major reliable approach to CCP in Smart Forestry 5.0, ensuring extremely accurate projections. The outcomes also imply that an increase in iterations leads to a fall in MSE values, corroborating the idea that more thorough training improves prediction accuracy. The study emphasises the value of machine learning-based optimisation techniques in raising the accuracy of climate forecast methods, with ESGBT-WoOA showing the best performance. The proposed ESGBT-WoOA for the CCP model on Smart Forestry 5.0 in terms of MSE is 61.65%, 54.05%, 46.88%, and 34.62% advanced than site-based approach, GM-AI, NSGA-II, and WSN-IoT, respectively.

4.3 MAE Analysis

Five different machine learning-oriented CCP methods used in Smart Forestry 5.0 are compared using the MAE analysis displayed in Fig. 8. The MAE among predicted as well as actual values is measured; smaller values indicate more accurate predictions. The site-based technique has the highest MAE values, ranging from 0.097 to 0.104, as can be seen in the table and graph. This suggests that this methodology is prone to notable deviations from actual climatic constraints. Despite providing a basic framework for climate prediction, this technique lacks the precision as well as efficacy related to more advanced techniques. With MAE values ranging from 0.084 to 0.091, the GM-AI strategy, which uses AI to drive optimisation, exhibits a little improvement

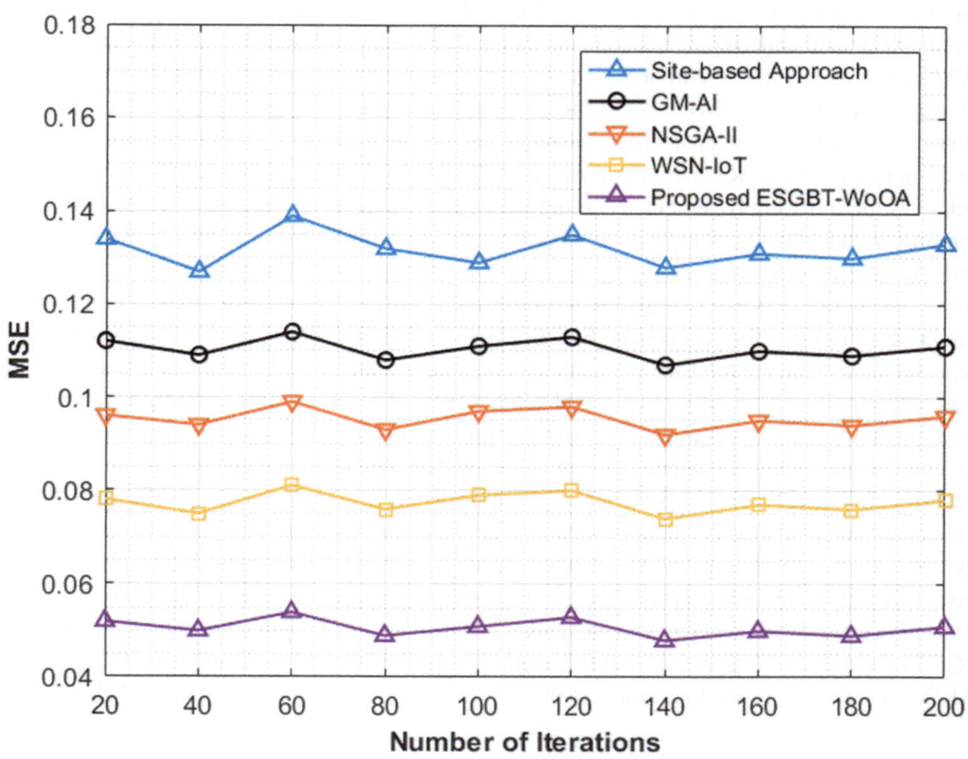

Fig. 7 MSE analysis

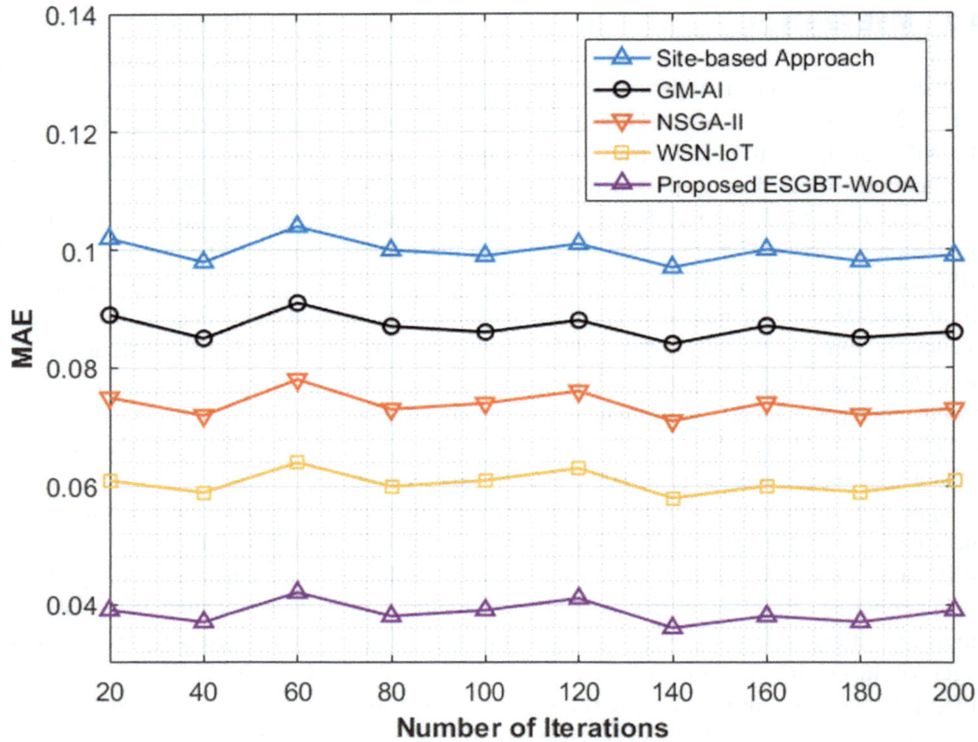

Fig. 8 MAE analysis

over the site-based technique. However, it is still insufficient to achieve maximum precision. With MAE values falling to among 0.071 and 0.078, the NSGA-II approach, an evolutionary optimisation scheme for multiple goals, improves prediction accuracy as well as shows a notable improvement in lowering absolute errors. With MAE values ranging from 0.058 to 0.064, the WSN-IoT strategy, which blends WSN as well as IoT methodologies, continues the trend of improved accuracy. IoT's real-time information collection capabilities reduce projections' absolute errors related to climate variation. However, the proposed ESGBT-WoOA achieves the most improvement. With the least MAE values, ranging from 0.036 to 0.042, this method outperforms entire remaining methods, demonstrating its improved ability to lower prediction errors. Compared to traditional and AI-tuned approaches, the proposed ESGBT-WoOA method represents the major reliable as well as accurate prediction technique, significantly reducing absolute errors. These findings highlight the advantages of using gradient boosting in conjunction with metaheuristic optimisation for Smart Forestry 5.0's CCP. The importance of training as well as optimisation in raising prediction accuracy is again shown by the steady decline in MAE values with more iterations. The proposed ESGBT-WoOA for the CCP model on Smart Forestry 5.0 with respect to MAE is 60.61%, 54.65%, 46.57%, and 36.07% higher than site-based approach, GM-AI, NSGA-II, and WSN-IoT, respectively.

4.4 MAPE Analysis

The prediction accuracy associated with several machine learning-oriented optimisation methods for predicting climate variation in Smart Forestry 5.0 is thoroughly assessed by the MAPE study. As MAPE determines the percentage difference among expected as well as actual values—smaller counts denoting higher prediction accuracy—it represents a crucial metric. The MAPE values across 200 rounds for five different approaches are shown in Fig. 9, highlighting significant variations in predicted accuracy. With MAPE values ranging from 5.398 to 5.521, the site-based technique has the highest percentage error, indicating difficulties maintaining forecast accuracy. This suggests that handling complex environmental information related to climate projections becomes more difficult using the traditional site-specific technique. With MAPE values ranging from 4.712 to 4.978, GM-AI, an AI-powered optimisation technique, shows a little improvement in its capacity to capture climatic variations. The prediction error is further reduced via the NSGA-II approach, an evolutionary algorithm for multiple objectives, which achieves MAPE values between 3.798 and 4.078. Prediction accuracy increases as a result of NSGA-II's enhanced optimisation capabilities. MAPE values drop to among 2.712 and 2.978 in the WSN-IoT strategy, which integrates WSN with IoT, demonstrating the benefits of using

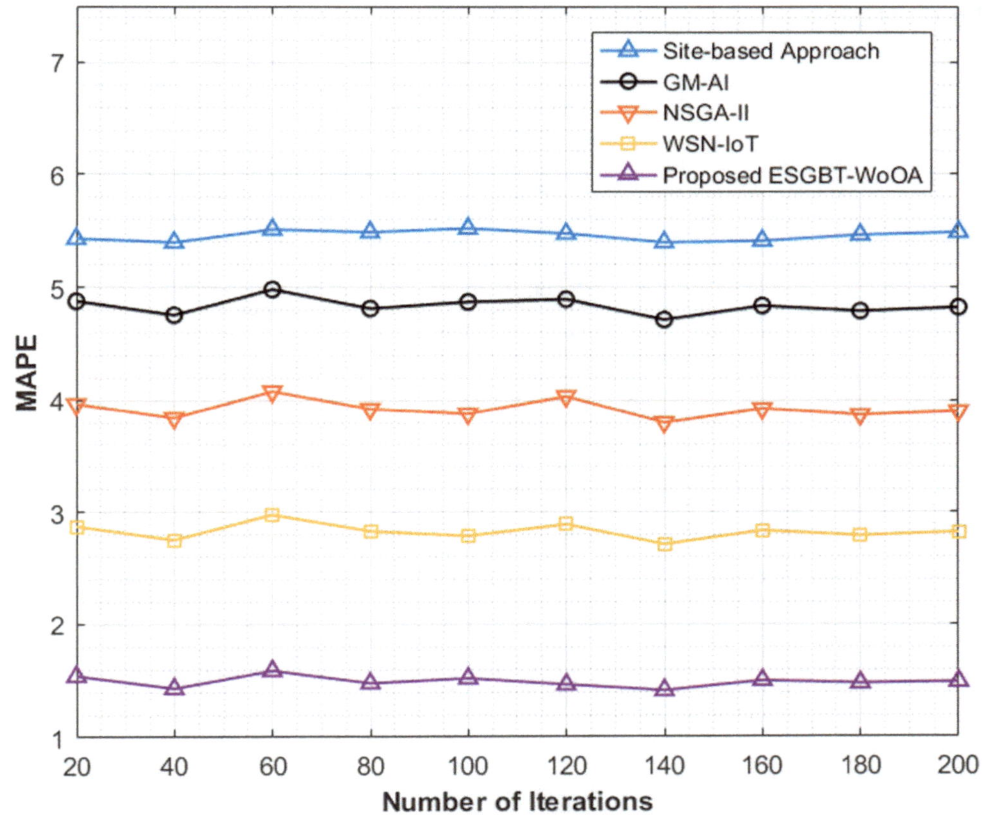

Fig. 9 MAPE analysis

real-time sensors for information collection. The proposed ESGBT-WoOA achieves the least MAPE values of all the approaches, ranging from 1.412 to 1.589. This demonstrates how well ESGBT-WoOA reduces prediction errors as well as improves prediction accuracy. As the major successful method for CCP in Smart Forestry 5.0, the significant drop in MAPE validates the effectiveness linked in combining gradient boosting with metaheuristic optimisation. By significantly reducing percentage errors as well as outperforming traditional and AI-oriented methods, the proposed ESGBT-WoOA method proves to be the major reliable approach for accurate CCP. The proposed ESGBT-WoOA for the CCP model on Smart Forestry 5.0 in terms of MAPE is 72.87%, 69.13%, 61.81%, and 47.25% more than site-based approach, GM-AI, NSGA-II, and WSN-IoT, respectively.

4.5 RMSE Analysis

The RMSE analysis looks at the accuracy as well as reliability related to the many methods used in Smart Forestry 5.0 to anticipate climate variation. The standard deviation associated with the prediction errors is measured using the widely used RMSE error measure, in which least values indicate better method accuracy. The RMSE results for five different approaches over 200 iterations are shown in Fig. 10, which show clear effective variations. The site-based approach had the highest error of all the methods, with RMSE values ranging from 0.354 to 0.372. This suggests that the smart forestry model's climate variation components are difficult to predict using traditional methods. When compared to the site-based technique, GM-AI, which uses an AI-oriented approach, exhibits improvements with RMSE values among 0.320 and 0.342, showing less prediction errors. The NSGA-II method, a multi-objective optimisation approach, yields RMSE values between 0.294 and 0.315, indicating further improvements. More accurate climate projections result from NSGA-II's increased processing capacity. The benefits of real-time environmental information collection as well as processing are shown by the WSN-IoT methodology, which uses WSNs in conjunction with IoT technologies to reduce RMSE values to a range of 0.262 to 0.283. The proposed ESGBT-WoOA method achieves the least RMSE values, ranging from 0.210 to 0.232, significantly outperforming entire existing methods. According to this, ESGBT-WoOA represents the major reliable method for Smart Forestry 5.0 applications as it lowers prediction errors as well as increases prediction precision. The proposed ESGBT-WoOA method outperforms conventional as well as AI-oriented methods in terms of prediction accuracy. Its effectiveness in increasing the accuracy of CCP is confirmed by the significant

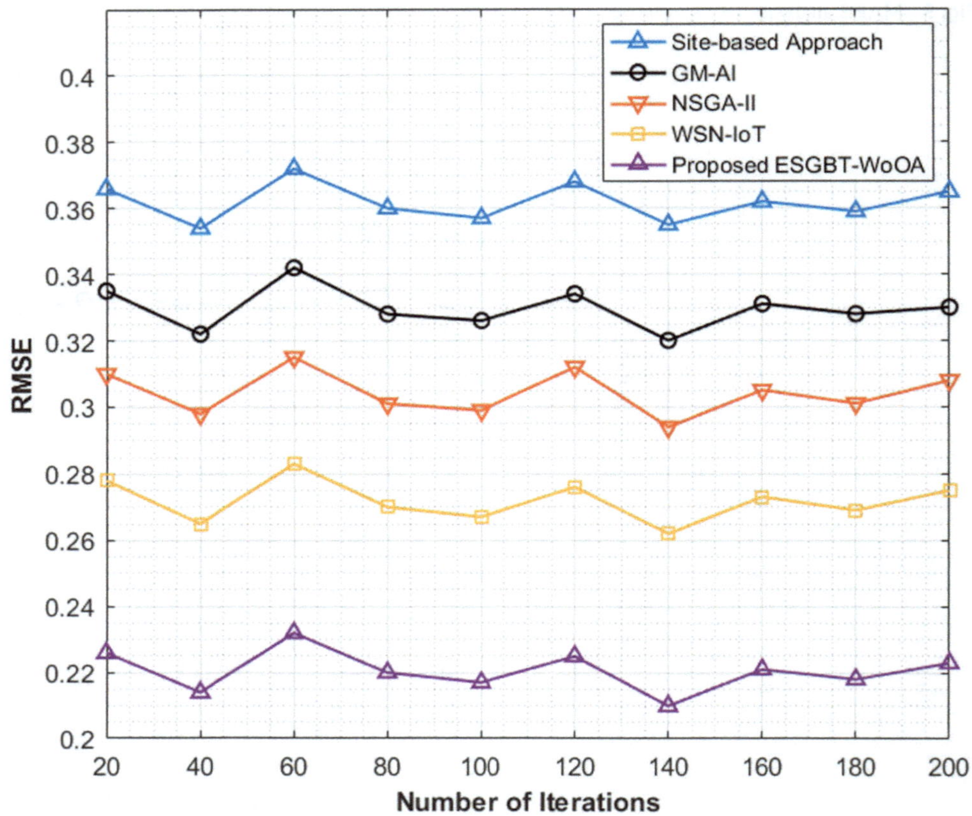

Fig. 10 RMSE analysis

reduction in RMSE, making it the major appropriate as well as successful approach for Smart Forestry 5.0. The proposed ESGBT-WoOA for the CCP model on smart forestry 5.0 with respect to RMSE is 38.90%, 32.42%, 27.60%, and 18.91%greater than site-based approach, GM-AI, NSGA-II, and WSN-IoT, respectively.

4.6 SMAPE Analysis

The SMAPE describes a crucial metric for evaluating the accuracy related to CCP methods because it computes the percentage error while equally accounting for overestimations as well as underestimations. Figure 11 demonstrates that a prediction method with a lower SMAPE value is more accurate as well as reliable. The site-based technique's relatively poor prediction accuracy is reflected in its greatest SMAPE values, which range from 4.692 to 4.823. This suggests that because traditional climate modelling techniques have trouble generalising over different iterations, they may not be the best fit for Smart Forestry 5.0 applications. With SMAPE values between 3.842 and 4.032, the GM-AI method exhibits a notable improvement over the site-based approach. The benefits of AI-oriented methods in improving forecasts as well as reducing prediction errors are shown by this decline. The NSGA-II method shows further improvement, with SMAPE values ranging from 2.989 to 3.187. The multi-objective optimisation capability of the NSGA-II improves climate forecast accuracy as well as adaptability. The WSN-IoT architecture achieves an even lower SMAPE, ranging from 2.315 to 2.498, by leveraging WSNs as well as IoT information collecting. This decline demonstrates how improving method accuracy may be achieved by obtaining real-time environmental information. The recommended ESGBT-WoOA method achieves the least SMAPE values of entire methods, ranging from 1.721 to 1.932. This significant drop demonstrates how ESGBT-WoOA effectively lowers percentage errors throughout several rounds, significantly increasing prediction accuracy. ESGBT-WoOA represents the major accurate as well as efficient approach to CCP in Smart Forestry 5.0, outperforming entire remaining methods in terms of predictive accuracy. Its remarkable effectiveness highlights its potential to revolutionise environmental monitoring as well as decision-making procedures. The proposed ESGBT-WoOA for the CCP model on Smart Forestry 5.0 in terms of SMAPE is 61.66%, 53.56%, 40.52%, and 23.95% better than dite-based approach, GM-AI, NSGA-II, and WSN-IoT, respectively.

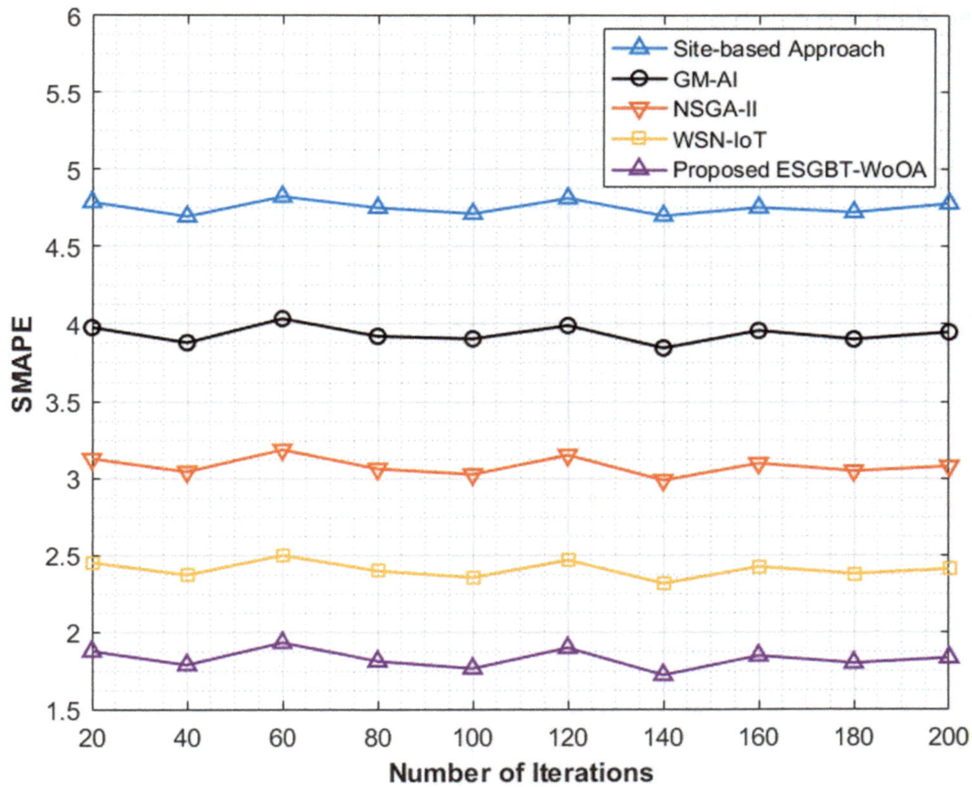

Fig. 11 SMAPE analysis

4.7 L1 Norm Analysis

An important statistic for assessing the error magnitude in CCP methods represents the L1 Norm, often known as the Manhattan Distance. As Fig. 12 demonstrates, a lower L1 Norm value indicates more prediction accuracy as well as resilience, resulting it an essential parameter for evaluating the effectiveness associated with different methods used in Smart Forestry 5.0. The site-based approach's restricted prediction accuracy is demonstrated by its maximum L1 Norm values, which vary from 1.198 to 1.312. This suggests that the complex environmental dynamics needed for precise climate variation projections are challenging for standard site-specific techniques to take into consideration. The L1 Norm values of the GM-AI method, which range from 1.032 to 1.178, demonstrate a notable improvement. This decline demonstrates the benefits of using AI-oriented techniques in climate modelling, which enable more efficacy as well as agility in forecasting climate variation. The NSGA-II method shows further improvement, with L1 Norm values ranging from 0.899 to 1.038. Multi-objective evolutionary optimisation, which helps to improve climate predictions by efficiently controlling a variety of environmental elements, is responsible for the method's increased precision. The L1 Norm, which ranges from 0.731 to 0.889, is also lowered by the WSN-IoT methodology, which uses WSNs in conjunction with IoT-enabled real-time information collecting. Prediction accuracy as well as reliability are improved by using sensor-oriented real-time climatic data. The proposed ESGBT-WoOA method exhibits the lowest L1 Norm values of entire methods, ranging from 0.454 to 0.712. This significant decline highlights how effective gradient boosting in conjunction with optimisation algorithms is in improving climate variation projections. The method's exceptional effectiveness, reliability, as well as adaptability in Smart Forestry 5.0 are demonstrated by its capability to successfully reduce error levels via several rounds. ESGBT-WoOA outperforms entire remaining methods, making it the major appropriate as well as successful method for predicting climate variation in Smart Forestry 5.0. The proposed ESGBT-WoOA for the CCP model on Smart Forestry 5.0 with respect to L1 Norm is 64.42%, 59.72%, 53.10%, and 44.57%superiorto site-based approach, GM-AI, NSGA-II, and WSN-IoT, respectively.

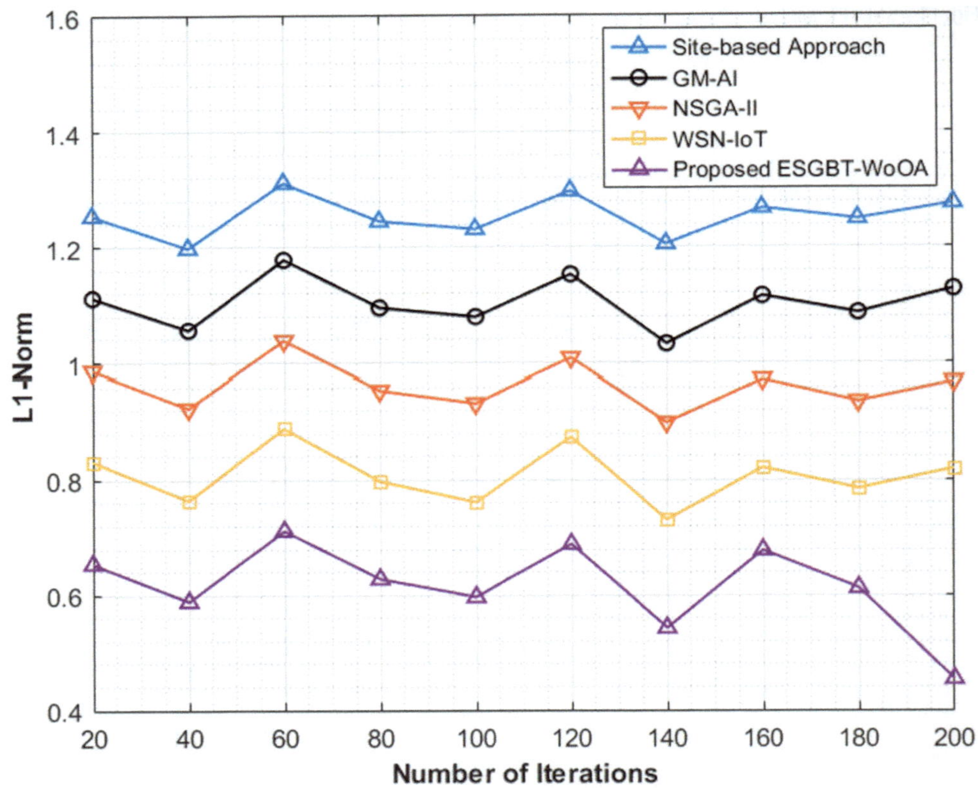

Fig. 12 L1 norm analysis

5 Conclusion

This chapter used a machine learning-based optimisation technique to apply the CCP model to Smart Forestry 5.0. The CMIP6 was the original source of data on how forests were affected by climate change. The gathered data was then processed using the Kalman filtering method. The HHT approach was used to extract characteristics from these pre-processed data. In the end, a new machine learning model called ESGBT was used to forecast these attributes, and WoOA was an optimisation technique that was used to alter the parameters of SGBT. The ultimate fitness function was defined as minimising error and maximising prediction accuracy. According to the findings, the proposed ESGBT-WoOA for the CCP model in Smart Forestry 5.0 performed better than other approaches in terms of a number of error measures. For the CCP model in smart forestry, the proposed ESGBT-WoOA performed better than existing techniques by 61.65% in error reduction and 13.06% in prediction accuracy, respectively.

5.1 Significance of the Study

The significance of this research lies in its innovative approach to CCP in Smart Forestry 5.0 using optimisation driven by machine learning. By employing advanced computer techniques, the chapter enhances the accuracy associated with climate projections, enabling proactive decisions in forest management. Combining CMIP6 climate information, which has been pre-processed using the Kalman filtering technique, ensures that the information is reliable as well as well-maintained for further research. Key climatic factors affecting forest ecosystems can be identified using the HHT for feature extraction. The study presents ESGBT, a complex machine learning method that has been optimised using the WoOA. This approach outperforms traditional methods in terms of prediction accuracy as well as error reduction. In order to address the consequences related to climate variation, the chapter emphasises the importance of multidisciplinary techniques that combine AI, climate science, as well as sustainable forestry. The findings support the advancement of Smart Forestry 5.0 by improving resource management as well as forest conservation's resilience, sustainability, as well as efficacy. The integration of machine learning and optimisation within Smart Forestry 5.0 is no longer a purely theoretical construct but is actively being implemented across several real-world scenarios. For instance, in Canada's boreal forests, AI models are being used to monitor forest health and predict wildfire risks by analysing satellite imagery and IoT sensor data in real time. In Finland, machine learning algorithms combined with optimisation techniques are deployed to forecast timber yield and plan sustainable logging schedules based on dynamic environmental conditions. Similarly, in tropical regions such as the Amazon, convolutional neural networks (CNNs), and gradient boosting frameworks are used to detect illegal deforestation and assess biodiversity threats. These implementations demonstrate how advanced predictive models and optimisation algorithms are driving proactive decisions, improving resource allocation, and promoting resilience against climate-induced threats. By showcasing such real-world applications, the relevance and practicality of the proposed ESGBT-WoOA framework in this chapter are further reinforced, highlighting its potential to transform forestry management across diverse ecological settings.

5.2 Novelty of the Proposed Work

In Smart Forestry 5.0, the proposed ESGBT-WoOA offers several improvements that significantly improve CCP. The following describes this chapter's unique features:

- **Advanced Machine Learning Framework:** The ESGBT-WoOA combines the WoOA with an inventive combination of ESGBT. By modifying hyperparameters in real time, this combination approach enhances model training and ensures robust adaptation to shifting climatic conditions. Unlike traditional techniques, ESGBT allows for real-time learning through streaming information, continuously enhancing predictions as new environmental information becomes available.
- **Smart Hyperparameter Optimisation:** The ESGBT method can only be improved with the help of the WoOA. In order to prevent overfitting or underfitting, it flexibly adjusts crucial parameters including tree depth, learning rate, as well as the count of estimators. WoOA ensures that forecasts stay extremely accurate while lowering computing load by utilising an adaptive learning technique.
- **Effective HHT Feature Extraction:** HHT is used to capture important non-linear as well as non-stationary climatic trends, as opposed to conventional Fourier or wavelet transforms. This method improves the detection of complex climatic patterns, which results in more accurate forecasts. Feature extraction is more detailed when climatic information is broken down into Intrinsic Mode Functions (IMFs).
- **Robust Pre-processing with Kalman Filtering:** Kalman filtering reduces noise as well as inconsistencies in climate variation information obtained from CMIP6 to increase the dataset's reliability. This increases prediction precision by ensuring that the ESGBT-WoOA method runs on clean, high-quality input information.
- **Outstanding Effectiveness in CCP:** The proposed ESGBT-WoOA achieves: 13.06% improved prediction accuracy as well as 61.65% lower error rates compared to traditional techniques including site-based approaches, GM-AI, NSGA-II, as well as WSN-IoT. This significant effectiveness improvement demonstrates how well gradient boosting and optimisation-oriented hyperparameter adjustment work together.

In Smart Forestry 5.0, the ESGBT-WoOA provides a novel approach to CCP. It is a reliable, efficient, as well as scalable solution for sustainable forest management because of its capacity to evaluate vast amounts of real-time climatic data while adapting to shifting environmental conditions.

5.3 Challenges and Limitations

The challenges as well as limitations of the introduced CCP model on Smart Forestry 5.0 can be described as below:

1. **Computational Complexity:** The CCP method for Smart Forestry 5.0 requires a significant amount of processing power due to the mix of optimisation techniques as well as machine learning. Real-time execution is made more difficult by the need for high-effectiveness computer systems to handle large datasets from IoT devices, CMIP6, as well as satellite photos.
2. **Exploration–Exploitation Trade-Off:** The CCP system, which relies on gradient boosting as well as reinforcement learning approaches, faces a trade-off among exploitation (increasing forecasts utilising available data) as well as exploration (finding novel climatic patterns). Insufficient equilibrium between these two factors might lead to imprecise forecasts.
3. **Challenges with Data Dependency as well as Quality:** The accuracy associated with climate projections depends on the completeness as well as quality of both historical as well as current information. Inconsistent or missing information from climate databases as well as IoT devices might lead to inaccurate predictions.
4. **High Memory as well as Processing Requirements:** The training as well as hyperparameter optimisation of the ESGBT method require high memory as well as computing resources. This could limit its application in contexts with low resources or on edge devices.
5. **Limited Knowledge Related to Forecasts:** Because machine learning algorithms are opaque, policymakers as well as forest managers find it difficult to understand the reasoning behind certain climate projections. This ambiguity may cause people to lose faith in AI choices.
6. **Handling Uncommon Climate Events:** While the CCP method is good at predicting general climate trends, it may struggle to predict uncommon but important events such as

droughts, wildfires, as well as sudden temperature shifts. It is still difficult to balance forecast accuracy for both typical as well as severe climatic fluctuations.

Future studies can increase the effectiveness as well as reliability of CCP in Smart Forestry 5.0 by addressing these challenges as well as limitations.

5.4 Future Scope

The promising results of this chapter open up a lot of possibilities for further research as well as development. There are several chances to improve CCP in Smart Forestry 5.0 by combining AI, optimisation techniques, as well as real-time information processing. These are significant possible study topics.

Self-Healing IoT Networks for Climate Information Collection: The reliability associated with sensor networks represents a major issue in smart forestry 5.0. Using AI-powered defect diagnosis to create self-repairing IoT frameworks may improve real-time environmental monitoring as well as increase information accuracy. This development might lead to more precise techniques for managing as well as conserving forests.

Adversarial Learning to Enhance Security in Climate Methods: Adversarial machine learning techniques can be used to address cybersecurity concerns in IoT-oriented climate prediction methods. Future studies can look into ways to keep forecast methods reliable as well as shield climate records from malevolent attacks.

Integration with 6G as well as Beyond for Smart Forestry: With ultra-fast information transfer as well as real-time climate monitoring, future wireless technology advancements, like 6G networks, have the feasibility to revolutionise smart forestry. Research may focus on improving information collection as well as processing using sophisticated communication frameworks.

Most accurate climate projections, proactive forest care, as well as sustainable environmental preservation will all be made possible by these advancements.

5.5 Final Remarks

This study demonstrated the CCP method's efficacy in Smart Forestry 5.0 using an advanced machine learning-based optimisation technique. The CMIP6 dataset was used in the study as the primary source of information on how climate fluctuation affects forests. The Kalman filtering technique was used to pre-process the collected information, ensuring information homogeneity as well as noise reduction. Significant climatic trends were found by feature extraction using the HHT approach. The WoOA was used to optimise the parameters of the ESGBT algorithm, which was suggested for predictive modelling. This method's ultimate objective was to decrease error and increase forecast precision. The results show that in smart forestry scenarios, the proposed ESGBT-WoOA system significantly outperforms existing methods in terms of CCP accuracy. Compared to conventional approaches, the proposed technique achieved a 13.06% improvement in predicted accuracy as well as a 61.65% reduction in error. These findings demonstrate how AI-powered optimisation may address the complex as well as dynamic problems associated with climate variation in forestry. Future studies can enhance these techniques by adding quantum computing, adaptive learning techniques, as well as real-time IoT information for increased effectiveness.

Despite the strong performance of the proposed ESGBT-WoOA model, certain limitations were observed during the simulation phase. First, the relatively small population size (10) and fixed iteration count (200 rounds) were selected to balance computational feasibility and result quality. However, higher population sizes and more iterations could potentially yield more stable and generalised models, albeit at the cost of greater computational overhead. Second, while the CMIP6 dataset offers comprehensive climate projections, its granularity may limit resolution in localised forest ecosystems, especially in heterogeneous terrains or microclimatic zones. Additionally, real-time data integration from IoT devices was simulated rather than deployed in an actual hardware environment, which may not fully reflect delays or communication errors that occur in practical scenarios. The model's performance may also be affected by seasonal data imbalance, such as an overrepresentation of common weather patterns versus rare anomalies. Lastly, while MATLAB was chosen for its simulation convenience, it may not represent the most efficient platform for deploying large-scale, real-time systems in forestry settings. Addressing these limitations in future work could improve the generalisability, robustness, and deployment readiness of the proposed Smart Forestry 5.0 framework.

Data Availability Statement The datasets generated and/or analysed during the current study are available in the CMIP6 repository.

References

Arivumani, S. S., & Nagarajan, M. (2024). Adaptive convolutional-LSTM neural network with NADAM optimization for intrusion detection in underwater IoT wireless sensor networks. *Engineering Research Express, 6*(3), 035243.

Aun, K., Kukumagi, M., Varik, M., Becker, H., Aosaar, J., Uri, M., Morozov, G., Buht, M., Uri, V. (2021). Short-term effect of thinning on the carbon budget of young and middle-aged Scots pine (Pinus sylvestris L.) stands. *Forest Ecology and Management, 492.*

Bowditch, E., Santopuoli, G., Binder, F., del Río, M., La Porta, N., Kluvankova, T., Lesinski, J., Motta, R., Pach, M., Panzacchi, P., Pretzsch, H., Temperli, C., Tonon, G., Smith, M., Velikova, V., Weatherall, A., Tognetti, R. (2020). What is climate-smart forestry? A definition from a multinational collaborative process focused on mountain regions of Europe. *Ecosystem Services, 43.*

Brodribb, T. J., Powers, J., Cochard, H., & Choat, B. (2020). Hanging by a thread? Forests and drought. *Science, 368*, 261–266.

Cesco, S., Ascoli, D., Bailoni, L., Bischetti, GB., Buzzini, P., Cairoli, M., Celi, L., Corti, G., Marchetti, M., Mugnozza, G. S., Orlandini, S., Porceddu, A., Gigliotti, G., & Mazzetto, F. (2024) Smart management of emergencies in the agricultural, forestry, and animal production domain: Tackling evolving risks in the climate change era. *International Journal of Disaster Risk Reduction, 114.*

Chen, J., Feng, K., Hannula, S.E., Kuzyakov, Y., Li, Y., Xu, H. (2021). Interkingdom plant microbial ecological networks under selective and clear cutting of tropical rainforest. *Forest Ecology and Management, 491.*

Clemmensen, K. E., Durling, M. B., Michelsen, A., Hallin, S., Finlay, R. D., & Lindahl, B. D. (2021). A tipping point in carbon storage when forest expands into tundra is related to mycorrhizal recycling of nitrogen. *Ecology Letters, 24*(6), 1193–1204.

Cosyns, H., Joa, B., Mikoleit, R., Krumm, F., Schuck, A., Winkel, G., & Schulz, T. (2020). Resolving the trade-off between production and biodiversity conservation in integrated forest management: Comparing tree selection practices of foresters and conservationalists. *Biodiversity Conservation, 29*, 3717–3737.

De Vries, W., de Jong, A., Kros, J., & Spijker, J. (2021). The use of soil nutrient balances in deriving forest biomass harvesting guidelines specific to region, tree species and soil type in The Netherlands. *Forest Ecology and Management, 479*, 118591.

Forsman, A., Sunde, J., Salis, R., & Franzen, M. (2024). Latitudinal gradients of biodiversity and ecosystem services in protected and non-protected oak forest areas can inform climate smart conservation. *Geography and Sustainability, 5*(4), 647–659.

Gao, D., Joseph, J., Werner, R.A., Brunner, I., Zürcher, A., Hug, C., Wang, A., Zhao, C., Bai, E., Meusburger, K., & Gessler, A. (2021). Drought alters the carbon footprint of trees in soils—tracking the spatio-temporal fate of 13C-labelled assimilates in the soil of an old-growth pine forest. *Global Change Biology, 27*(11), 2491–2506.

Geoff Wang, G., Lu, D., Gao, T., Zhang, J., Sun, Y., Teng, D., Yu, F., & Zhu, J. (2024). Climate-smart forestry: An AI-enabled sustainable forest management solution for climate change adaptation and mitigation. *Journal of Forestry Research, 36*(7).

Georgiev, G. D., Hristov, G., Zahariev, P., & Kinaneva, D. (2020). Forest monitoring system for early fire detection based on convolutional neural network and UAV imagery. In 2020 28th National Conference with International Participation (TELECOM), Sofia, Bulgaria, pp. 57–60. https://doi.org/10.1109/TELECOM50385.2020.9299566

Gong, C., Tan, Q., Liu, G., & Xu, M. (2021). Forest thinning increases soil carbon stocks in China. *Forest Ecology and Management, 482*, Article 118812.

Holzinger, A., Schweier, J., Gollob, C., Nothdurft, A., Hasenauer, H., Kirisits, T., Häggström, C., Visser, R., Cavalli, R., Spinelli, R., & Stampfer, K. (2024). From industry 5.0 to forestry 5.0: bridging the gap with human-centered artificial intelligence. *Current Forestry Reports, 10*, 442–455.

Holzinger, A., Schweier, J., Gollob, C., et al. (2024). From industry 5.0 to forestry 5.0: bridging the gap with human-centered artificial intelligence. *Current Forestry Reports, 10*, 442–455. https://doi.org/10.1007/s40725-024-00231-7

Janani, S., Ramaswamy, M., Samuel Manoharan, J. (2018). Spectrum sensing based heed routing performance enhancement strategy for cognitive radio sensor networks. In G. Venkataramani, K. Sankaranarayanan, S. Mukherjee, K. Arputharaj, S. Sankara Narayanan (Eds.), Smart secure systems—IoT and analytics perspective. ICIIT 2017. Communications in computer and information science, vol. 808. Springer, Singapore.

Kašanin, M., Antić, N., Stojadinovic, S., & Živanović, N. (2024). Soil sensitivity index as an indicator for climate-smart urban forestrysoil sensitivity index as an indicator for climate-smart urban forestry. *Geopauta, 8.*

Lull, C., Bautista, I., Lidon, A., del Campo, A.D., Gonzalez-Sanchis, M., García-Prats, A. (2020). Temporal effects of thinning on soil organic carbon pools, basal respiration and enzyme activities in a Mediterranean Holm oak forest. *Forest Ecology and Management, 464.*

Mahmood, T., Asif, M., & Raza, Z. H. (2024). Smart forestry: The role of AI and bioengineering in revolutionizing timber production and biodiversity protection *15*, 1176–1202.

Manoharan, J. S., et al. (2023). A novel hybrid machine learning approach for traffic sign detection using CNN-GRNN. *Journal of Intelligent and Fuzzy Systems, 44*, 1283–1303.

Mayer, M., Prescott, C. E., Abaker, W. E., Augusto, L., Cecillon, L., Ferreira, G. W., James, J., Jandl, R., Katzensteiner, K., & Laclau, J.-P. (2020). Influence of forest management activities on soil organic carbon stocks: A knowledge synthesis. *Forest Ecology and Management, 466*, 118127.

O'Hara, J. K. (2024). U.S. climate-smart agricultural and forestry incentives: From offsets to clean energy credits. *Agricultural and Resource Economics Review, 53*(3), 1–16.

Riutta, T., Kho, L. K., Teh, Y. A., Ewers, R., Majalap, N., & Malhi, Y. (2021). Major and persistent shifts in below-ground carbon dynamics and soil respiration following logging in tropical forests. *Global Change Biology, 27*, 2225–2240.

Soundararaj, A. J. (2024). Godfrey winster sathianesan, task offloading scheme in mobile augmented reality using hybrid monte carlo tree search (HMCTS). *Alexandria Engineering Journal, 108*, 611–625.

Sterck, F. F., Vos, M. A., Hannula, S. E. S., de Goede S. S., de Vries, W. W., den Ouden, J. J, Nabuurs, G. J., van der Putten, W. H., & Veen, C. G. F. (2021). Optimizing stand density for climate-smart forestry: A way forward towards resilient forests with enhanced carbon storage under extreme climate events. *Soil Biology and Biochemistry, 162.*

Wang, G. G., Lu, D., Gao, T., et al. (2025). Climate-smart forestry: An AI-enabled sustainable forest management solution for climate change adaptation and mitigation. *Journal of Forest Research, 36*, 7. https://doi.org/10.1007/s11676-024-01802-x

Xu, W., Shrestha, A., Wang, G., & Wang, T. (2024). Site-based climate-smart tree species selection for forestation under climate change. *Climate Smart Agriculture, 1*(2), 100019.

Smart Approaches to Forestry: Techniques and Tools for Precision Resource Management

Jahanara Begum, Gattu Shravani, T. Saravanan, R. Hemavathi, and Gayatri Parasa

Abstract

Smart forestry (SF) is a novel approach that integrates progressive approaches such as big data, artificial intelligence (AI), internet of things (IoT), and remote sensing (RS) to enhance forest resource management (FRM) while reducing the environmental impact. The integration of these advanced technologies into traditional forestry enables to make predictions and optimal decisions automatically for sustainable forestry management. The effectiveness of these approaches in this field is dependent upon the accessibility of high-quality data, illustrating the significant role played by sensing techniques in the smart revolution of forestry. But, to achieve the complete efficiency of SF demands incorporation of different approaches including IoT, AI, big data, etc., throughout the process chain for precise resource management in forestry. This document explores the smart approaches used in digital transformation of forestry, particularly the technologies used in resource management. These advanced technologies allow to gather real-time data from forest, automatic data analysis and predictive decision-making, which reduces manual burden and provides better resource management. In smart forestry, the remote sensing technologies enables to identify and categorize species, monitors tree health, heat stress, water stress, and detects illegal logging. In addition, AI-based model integrated with data acquired from sensing technologies provide automatic decision-making regarding forest fire prediction, climate forecasting, and rainfall analysis in the region. This predictive technology helps in preserving forest resources from natural calamities and man-made disasters through early warning and timely intervention. Moreover, the sensing approaches provide continuous monitoring, fungi prediction, pest control and disease classification, protecting forest resources from damage. This document explores the digital transformation of forestry by analyzing these smart applications and assessing their effectiveness in improving forest management.

Keywords

Smart forestry · Artificial intelligence (AI) · Remote sensing · Resource management · Robotic technology · Smart forestry

1 Introduction

Forest contributes a significant role in supporting life on our planet and acts as a carbon sink, eliminating pollution and greenhouse gases in the environment. Recent study on forest conservation reported that around 1.6 billion people's lives depend on forest and more than 300 million people rely directly on forest resources as their source for survival (Jing et al., 2023). The study also stated that more than three-quarters of the world's terrestrial biodiversity including plants, fungi, bacteria, and animals live in forest and they are essential in

J. Begum (✉)
Department of Computer Science and Engineering, Hyderabad Institute of Technology and Management, Hyderabad, India
e-mail: zaib.jaha@gmail.com

G. Shravani
Department of CSE (Cys, DS, AI & DS), VNRVJIET, Bachupally, Hyderabad, Telangana, India
e-mail: gattushravani513@gmail.com

T. Saravanan
Department of Information Technology, P.S.R. Engineering College, Sivakasi, Tamilnadu, India
e-mail: saravanan.t@psr.edu.in

R. Hemavathi
Department of Civil Engineering, Sri Krishna College of Engineering and Technology, Kuniyamuthur, Coimbatore, Tamilnadu, India
e-mail: hemavathi859@gmail.com

G. Parasa
Department of CSE—AIML, KG Reddy College of Engineering and Technology, Hyderabad, Telangana, India
e-mail: drgayathriparasa1989@gmail.com

M. Ashwin et al. (eds.), *Smart Forestry 5.0 for Powered AI Sustainable Environment and Climate*,
Sustainable Artificial Intelligence-Powered Applications, https://doi.org/10.1007/978-3-031-99086-1_6

regulating ecosystem (Wang et al., 2024). Also, the report published by Food and Agricultural Organization (FAO) in 2020 stated that around 80% amphibian species, 75% birds, and 68% mammals live in forest (Mahmood et al., 2024). Moreover, the forest performs a range of functions such as carbon dioxide reduction, pollution control, greenhouse gas minimization, etc., that positively influence human life and environment (Ehrlich-Sommer et al., 2024). Furthermore, the forest is critical for supporting well-being of biodiversity, human wealth, air and water quality, wildlife habitats, etc. The recent climate change dispute illustrated the importance of forest in global nutrient cycling and stabilizing biodiversity (Wang et al., 2025). The report named "State of the World's Forest" released by the United Nation's FAO demonstrated the adverse impacts of fragmentation, deforestation, urbanization and agricultural extension. These activities affected more than 100 million hectares of forest area (Holzinger et al., 2024). In addition, factors such as pests, drought, plant diseases, forest fires, invasive species and heavy weather conditions.

FAO's study also demonstrated that during the period 2010–2020, the forest area has significantly reduced from 7.8 to 4.7 million hectares. This reduction in forest area creates a need for national and international efforts to fight against deforestation, illegal hunting and logging by introducing techniques for sustainable forest resource management (Holzinger et al., 2022). Over the past years, various international practices such as 17 sustainable development goals (SDGs), Biodiversity Aichi Targets, etc., are introduced to improve forest resources. However, to achieve these goals, smart approaches need to be implemented for sustainable forest resource management (Hoppen et al., 2024). Therefore, smart forestry was introduced, which combines technologies such as information technology, geographic data, RS, unmanned aerial vehicles (UAVs), AI, IoT, etc. This technology protects forest from natural as well as man-made disaster through continuous monitoring and management (Uddin et al., 2024). The typical forest resource management involves periodic assessment by forest officers, which is not effective and needs more manpower for precise management (Torresan et al., 2021). On the other hand, smart forestry allows finding and tracking of forest area continuously, which reduces illegal hunting and logging. This chapter works on providing overview of available smart approaches for sustainable forest resource management. It also interprets the working of each technology and analyzes its advantages and disadvantages.

2 Forest Management and Smart Forestry

Forest management or forest resource management, a subfield of forestry concerned with overall legal, economic, administrative, technical, and social aspects related to forest protection and resource management (Venanzi et al., 2023). This involves regulation of wildlife, air quality, water resources, wood products, plants, animals, mammals, timber, and other resource values. Typically, management techniques aim for forest conservation, fire prediction and prevention, effective construction roads and pathways through forest, replanting different species, etc., (Wang et al., 2024). Traditional forestry management is depicted in Fig. 1.

In recent years, information technology has become the driving force for human development and civilization. Also, it has become of the rapidly growing and most influencing technologies in the world and it is being applied in various fields including forestry for optimal forest conservation. Smart forestry defines deep integration of advanced IT technologies

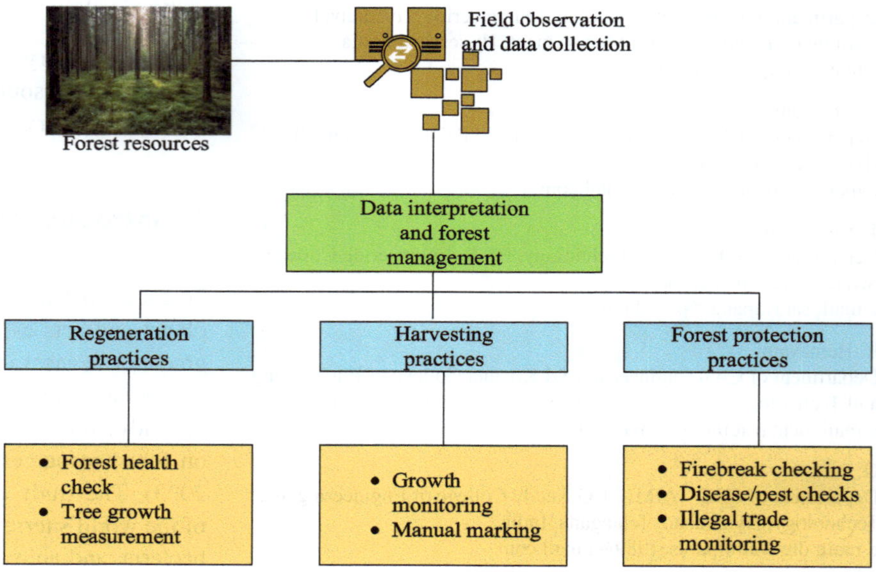

Fig. 1 Traditional forestry management

Table 1 Comparison of traditional forestry and smart forestry

Features	Traditional forestry	Smart forestry
Definition	Forest resources are managed through field observation	Management of forest resources using IoT, big data, etc.
Technology used for data collection	Manual field observation through survey	IoT, satellite imaging (SI), remote sensing
Coverage area	Limited	Large area including rural and remote places
Management strategy	Knowledge expertise of the forest workers	Artificial intelligence, machine learning, deep learning
Cost	High due to requirement of high manpower	Low, as it offers automatic prediction
Accuracy	Prone to human errors	High and offers continuous monitoring
Applications	Basic applications like timber extraction, fire line creation, forest patrolling	Fire prediction and prevention, illegal logging detection, real-time monitoring

such as IoT, cloud computing (CC), AI, RS, big data, mobile internet, etc., for optimally managing forestry production and its resources (Hoppen et al., 2024). The main objective of smart forestry is to ensure automatic forest resource monitoring, fire detection and control, optimal decision making, pest control, disease control, and wildlife protection with the help of advanced technologies. Table 1 provides the comparison between traditional forestry and smart forestry.

3 Smart Approaches to Forestry

Application of advanced technologies in forestry is analyzed in this section.

3.1 Remote Sensing-Based Smart Forestry

In forest research, digital transformation indicates a vital transformation towards utilizing modern technologies for forest resource conservation and sustainable management. This transformation needs large volume of data, creating a need for remote sensing technology (Karunathilake & Le, 2023). Remote sensing (RS) represents the process of capturing earth's surface precisely using sensor instruments. These sensor devices are typically modeled to accumulate information from the surroundings that can be forwarded into server for forestry decision-making. This sensing technology provides a collective knowledge base for the forestry community. Light detection and ranging (LiDAR) is one of the crucial sensor devices that performs tasks such as mapping forest topography, observing forest health, and evaluating biomass. Also, this sensor offers high-resolution 3-dimensional (3D) data, which assist in accurately measuring canopy structure, tree height, and terrain attributes, providing an optimal way for sustainable timber harvesting (Karunathilake & Le, 2023). Other widely used sensing device is soil pH sensor, which plays pivotal role in gathering information related to water content and nutrient content in forest soils, allowing the forest workers to make optimal decisions in irrigation practices and drought management. Similar to soil pH sensor, temperature sensor detects the temperature and humidity in the forest regions and helps in predicting forest fire and rainfall more precisely. Acoustic sensors are deployed in RS technology for observing biodiversity and assessing wildlife health. It also helps in conversation efforts by identifying and recording sounds of various animal and bird species. In addition to this, RS in smart forestry also deploys several sensors including moisture and humidity sensor, RGB camera, etc., to collect data precisely from the forest region. In recent times, drones and aerial sensors are integrated to provide information related to forest health, tree density and pest or disease outbreaks. It acts as a versatile platform and gathers data rapidly with high precision. The digital transformation of forestry not solely depends on these sensing devices but it relies on effective integration of all devices into a cohesive system, which offer enrich knowledge about forest needed for sustainable forest resource management (Holzinger et al., 2022). Table 2 lists different sensor devices used in forestry along with its advantages and disadvantages.

3.2 Satellite Imaging-Based Smart Forestry

Over the past decades, satellites imaging technology is applied in forestry to observe deforestation, forest health and FRM. Currently, many satellites used in forest research is increasing progressively and these satellites are often furnished with imaging devices, generating high-resolution, multispectral, and hyperspectral data. Also, this imaging technology provides wide data varieties using thermal and radar imaging devices, offering information related to atmosphere's biological composition. In forestry, this incorporation takes benefits of monitoring huge areas without the need for manual observation (Octavia et al., 2022). Also, it eliminates the need for gathering forest data manually and provides long-term surveillance of forest region in an inexpensive manner. Typically, the data generated using satellite imaging technology are used for assessing forest health, biomass assessment, and inventory analysis. Although this technology offers numerous advantages, it has several limitations. The major drawback of

Table 2 Sensing devices used in smart forestry

Device type	Application	Advantages	Disadvantages
Global navigation satellite systems (GNSS) and Wi-Fi	Localization	Precise position estimation	Not effective in covering terrain because of signal inconsistency
Acoustic sensors	Animal sound monitoring	Provide insight related to health of animal and bird species	Noise interference
Soil pH	Estimates water and nutrient content in soil	Provide insight related to nutrient uptake efficiency of soil	Cannot match with accuracy achieved as in-laboratory measurement
Temperature sensors	Temperature prediction	Offer information related heat variations. Also, it can be interconnected to different data like humidity, etc., allowing to determine temperature in different media	Very narrow application in forestry when not interconnected with other data. An optimal design is needed continuous monitoring
RGB camera	Produces images of forest surroundings	Provide insights regarding forest road condition, wildlife observation and navigational assistance	Interpreting pure image data is a complex task. Huge amount of data produced when high-quality photo/video is deployed
Moisture and humidity sensors	Determines water content in water and air	Generates climate-related information, which is interconnected with other data like pH	An optimal design is needed continuous monitoring. Implementation setup is intricate
Thermal imaging camera	Estimates heat stress in forest region	Provide insights of thermal grades. Assist in phenotyping and forest fire prediction	Huge data are produced and handling those image needs technical expertise
LiDAR sensor	Generate topographical information	Useful in biomass evaluation, forest structure assessment, etc.	Highly expensive. Face issues with rain on the transparent window
Multispectral cameras	Generate multiple spectra	Simple experimental setup. Produce massive range of information correlative to multiple forest sources	Accumulating numerous wavelengths creates error

this technique is that data was gathered from huge distance from source, making it less efficient in case of assessing ground level of forestry. On the other hand, weather is one of the most influencing factor is satellite imaging. Poor weather conditions hinder the utilization of optical sensor. In addition, adverse weather conditions like thick cloud completely stops data generation for a certain period of time, creating issues in real-time monitoring. Also, it is important to acknowledge that this technology cannot capture data beneath the tree canopy. These inherent challenges can be addressed by integrating satellite information with other data sources for better forestry management. Figure 2 displays the sensor-based forestry.

3.3 Unmanned Aerial Vehicles

UAVs or drones have become an important technology currently used in forestry management. Similar to traditional sensing techniques used in RS, UAVs offer comprehensive insights related to forest health, wildlife, tree intensity, biodiversity loss, etc. UAVs are not sensors, but they act as a platform for transferring the information captured by different sensors and are critical in forest information collection. Evolution of drones has transformed forestry management systems, as it is inexpensive compared to conventional aircrafts like planes and helicopters. This cost-effective nature of UAVs makes it more accessible in numerous forestry applications like fire prediction, tree health analysis, forest health assessment, wildlife observation, etc. In addition, this technology provides more flexibility and accessibility, making it effective for functioning in diverse environments (Shivaprakash et al., 2022). Also, drones have the potential of reaching highly challenged and risky regions, aids in assessing forest ecosystem more effectively and precisely. This efficiency of UAV has boosted smart forestry, allowing to analyze complex terrain regions and sensitive ecological zones effectively. Notable advantage of this technology is its capacity to produce images with better resolution by hovering at inferior altitudes compared to other aircrafts. This unique feature of drones provides comprehensive data, which is useful in providing detailed assessment and monitoring of forest regions in different fields of forestry with high precision. In smart forestry, these drones are very useful in identifying species and also assist in controlling forest fire in its initial stages. The data from the drones are typically transferred to cloud servers for optimal decision-making where artificial intelligence was deployed to automatically predict forest fire and heavy rainfall. Recently, drones are equipped with AI models, allowing

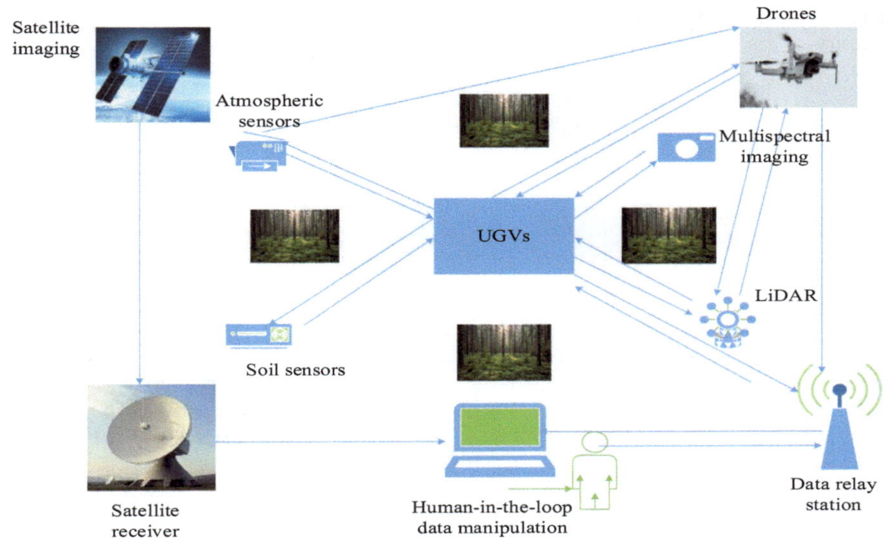

Fig. 2 Sensor-based forestry

automatically mapping tree species and performing classification (Nie et al., 2022).

3.4 IoT-Based Smart Forestry

Recently, IoT has emerged as a booming technology and has revolutionized entire world. In forest research, the integration of IoT has transformed the conventional practices and provides more sustainable and effective way to manage forest resources. The IoT technology makes forest management more efficient and less labor-intensive. The term "IoT" represents the interconnection of physical devices such as sensors, allowing collecting information from surrounding and transferring the accumulated data with each other devices (Ntawuruhunga et al., 2023). Through sensors and IoT nodes, forest workers can continuously monitor large forest regions accuracy, enabling them to evaluate tree health, soil conditions, wildlife, and other ecological conditions. This revolution from manual forestry to IoT-based forestry provides proactive decision-making and improves sustainability of FRM. This technology uses sensors (defined in Table 2) and produces continuous and constant stream of real-time data regarding tree health, pest, diseases, different species, temperature, heat stress, etc., allowing foresters to monitor wildlife and ecological conditions continuously remotely. These insights provided by IoT helps in detecting changes in the forest ecosystem and enable to take appropriate action to protect endangered species and biodiversity from external threats.

It provides real-time observation, which is crucial for ensuring forest ecosystem stability. Unlike conventional monitoring practices, which depends on labors and periodic surveys are interchanged with sensor devices. This reduces the labor cost and offers automatic health monitoring. These sensing instruments provide extensive insights into real-time forest conditions, identifying early disease signs, different pests, and other environmental stress. Additionally, the IoT technology large coverage of forest regions compared to manual assessments, making it flexible and accessible in real-world scenarios. Currently, the IoT was integrated with AI models for automatic prediction and classification of diseases, and appropriate fertilizers are sprayed through UAVs to control the spreading of diseases. This real-time data collection also helps in predicting rainfall in the region, which assists in optimal decision-making related to irrigation practices and water storage. Figure 3 presents the outline of IoT-based forestry.

3.5 Artificial Intelligence

AI enables the system to perform a task as similar to that of human. This technology allows system to analyze the data and make predictions or recommendation based on objective. AI is the subfield of computer science committed to developing machines with the potential of performing tasks that generally demand human intelligence (Wang & Azam, 2024). In other words, AI defines the process of creating algorithms and computational protocols that allow computer systems to study from data, analyze their interconnections, solve problems and take best decisions. Currently, AI technology is used in numerous applications such as object identification, medical image processing, robotics, natural language processing and computer vision. In forestry management, AI analyzes large data gathered from sensing equipment like cameras, IoT devices, UAVs, etc., and make prediction based on its core objective. It examines complex environmental

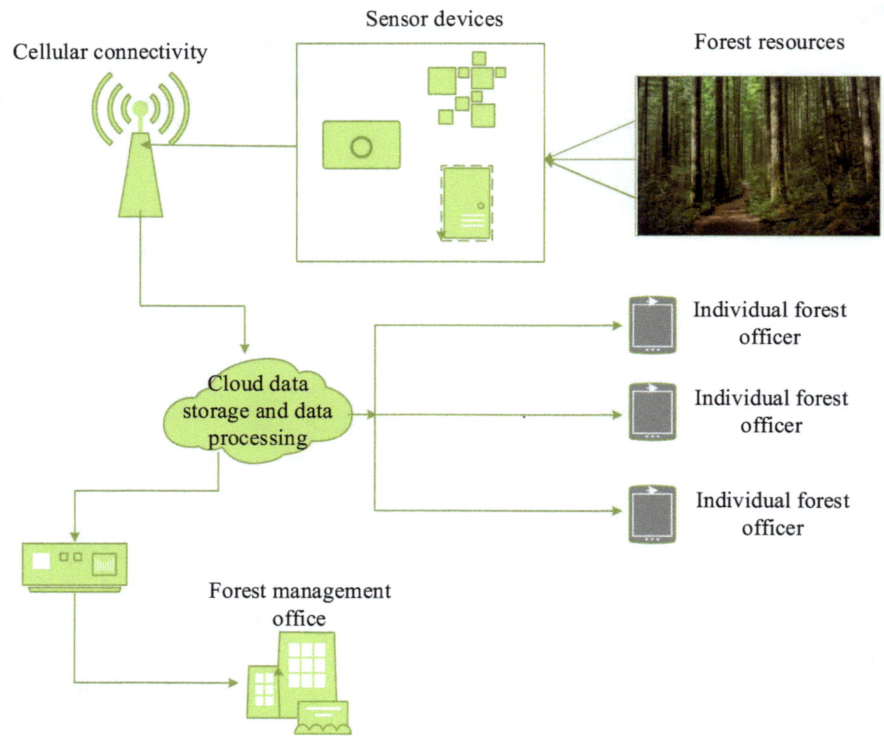

Fig. 3 Outline of IoT-based forestry

data, and learns intricate patterns within data through either supervised or unsupervised learning.

3.5.1 Machine Learning

ML is an AI framework that enables system to mimic human's learning tactics, to do operations independently, and to enhance their outcomes by analyzing more real-time data and upgrading their knowledge based on experience. The learning system of ML model is divided as: decision process, error, and optimization. In decision process module, the ML model is fed with some data (either labeled or unlabeled data), and it is trained to make predictions similar to the pattern in the data. An error determines model's accuracy and optimization module enables system to better fit into training sequence and weights are modified to minimize deviation between the actual and estimated outcome. In forestry, the algorithms such as support vector machine (SVM), k-nearest neighbor (KNN), decision tree (DT), and gradient boosting machine (GBM) are used as base model to train the system for making real-time predictions regarding species types, diseases, pests, etc.

3.5.2 Deep Learning

DL, a subfield of ML, which utilizes artificial neural networks (ANN) to study from data to meet its core objective. ANNs are inspired by the human, and they are deployed for solving a wide multiple problems which involves natural language processing, and speech/image detection. ANNs are comprised of multiple connected nodes, which learn specific data attribute. After learning, the weights are refined to increase detection accuracy, and this adjustment phase is termed as training. Once the network was trained, it makes estimates with unknown sample. Most commonly used DL algorithm include convolutional neural network (CNN), recurrent neural network (RNN), long short-term memory (LSTM), etc., and they showed promising results in making prediction and classification tasks in forestry.

3.5.3 Collaborative AI

Federated learning (FL) is an AI model that collaboratively learns the data patterns across distributed datasets without compromising data privacy. With FL strategy, organizations can utilize maximum potential of big data while also navigating privacy regulations. In traditional AI methods, all training data are aggregated in a central location for model training, while in FL system learns in a decentralized manner and only model updates are returned to global server. Thus, by keeping data localized, this strategy reduces the attack surface for potential hackers. Also, it reduces the need to move large databases, saving bandwidth and minimizes latency in model training. In forest research, FL is useful in protecting sensitive information gathered using sensing devices.

3.5.4 Reinforcement Learning

RL is a ML algorithm that illustrates how agents provide real-time solutions by trial-and-error process to increase cumulative rewards. It enables system to study by cooperating with an environment and by accepting response through actions. This response either arises as rewards or penalties. This concept of this methodology revolves around an agent's interaction with the environment to reach its objective. Here, the agent performs actions and receives feedback to optimize its decision-making process. In forestry, the integration of RL helps in decision-making regarding implementing irrigation practices, rainwater conservation, fire control, etc., offering sustainable forest resource management.

3.5.5 Transformer Models

This type of neural network learns the patterns by stalking interconnections in sequential dat. They utilize self-attention mechanism that enables it to understand the relation between different sections of input sequences and make predictions based on the learned content. Before transformers, training neural networks is a complex task, making with large, making it expensive and time-consuming. By estimating configurations between components statistically, they mitigate the necessity of trillion images and petabytes of text data on the web, reducing the neural network simple and effective. Table 3 analyzes the different AI models, its application area, advantages and disadvantages.

3.6 Robotic Technology

Robotics serves as the basic system to collect extensive data through modern forestry technology. The robotic base needs to operate as an unmanned ground vehicle (UGV) to execute multiple simultaneous tasks under harsh environmental conditions. Thus, it needs autonomous navigation features to independently explore complex unstructured forest terrains effectively during deployment. Effective perception and mobility of robots on forest terrains demand integration of it with LiDAR, RGB-D cameras and ultrasonic sensors. The current robotic systems have made progress but they encounter major obstacles in distinguishing between high grass and solid structures such as walls which requires human operator intervention (Dave et al., 2025). This limitation emphasizes the necessity of human supervision during vital operational phases. Hence, the studies focused on completely automating robots. The complete automatic robots need extended autonomous operation capability while remaining disconnected from human support groups. This automation necessitates a resilient power supply system to maintain all onboard sensors and calculation units. Therefore, researchers start investigating future energy solutions which include solar-charging systems together with hybrid battery solutions and mission-oriented power management methods to advance forestry robots' operational time and system efficiency. The combination of autonomous delivery systems and energy-saving system design will improve forestry operations through increased production speed and enhanced safety together with sustainable practices.

4 Application of Smart Approaches in Forestry

The real-world applications of smart approaches including forest species identification, fungi detection, forest fire prediction, illegal logging detection, and tree height assessment.

4.1 Forest Species Detection

Involving species habitats analysis within a forest environment is critical for assessing losses due to forest fires, droughts, and climate change (Ali et al., 2025). For human

Table 3 Analysis of different AI models

AI type	Algorithms	Application	Advantages	Disadvantages
Machine learning	SVM, DT, GBM, KNN	· Forest fire prediction · Disease classification · Species detection	· Learns patterns and structured data well · High interpretability	Cannot predict complex patterns and needs large labeled data
Deep learning	CNN, RNN, LSTM	· Tree species and disease detection · Detects illegal logging · Provide real-time surveillance · Biomass estimation	· High accuracy compared to ML models · Sequence data processing	Limited generalization and risk of overfitting is high
Collaborative AI	Federated learning	· Secure data transmission in smart forestry	· Protects sensitive information from threats · Collaborative learning	High computational complex and risk of false predictions
Reinforcement learning	Deep Q-network, Q-learning	· Resource management · Resource optimization	· High adaptability in dynamic environments	Design complexity and needs large volume of data
Transformer AI	BERT, vision transformer	· Disease prediction and species classification	· High interpretability	Data dependent

researchers, ecological monitoring in forests is a laborious business due to the long sittings in the field for recording birds. This becomes even more complicated if there are many species to be studied that will require several experts to be recording sounds. A bird monitoring framework was modeled with the ability to listen to many birds simultaneously. The bird's activities including arrival, departure, their locations, and nest pattern deviations are recorded in real-time by the system. So far, species like frogs, crickets, grasshoppers, and marine mammals have been trailed. The major differences come from the natural and human-made noise in the forest settings. Besides, the population of bats is monitored through their wing sounds.

4.2 Tree Health Monitoring

The health of trees is usually detected by evaluating the alterations of some essential metabolic materials and cellular functions for a timely detection of stress. Stress is evaluated from nitrogen interaction with such systems in plants and soils, the applications of ammonium nitrate and ammonium sulfate in amounts deemed excessive, contain sulfur and nitrogen, therefore the correlation between nitrogen assimilation and photosynthesis is analyzed. Stress application leads to anthocyanin production by trees. Metabolic stress can thus be a different indicator among the diverse tree species (Liu et al., 2021). One can thereby identify how different disruptive deviations in single metabolites and their interconnections can be harnessed to engineer resilient biosynthetic genes for growing healthy plants.

4.3 Illegal Logging Detection

It is a primary in forest ecosystems degradation and ultimately leads to biodiversity losses. The illegal logging sensing deploy microphone devices capable of detecting logging-related sounds such as saws, axes, and transport. Such information can then be communicated to relevant enforcement agencies in real time through wireless communication links.

4.4 Forest Fire Prediction

Satellite technology can detect and sense the forest fires using IR sensors mainly for measuring the infrared radiation inhibited by wildfires. Results of RS of the 2019 Amazon rainforest fires have established that these fires emanated from cleared forest lands meant for agriculture and planting purposes (Nitoslawski et al., 2021). Extreme land utilization and growing deforestation are the popular ecological problems not brought by natural circumstances but manipulated by humans as major causations of these fires.

4.5 Invasive Species and Fungi Sensing

Major invader non-native exotic species are indeed an entire battle for forest management. These species are the actual culprits of major ecological revolutions about which advanced sensing and monitoring will have to be applied to reduce, eradicate and be minimized concentrate on the effects and propagation of invasive species. Mostly, it is timber, firewood, and forest transport vehicles that carry them (Kumar et al., 2022). Fungi of all types also minutely contribute to the environment for auguring the acquisition of exotic species, which then predate over their native counterparts on these wood substrates.

Major invader non-native exotic species are indeed an entire battle for forest management. These species are the actual culprits of major ecological revolutions about which advanced sensing and monitoring will have to be applied to reduce, eradicate and be minimized concentrate on the effects and propagation of invasive species. Mostly, it is timber, firewood, and forest transport vehicles that carry them. Fungi of all types also minutely contribute to the environment for auguring the acquisition of exotic species, which then predate over their native counterparts on these wood substrates. These also protect against biofuel conversion for bioenergy. Identifying fungal species can involve several approaches, including molecular and microscopic examination. Gene sequencing may also inhibit soil-borne fungi. Scanning electron microscopy (SEM) can be used for wood decay study. Computer tomography (CT) can track fungal decay and predicts moisture where data is then accessed deep inside the wood structure.

4.6 Tree Height Detection

Air-borne LiDAR offers precise assessment of vegetation height. This method fires a laser beam from above the vegetation canopy to the ground. The reflected energy is used to detect vegetation height, density, and structure. In addition, the vegetation change tracker is able to generate optimal forest estimates through SI. The tracker operates autonomously using Landsat data for mapping (Le et al., 2023). The resultant maps can be used to make clear distinctions between forested and non-forested areas. At the moment, false indicators of forest disturbance are among the main challenges in FRM, especially in wetland and agricultural landscapes. This tracker will further help identify these false positives by way of enhanced mapping. Time Sync employs a similar unofficial database to keep tabs on vegetation status in a temporal

Table 4 Real-world applications of advanced technologies in forest resource management

Technology used	Location	Application	Benefits	Outcomes
IoT and artificial intelligence (AI) (Górriz-Mifsud et al., 2022)	Catalonia, Spain	Forest and climate monitoring and prediction	Pest and disease prediction, and portable to climate variation detection	Improved monitoring of forest biomass usage and wildfire management
AI, cloud technology, and remote sensing (LiDAR) (Wang et al., 2024)	Qingyuan, China	3D forest reconstruction, AI-based prediction designs, digital twin for virtual simulations	Real-time forest carbon–water flux monitoring and carbon sink analysis	Improved forest resilience, and helps in maintaining carbon neutrality
Satellite imaging and RS technology (Dang et al., 2021)	South-west coastal region, Mekong Delta, Vietnam	Wetland prediction and sea level assessment	Accurately identifies wetland with 86.7% accuracy and forecasted sea level rise with 93.3% accuracy	Assist in preserving mangroves, rice fields, and forested wetlands
Unmanned aerial vehicle (UAV) (Mohan et al., 2021)	United States, Australia, Spain, and Thailand	Mangrove restoration, and post-wildfire forest refurbishment	Time and labor efficient, cost effective, and can function in remote regions	Precise seeding, species detection and forest surveying
Machine learning, Robotic technology (Chisika & Yeom, 2023)	Nairobi, Kenya, and Kampala, Uganda	Vegetative resource assessment, land cover mapping, and monitoring urban forest resources	Manpower reduction, enhances ecological planning, and intelligent decision making	Identifies forest fires, and other wildlife threats accuracy and faster compared to manual method

manner through timeseries modeling. The drought periods were found to be highly correlated with decrease in fresh canopy. It is the effect of duration and severity of the short-term summer drought, among others like geography variation, structure and composition, soil and topography and habitat outbreak of pests and diseases. Table 4 summarizes the real-world case studies related to forest resource management using advanced technologies.

5 Challenges and Future Work

Here, the document analyzed some of the most prominent advanced technologies that can be used in the forestry domain. This is a good foundation for establishing the much-needed data generation. There are still challenges to be worked out, though. To provide an AI benefit, one must deploy a sufficiently large number of sensors holding their data at one point. Existing data can be used, but most often the datasets from different locations and times of acquisition are not connected, meaning they are not synchronized. The two are critical in helping an AI make a meaningful connection to different data sets. Everything is done with respect to one location and one time, much as what the modern weather station does. Sensor technologies may be the way to go towards enabling this very widespread generation of data for that automated collection. The very advanced setups such as imaging technologies add value only when enough ground really is covered and digitized in this way.

Systems may be simpler to automate for data collection. We've, as indicated in above, a robotic backbone system in mind to achieve this, but there's riper for exploration. For instance, affordable and standardized LiDAR scanners could be provided to existing forest equipment that already are present in the forests of today so as to create larger amounts of data. Such moves could hugely boost the digital transformation of the forest industry. And of course, placing all these systems under a large umbrella would be the entry of the industry, for post-research-convenient and even more for practical consideration. As to the capital and operational costs incurred, they will be key during development because they will determine the economic value of the digital transformation, not only in research but even more for the industry itself. Our future goal is to establish a completely autonomous robotic data collection system that can traverse the vast forest road networks on its own and start generating data. In that context, we are planning to start building up the baseline system for robotic data generation platform and low-cost sensor setups that can be integrated into equipment that is already moving through forests on a regular basis. The immediate usability for different forest experts can first be achieved by feeding the databases with high-quality data at short intervals.

Having comparable data from more sites may further facilitate some of the work of individual scientists, who would thus have access to more data than they could gather into their experiments. There is an ability to check if certain models working for a particular area might be used for others by checking against results from data already in a database. We would like to stress that these databases above would be very simple at first, but that the strength behind such an automated data collections is building exponentially over time.

The moment that these data become available any research question that is dependent on data which can be acquired as described within this manuscript can be tackled using models. This is where human-in-the-loop AI, as it were, will become a very invaluable instrument for the forest expert. The experts should be in a position to help answer their own research question by getting to see new types of links that have not so far been known or explored. Computing such complicated relationships between various data category and feeding them to the expert and additionally seeking more input from the expert is an excellent tool of AI in forestry for extremely difficult and complicated questions. This organized data flow can be used to train AI models which develop with new inputs leading to more robust and generalized models. The development of interoperability standards at an early stage enables the combination of data from various systems and manufacturers and platforms into one unified framework, which helps in appropriate decision-making. On the other hand, the application of edge computing reduces both data transmission expenses and energy needs which operate in forested areas without grid connectivity. In addition, storing lightweight AI models as part of robotic hardware and sensor components enables real-time data filtering of important information so central storage systems can become more efficient and scalable, addressing issues like limited internet access and restricted power capabilities.

6 Conclusion

Smart technologies including IoT, RS, AI, and SI have revolutionized the forest management practices. AI plays a crucial in the enduring digital revolution, providing innovative solutions across different fields. But, it is significant to spot that the AI technique's success relies on the consistency and quantity of data deployed for training.

This emphasizes the pivotal role of integration of sensing technologies with AI in forestry management where sensing devices provide wide range of data and the AI model make automatic predictions. This comprehensive integrated approach will maximize AI model's efficiency in digital revolution while eliminating threats and ensuring confidentiality. Also, these technologies are currently used in various industries can be applied in forestry applications for real-time monitoring, disease control, and forest health evaluation. In addition, robotic technologies are studied for their potential use in automatic tree planting operations as well as precise spraying and forest patrol services in challenging areas and sample retrieval tasks in distant regions. The combination of robotic systems with their AI-driven software for navigation decision-making functions results in better efficiency coupled with safety and scalability of forest management activities. The integration of robotic technology with sensor network and AI techniques enables to create an autonomous system, which provides effective management of forest resources along with monitoring functions.

The merging of sophisticated technological systems creates an opportunity that will establish a more sustainable period for forestry industry practices. Through this system society achieves the capability to make advanced commitments while spotting threats before they emerge and to fast react to changing environments and maximize the use of available resources. Smart forestry development needs sustained investigation together with technological progress and technological cooperation between experts from environmental science and governmental authorities to fulfill its potential for forest ecosystem preservation. Also, to enhance the practical relevance of smart forestry, the policymakers should develop standard protocols for addressing the prominent issues inherent in these technologies such as security, data privacy and other ethical concerns. In addition, engaging educational campaigns could help to create trust among the local communities and assist in combining smart technologies with traditional forestry practices, which provides an optimal way for sustainable forest resource management. On the other hand, framing regulatory protocols would facilitate seamless data sharing among research institutions that helps in developing novel solutions for better resource management.

References

Ali, G., Mijwil, M. M., Adamopoulos, I., & Ayad, J. (2025). Leveraging the internet of things, remote sensing, and artificial intelligence for sustainable forest management. *Babylonian Journal of Internet of Things, 2025*, 1–65.

Chisika, S. N., & Yeom, C. (2023). Smart urban forest management in east Africa: The case of Nairobi and Kampala Cities. *Sage Open, 13*(3), 21582440231194137.

Dang, A. T. N., Kumar, L., Reid, M., & Nguyen, H. (2021). Remote sensing approach for monitoring coastal wetland in the Mekong Delta, Vietnam: Change trends and their driving forces. *Remote Sensing, 13*(17), 3359.

Dave, R., Kaunert, C., & Singh, B. (2025). Wildlife and forest resource management with artificial intelligence. In *Machine learning and internet of things in fire ecology* (pp. 301–324). IGI Global Scientific Publishing.

Ehrlich-Sommer, F., Hoenigsberger, F., Gollob, C., Nothdurft, A., Stampfer, K., & Holzinger, A. (2024). Sensors for digital transformation in smart forestry. *Sensors, 24*(3), 798.

Górriz-Mifsud, E., Ameztegui, A., González, J. R., & Trasobares, A. (2022). Climate-smart forestry case study: Spain. In *Forest bioeconomy and climate change* (pp. 211–228). Springer International Publishing

Holzinger, A., Saranti, A., Angerschmid, A., Retzlaff, C. O., Gronauer, A., Pejakovic, V., Medel-Jimenez, F., Krexner, T., Gollob, C., & Stampfer, K. (2022). Digital transformation in smart farm and forest operations needs human-centered AI: Challenges and future directions. *Sensors, 22*(8), 3043.

Holzinger, A., Schweier, J., Gollob, C., Nothdurft, A., Hasenauer, H., Kirisits, T., Häggström, C., et al. (2024). "From industry 5.0

to forestry 5.0: Bridging the gap with human-centered artificial intelligence. *Current Forestry Reports, 10*(6), 442–455.

Hoppen, M., Chen, J., Kemmerer, J., Baier, S., Bektas, A. R., Schreiber, L. J., Mayer, D. G., Kaulen, A., Ziesak, M., & Rossmann, J. (2024). Smart forestry—A forestry 4.0 approach to intelligent and fully integrated timber harvesting. *International Journal of Forest Engineering, 35*(2), 137–152.

Jing, W., Kuang, Z., Scherer, R., & Woźniak, M. (2023). Big data and artificial intelligence technologies for smart forestry. *Frontiers in Plant Science, 14*, 1149740.

Karunathilake, E. M. B. M., Le, A. T., Heo, S., Chung, Y. S., & Mansoor, S. (2023). The path to smart farming: Innovations and opportunities in precision agriculture. *Agriculture, 13*(8), 1593.

Kumar, S., Meena, R. S., Sheoran, S., Jangir, C. K., Jhariya, M. K., Banerjee, A., & Raj, A. (2022). Remote sensing for agriculture and resource management. In *Natural resources conservation and advances for sustainability* (pp. 91–135). Elsevier.

Le, T. S., Harper, R., & Dell, B. (2023). Application of remote sensing in detecting and monitoring water stress in forests. *Remote Sensing, 15*(13), 3360.

Liu, T., Sun, Y., Wang, C., Zhang, Y., Qiu, Z., Gong, W., Lei, S., Tong, X., & Duan, X. (2021). Unmanned aerial vehicle and artificial intelligence revolutionizing efficient and precision sustainable forest management. *Journal of Cleaner Production, 311*, 127546.

Mahmood, T., Asif, M., & Raza, Z. H. (2024). Smart forestry: The role of AI and bioengineering in revolutionizing timber production and biodiversity protection. *Revista de Inteligencia Artificial en Medicina, 15*(1), 1176–1202.

Mohan, M., Richardson, G., Gopan, G., Aghai, M. M., Bajaj, S., Pabodha Galgamuwa, G. A., Vastaranta, M., et al. (2021). UAV-supported forest regeneration: Current trends, challenges and implications. *Remote Sensing, 13*(13), 2596.

Nie, J., Wang, Y., Li, Y., & Chao, X. (2022). Artificial intelligence and digital twins in sustainable agriculture and forestry: A survey. *Turkish Journal of Agriculture and Forestry, 46*(5), 642–661.

Nitoslawski, S. A., Wong-Stevens, K., Steenberg, J. W. N., Witherspoon, K., Nesbitt, L., & Konijnendijk Van Den Bosch, C. C. (2021). The digital forest: Mapping a decade of knowledge on technological applications for forest ecosystems. *Earth's Future, 9*(8), e2021EF002123.

Ntawuruhunga, D., Ngowi, E. E., Mangi, H. O., Salanga, R. J., & Shikuku, K. M. (2023). Climate-smart agroforestry systems and practices: A systematic review of what works, what doesn't work, and why. *Forest Policy and Economics, 150*, 102937.

Octavia, D., Sri Suharti, M., Wayan Susi Dharmawan, I., Nugroho, H. Y. S. H. Supriyanto, B., Rohadi, D., et al. (2022). Mainstreaming smart agroforestry for social forestry implementation to support sustainable development goals in Indonesia: A review. *Sustainability, 14*(15), 9313.

Shivaprakash, K. N., Swami, N., Mysorekar, S., Arora, R., Gangadharan, A., Vohra, K., Jadeyegowda, M., & Kiesecker, J. M. (2022). Potential for artificial intelligence (AI) and machine learning (ML) applications in biodiversity conservation, managing forests, and related services in India. *Sustainability, 14*(12), 7154.

Torresan, C., Garzón, M. B., O'grady, M., Robson, T. M., Picchi, G., Panzacchi, P., Tomelleri, E., et al. (2021). A new generation of sensors and monitoring tools to support climate-smart forestry practices. *Canadian Journal of Forest Research, 51*(12), 1751–1765.

Uddin, M., Chowdhury, A., & Kabir, M. A. (2024). Legal and ethical aspects of deploying artificial intelligence in climate-smart agriculture. *AI & Society, 39*(1), 221–234.

Venanzi, R., Latterini, F., Civitarese, V., & Picchio, R. (2023). Recent applications of smart technologies for monitoring the sustainability of forest operations. *Forests, 14*(7), 1503.

Wang, G. G., Lu, D., Gao, T., Zhang, J., Sun, Y., Teng, D., Yu, F., Zhu, J. (2024). Climate-smart forestry: an AI-enabled sustainable forest management solution for climate change adaptation and mitigation. *Journal of Forestry Research, 36*(1), 7.

Wang, J., & Azam, W. (2024). Natural resource scarcity, fossil fuel energy consumption, and total greenhouse gas emissions in top emitting countries. *Geoscience Frontiers, 15*(2), 101757.

Wang, T., Zuo, Y., Manda, T., Hwarari, D., & Yang, L. (2025). Harnessing artificial intelligence, machine learning and deep learning for sustainable forestry management and conservation: Transformative potential and future perspectives. *Plants, 14*(7), 998.

Integrating Biodiversity and Ecosystem Services: Pathways to Sustainable Management

R. Ganesh Babu, J. Seetha, Patil Mounica, T. Prabhakara Rao, and Subhash Bondhala

Abstract

Biodiversity refers to the vast array of life on our planet, delivering essential services that boost human health and well-being. Ecosystems, comprising living organisms and their physical environment, interact to provide numerous vital benefits to humans. However, the functioning of biodiversity models and ecosystem service models tends to be separate from each other. Integrating them creates optimized pathways needed for sustainable environment management. This chapter provides a detailed analysis on need for this integration and analyzes the strategies available for making integration possible. This chapter looks into the ideas and applications of biodiversity and the ecosystem services, and explains why it is important to integrate them. In addition, the chapter explores the need for integration by examining biodiversity and ecosystem service losses with recent case studies conducted in Amazon rainforest and Mekong Delta wetlands. The comprehensive analysis of the integration approaches and its challenges helps the future studies to create pathways for managing the environmental resources optimally. This chapter outlines the significance of this integration and analyzes strategies introduced in previous works to achieve to integrate ecosystem and the biodiversity services. In addition, a detailed investigation of the biodiversity and ecosystem services provides detailed understanding how these components interact to maintain the health of ecosystems and meet the needs of people. This holistic perspective improves the capacity to identify key areas in ecosystems and preserves them before degradation. This also helps in developing adaptive environmental policies based on science for improving sustainable management in various socio-ecological environments.

Keywords

Sustainability · Biodiversity · Ecosystem · Resource management · Sustainable practices

1 Introduction

Current global models assess the impact of human-induced stressors like climate change and land-use change on biodiversity and ecosystem services separately (Beillouin et al., 2021). This limits our understanding of the interconnectedness between biodiversity and ecosystem services under these stressors. Typically, projections of socioeconomic factors, such as population growth and land-use changes, are used to estimate their effects on biodiversity and ecosystem services in isolation (Heino et al., 2021). Therefore, models that only look at biodiversity or ecosystem services might not capture how both can affect each other at the same time. While some studies have incorporated aspects of nature or ecosystems (such as total biomass or specific species) into projections, they often overlook the crucial role of biodiversity in maintaining ecological functions that underpin ecosystem services

R. G. Babu (✉)
Department of Electronics and Communication Engineering, Koneru Lakshmaiah Education Foundation, Vaddeswaram, Guntur, Andhra Pradesh, India
e-mail: ganeshbaburajendran@gmail.com

J. Seetha
Department of Computer Science and Business Systems, Panimalar Engineering College, Chennai, India
e-mail: seetha.csbs@panimalar.ac.in

P. Mounica
EEE Department, Vardhaman College of Engineering, Shamshabad, Telangana, India
e-mail: Mounica.p@vardhaman.org

T. P. Rao
Department of Computer Science and Engineering, Aditya University, Surampalem, AP, India
e-mail: prabhakar.tatapudi@gmai.com

S. Bondhala
Dominion Energy, Network Security Engineer, Richmond, USA
e-mail: subhashbondhala@gmail.com

M. Ashwin et al. (eds.), *Smart Forestry 5.0 for Powered AI Sustainable Environment and Climate*, Sustainable Artificial Intelligence-Powered Applications, https://doi.org/10.1007/978-3-031-99086-1_7

(Kleemann et al., 2022). Biodiversity is affected by changes in the environment, but it also helps maintain the strength and function of ecosystems. Consequently, global predictions about biodiversity, ecological functions, and ecosystem services may be overly optimistic, as they assume that ecosystems will continue to provide the same benefits to humans even as biodiversity is lost (González-García et al., 2022). This ideology does not consider that ecological health is linked to the ability of ecosystems to provide services like pollination, clean water, and regulation of the climate.

Despite ongoing biodiversity loss, efforts to integrate models of biodiversity and ecosystem function are increasing, but remain insufficient (Li et al., 2022). The existence of numerous models operating at different spatial and temporal scales poses a significant integration challenge (Pereira et al., 2024). However, it is hard to unite the information or create one method for analyzing it and making policies because of data diversity. Furthermore, certain models include parameters that overlook biodiversity, despite potentially being dependent on it. Thus, prior to the development of integration strategies, scientists require a synthetic understanding of ways how the conventional methodologies works (Provost et al., 2023). Before developing integration strategies, scientists require a comprehensive understanding of existing models, including their biodiversity attributes, ecosystem functions, inputs, outputs, and operational scales. Identifying areas of overlap, gaps, and potential synergies between models is crucial (Zeng et al., 2022). This understanding enables the scientific community to pinpoint opportunities for effective integration, ultimately informing sustainable management strategies. This chapter outlines the definition of biodiversity and ecosystem systems, explains their values and need for their integration to ensure sustainable management.

2 Conceptual Analysis of Bio-diversity and the Ecosystem Services

Biodiversity encompasses the variety of species, their differences, and the diversity within ecosystems. Ecosystem services provide essential benefits, including water, food, climate regulation, pollination, nutrient cycling, and opportunities for recreation and spiritual connection. The detailed definition for biodiversity and ecosystem services is provided in below subsections.

2.1 Definition of Biodiversity

Biodiversity has multiple definitions and measurement approaches. For many, species are the fundamental unit of diversity, with species count being a key metric. Measures of species richness can vary, reflecting relative abundance, ecological relationships, or evolutionary connections between species. For effective conservation and management, a more nuanced definition of biodiversity is required, incorporating genetic analysis and considering key features like function and resilience within ecosystem composition (Jones & Tingley, 2022). In practical terms, biodiversity encompasses a range of values, including balance, interactions, and wilderness. A widely accepted definition, adopted by the Convention on Biological Diversity (CBD), describes biodiversity as "the variability among living organisms from all sources, including diversity within species, between species, and of ecosystems." The CBD definition has several key features, notably emphasizing variability as its core aspect. This refers to the range of variation, rather than spatial or temporal aspects, allowing for multiple diversity measures to be applied (Berthon et al., 2021). This definition highlights variability at three levels: within species, among species, and within ecosystems. It also recognizes the role of species interactions in shaping biodiversity. However, it doesn't account for metrics based on abundance or quantity, which are important for ecosystem services. Despite some limitations, the CBD definition is widely used, has policy relevance, and is inclusive. Although it may not explicitly capture certain biodiversity metrics like compositional diversity and biomass, it can be broadly interpreted to support ecosystem service assessments. Given the science, it seems unnecessary to debate the definition further, as it's already broad, inclusive, and embedded in key intergovernmental agreements focused on conserving diversity and managing ecosystem services.

2.2 Ecosystem Services

Ecosystem services refer to the benefits humans gain from ecosystems, categorized into provisioning, regulating, cultural, and supporting services. Due to their interconnectedness and the foundational role of supporting services, these categories have undergone refinements. These refinements facilitate economic valuation by distinguishing between final ecosystem services that benefit humans and the underlying ecological processes, as well as clarifying the difference between ecosystem services and the goods they provide (Ayompe et al., 2021). Ecosystems' biotic and abiotic interactions drive ecological processes, ultimately providing final ecosystem services. Humans manage ecosystems for these services, but the goods they produce, like flour from wheat, often require additional processing and inputs beyond the ecosystem's contribution. For instance, an ecosystem managed for tree growth (a final ecosystem service) can yield various goods, including timber, fuelwood, carbon sequestration, or recreational opportunities (McPhearson et al., 2022). The value of these goods to humans depends on context and additional inputs. Final ecosystem services, which directly

Table 1 Role of biodiversity in ecosystem functions and services

Characteristic	Biodiversity as a regulator of ecosystem processes	Biodiversity as a final ES	Biodiversity as a good
Type of organism	• Microorganisms • Primary producers (carbon capture, produce biomass) • Top predators (regulate population) • Pollinators (stabilize non-agricultural ecosystems)	• Wild relatives of livestock and crops • Organisms with secondary compounds • Pollinators	• Large vertebrates, such as birds and mammals, along with showy flowering plants • Flagship or umbrella species • Phylo-genetically distinct species • Endangered species
Ecosystem management	• Managing inputs is crucial for productivity, nutrient cycling, and decomposition, but it can compromise the system's ability to provide other essential services • Maintaining diversity is crucial for building resilience and resistance to both predictable and unpredictable environmental changes	Effective management aims to preserve diverse species groups and habitats	Targeted management strategies focus on enhancing the viability of specific species or species groups
Human benefit	Clean air, healthy fertile soils, pest and disease regulation, clean water, climate regulation, and fiber and food production	Enhancing genetic diversity for valuable resources like new medicines, crop varieties, livestock breeds, and pollinators	Aesthetic enjoyment, education, recreation, cultural services, inspiration, and tourism

contribute to goods and values, can be spatially analyzed to map alternative management approaches, along with their economic costs and benefits. Ecosystems form a complex network, starting with ecological processes and culminating in final services that provide human benefits. Environmental management typically focuses on these final services, rather than underlying processes or end products (Table 1).

3 Implications for Biodiversity and Ecosystem Management

Biodiversity and ecosystem services are intricately linked, with biodiversity supporting key ecosystem processes, directly influencing service delivery, and often being the valued resource itself, with specific genes, species, or traits playing crucial roles in various ecosystem benefits (Pisani et al., 2021). Managing ecosystems effectively involves balancing multiple objectives, such as maintaining ecosystem processes and delivering specific services, while recognizing that diversity plays varying roles in different services, requiring adaptive and context-dependent approaches. Biodiversity likely plays a significant role in regulating and cultural ecosystem services, as well as long-term resilience of provisioning services, with stronger evidence linking biodiversity to ecosystem stability rather than service outputs (Croci et al., 2021). Biodiversity's impact may stem from key species or functional traits, and its importance relative to other factors like biomass is an area of ongoing research. Additionally, biodiversity provides an 'insurance' role, enhancing ecosystem resilience and reliability in the face of uncertainty.

Figure 1 presents the basic interconnection between biodiversity and ecosystem services. The functional trait diversity at the top of the food chain helps influence the interactions and teamwork of species in the soil. The movements under the surface affect how strong the soil substratum is, which leads to more stable soil and better nutrient recycling (Terêncio et al., 2021). It demonstrates that healthy soils make ecosystems function optimally and offer services like water distillation, pollination, and climate change controlling. Biodiversity is the main link that makes all the interactions between ecosystems possible. It is crucial in preserving both natural and human-made ecosystems by changing the structure of habitats, locations where species live, and how ecosystems function. This process demonstrates that biodiversity should be protected not only for nature's sake. It is necessary to protect the benefits nature gives, especially as the environment becomes more uncertain.

4 Valuing Biodiversity and Ecosystem Services

Quantifying biodiversity and ecosystem services is crucial for informing effective and sustainable management decisions. However, valuation is a difficult task because of the variety of value theories and the inconsistencies of the paradigms involved (Weiskopf et al., 2022). Currently, different paradigms are developed for evaluating the value of biodiversity and ecosystem services.

4.1 Valuation Strategies

Given that there are different theories of value, it is essential to valuation exercises that (i) acknowledging multiple, sometimes conflicting, frameworks for valuing biodiversity and ecosystem services and (ii) exhibit transparency by indi-

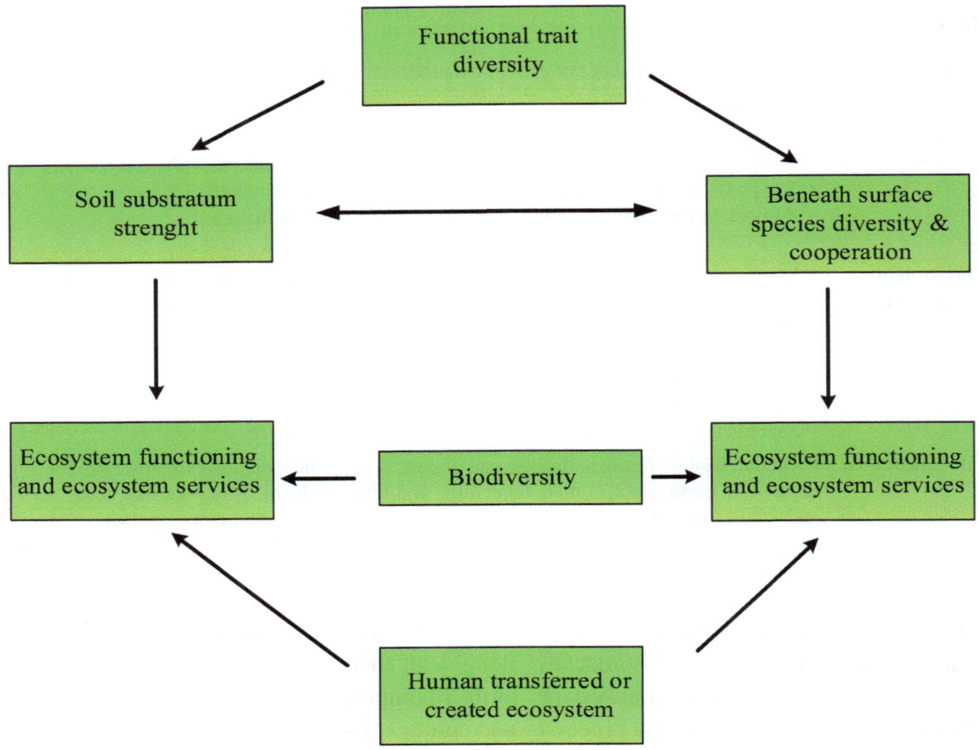

Fig. 1 Basic interconnection between biodiversity and ecosystem services

cating the selected valuation paradigm and the underlying assumptions. A thorough analysis of the way valuation is done reveals that there are two distinct paradigms that dominate. Biophysical valuation methods combine physical principles and ecological sciences, while preference-based methods are usually preferred in economic research (Yin et al., 2021).

4.1.1 Biophysical Valuation

The "cost of production" is the primary lens through which biophysical valuation methods view ecological systems. These approaches determine the biodiversity economic value and the ecosystem services by calculating the particular physical or ecological costs required to provide a given function, good or service. The biophysical valuation measures value by evaluating the observable and quantifiable costs considering labor hours, raw materials, energy use, or land area to maintain a particular ecological state or result (Xu et al., 2022). The approach highlights the material basis of ecosystems and focuses on the physical needs necessary to sustain ecosystem health. Biophysical valuation becomes particularly relevant when trying to capture the tangible and resource-intensive needs required to maintain ecosystems. It is a different way of doing things from standard economic valuation methods and is closer to the ecological principles. This method therefore challenges the anthropocentric foundation of preference-based approaches and emphasizes the importance of understanding ecosystems' functional capacity and resilience.

A study was conducted in Brazil's Atlantic Forest (Shennan-Farpón et al., 2022) to assess agroforestry product. This program calculated the cost of maintaining forest cover based on the tangible ecological benefits provided by forest watershed protection and carbon sequestration using adjusted market price strategy. In South Africa (Peacock et al., 2023), biophysical assessments have been applied to find out the resources that are needed to restore damaged catchments and increase water supply in the semi-arid parts of the country. Figure 2 presents the structure of biophysical and preference-based evaluation methods (Gómez-Baggethun & Groot, 2007, 2010).

4.1.2 Preference-Based Methods

Preference-based models are based on human behavioral models, which state that values are determined by individual preferences. The assumption behind this approach is that the monetary value of ecosystems can be compared not only with the value of other ecosystems but also with that of human-made or financial sectors (Noordwijk, 2021). Consequently, the values of ecosystems are frequently estimated and compared using financial metrics. It should be noted that these valuation methods based on different value theories and axioms do not go well together and cannot be used interchangeably. This disparity fuels ongoing debates about employing multiple metrics and value frameworks in environmental assessment. A case study was conducted in Lake Tana (Wubalem et al., 2023) using travel-cost methods to eval-

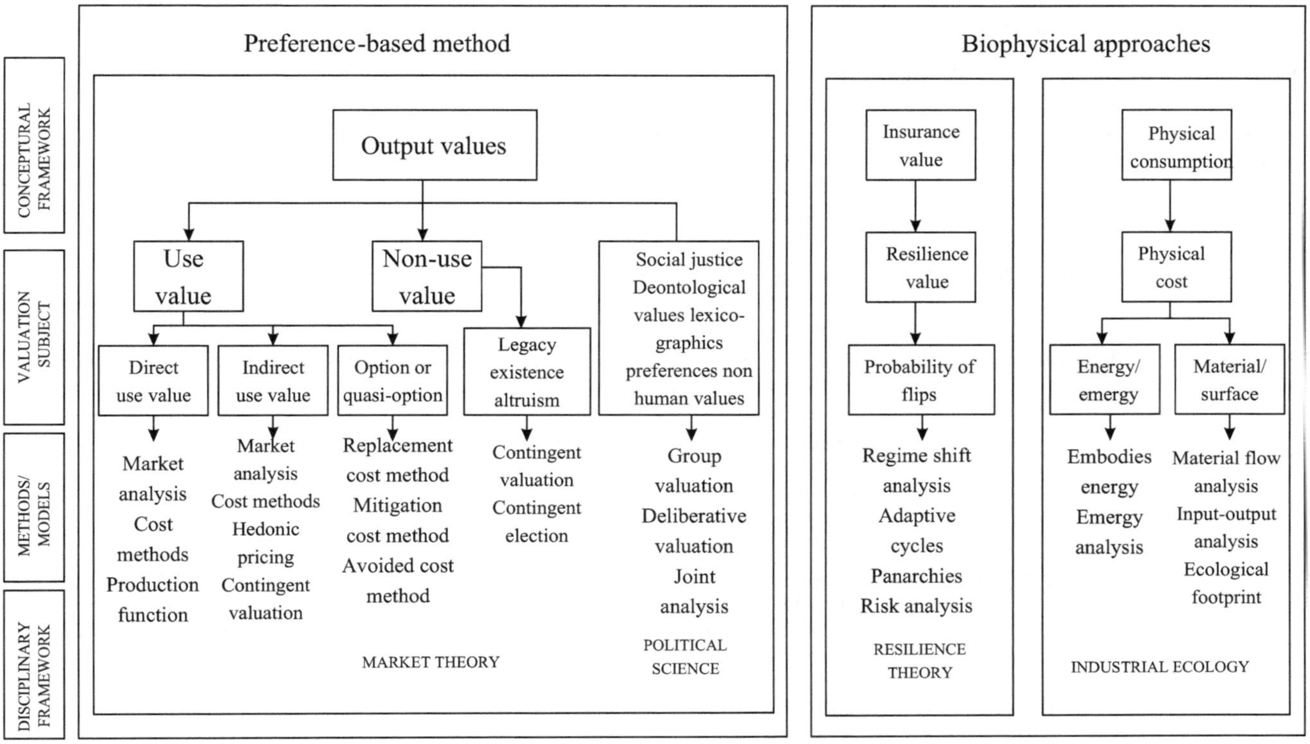

Fig. 2 Biophysical and preference-based evaluation methods

uate ecological benefits and to find out how much people are willing to spend for ensuring environmental management. Also, stated preference methods like contingent valuation were used after the Deepwater Horizon oil spill in the Gulf of Mexico to estimate how much people were willing to pay to help restore the marine environment, particularly the services that the ocean provides (Whitehead et al., 2023). Table 2 illustrates the summary of valuation methods.

5 Need for Integration of Biodiversity and the Ecosystem Services

This section devolves the answers for the questions regarding the necessity for integration of biodiversity and ecosystem services and what are the main causes influencing integration.

5.1 Decline in Biodiversity

Variety and differences in life forms, within species and across ecosystems and genes, are being lost at a rapid and alarming rate on a global scale (Loos et al., 2023). Such degradation is mainly caused by human activities such as habitat fragmentation, conversion of lands, deforestation, over exploitation of resources, environmental pollution, and increased the effects of climate change. The fragmentation and extensive loss of forests, wetlands, coral reefs and grasslands—ecosystems that are critical in sustaining high levels of biodiversity—is reducing the critical space and resources needed for many species to survive and reproduce. The agricultural growth, particularly which is based on monocultures and excessive chemical application, has significantly damaged the populations of native plants and animals. Reducing and standardizing ecosystems reduces their functional diversity, reducing their ability to resist and recover from disturbances. According to the current data, species are disappearing at the rate 100 to 1,000 times faster than a normal historical rate. This decline affects endangered species, as well as the unheralded contributions of pollinators, soil microbes, and amphibians that support the stability and productivity of ecosystems. In addition, degrading genetic biodiversity results in degradation of air quality, water quality, disease control, and sustainable food supply. This situation emphasizes the gap between the present development focus and the requirement for ecological sustainability, emphasizing the need for integrated conservation strategies that appreciate biodiversity as a treasure of the world and a vital lifeline.

To assess the adverse impact of biodiversity loss, a case study was conducted in Amazon Rainforest (Silva-Junior et al., 2023), which highlights the major damage caused by deforestation. The case study illustrate that over 500,000 km^2 (11.1%) of forest cover was lost between 2000 and 2020, driving a 20–30% decline in mammal and bird species rich-

Table 2 Summary of valuation methods

Valuation paradigms	Sub category of valuation method	Definition	Applications to ecosystem services	Real-world Examples
Biophysical method	Market prices with adjustments	Use market prices with adjustments for changes (For example: non-competitive practices, taxes, subsidies)	Crops, livestock, woodland	Agroforestry product assessment in Brazil's Atlantic Forest (Shennan-Farpón et al., 2022)
	Production function model	Assessing the value of an ecosystem service by its impact on production outputs	Maintenance of beneficial species, flood protection agricultural productivity	Ecological restoration in South Africa (Peacock et al., 2023)
Preference-based method	Revealed-preference methods	Analyzing spending on market goods tied to ecosystem services reveals their value through methods like expenditure avoidance, travel-cost analysis, and hedonic pricing, which assess substitutes, recreational activities, and environmental impacts on property values	Water quality, quiet and peace, amenity, recreation	Economic valuation of Lake Tana (Wubalem et al., 2023)
	Stated-preference methods	Survey-based methods like contingent valuation and choice modeling gauge people's willingness to pay for environmental improvements, either by direct valuation or by presenting trade-offs between environmental goods and prices	Species, water quality, air quality, non-use values, conservation,	Gulf of Mexico restoration post-oil spill (Whitehead et al., 2023)

ness and a 30–40% reduction in insect populations, with specific losses like 35% of beetle species in logged areas. It proves that when habitats are destroyed, species often suffer, as 15–20% of endemic species are now at risk of extinction. Using remote sensing and species counts to track biodiversity loss, research shows habitat fragmentation limits jaguar mobility by 25%, highlighting conservation challenges. This situation puts the forest ecosystem at risk, as pollinators and seed dispersers are necessary for forests to grow back. The findings suggest that if no preventive measures are done, the Amazon's ecosystem could fail. Also, it highlights that if 10% of damaged areas are restored, it could help bring back 15–20% of the species living there.

5.2 Ecosystem Service Losses

The delivery of benefits from nature to people is threatened by the continuous degradation of ecosystems. Ecosystem services, including food, climate regulation, pollination, and recreation, are declining or becoming more unpredictable in many regions (Liu et al., 2022). Changes in ecosystems is the result of land-use change, environmental pollution, unsustainable harvesting, and interference with natural cycles have resulted in serious impairments of these ecosystem services. For instance, the draining of wetlands caused by intensive exploitation of both groundwater and surface water reduces their water purification and flood control role. Not only does deforestation decrease the world's carbon storage, but it also disrupts normal rainfall patterns, promotes more severe erosion, and weakens local climate regulation. Pollinator population collapses may have negative effects on food availability, and mangroves and coral reefs loss exposes communities to greater storm and sea-level rise risks. The impacts of degradation of ecosystem services are felt most by the most vulnerable communities because they are mainly dependent on natural resources and hence are poorer and less resilient to environmental threats. The economic value of such losses is not typically accounted for in standard metrics, which tend to underestimate the actual financial cost of ecosystem destruction.

A case study in the Mekong Delta (Hecht et al., 2019) assessed ecosystem service losses resulting from upstream dam construction and agricultural expansion, highlighting impacts on downstream services. The data analysis depicts that floodwater storage space has gone down by about 20–25%, which means flood damage each year is now $200 million more, water treatment has gotten less effective by 15–20%, so it costs $50–80 million more to clean water every year, and fishery yields have dropped by 40–50%, which means the industry loses about $1.2–1.5 billion every year. These losses, measured using water models and money values, show how building 11 big dams has messed up the way water moves and caused wetlands to get damaged. This study used data from the Mekong River Commission and input from local people to show how the Delta affects both the environment and the economy, and also shows that the Delta provides half of Vietnam's fish for people to eat. The findings show that millions of people in these areas are at risk of not having enough food and struggling with money problems, especially in the communities further away from the main city. While restoration efforts like bringing back floodplains could help bring back about 10 to 15% of what the Mekong used to provide, getting different countries to work together is still a big challenge because they often have different goals and needs.

5.3 Interdependence Between Biodiversity and the Ecosystem Services

Biodiversity is the diversity and variability of the forms of life at ecosystem, species, and genetic levels. It is the basis upon which ecosystem services are formed. Humans get these services from nature, and they are closely connected to the way ecosystems function (Kattel, 2022; Qiu et al., 2022). Ecosystem services, broadly divided under cultural, provisioning, regulating, and the supporting services, are inextricably linked to the structure and functioning of biological diversity. For example, food and medicinal plant provisioning depends on species diversity; natural pest management and pollination are enabled by diverse bird and insect populations; and carbon sequestration is increased by diverse vegetation communities. Thus, biodiversity plays an important role in direct and indirect functioning of ecosystems. In the absence of healthy biodiversity, ecosystems become less resilient, less productive, and less able to deliver critical services. Because of this, the environment becomes more sensitive to changes and the regular flow of services is interrupted. In spite of this built-in connection, ecosystem services have too often been valued and managed separately, with little consideration for the biological diversity that supports them. This difference between policy and practice often results in inefficiency and a long-term deterioration of natural resources and the environment.

Typically, biodiversity is the base of all ecosystems, providing information regarding how they work in accordance to the changes. At the same time, ecosystems give living things the places and resources they need to live, change, and grow over time. This two-way interaction helps keep things like nutrient recycling, energy moving through the ecosystem, and keeping the climate steady so that the ecosystem can keep going on its own. For instance, in a wetland ecosystem, there are many different kinds of plants, and these plants help grow different kinds of microbes and animals, which then help clean the water, reduce flooding, and form the soil. These ecosystem functions, in return, help keep the environment healthy and support the different kinds of living things that live there. Disturbances in either component, like when species go extinct or habitats get damaged, can start a chain reaction that messes up the ecosystem and makes it harder for different kinds of living things to survive. Hence, biodiversity and ecosystems work together as one for keeping things in balance. A loss in one part can make the other weaker, not because they're directly connected, but because they both rely on each other for managing the environment resources.

6 Strategies for Integrating Biodiversity and Ecosystem Services:

Integrating biodiversity and ecosystem services requires two main approaches: directly linking biodiversity models to ecosystem service models or using ecosystem function models as intermediaries. To achieve successful integration, four crucial questions need to be answered, laying the groundwork for effective conservation and management strategies that balance both biodiversity and ecosystem services (Fang et al., 2023). First, the nature and reliability of the data needed for integrating biodiversity with ecosystem services should be clearly defined. This involves testing whether biodiversity model outputs are compatible with the input specifications of ecosystem service models and verifying the empirical evidence for the biodiversity–ecosystem function relationships. Second, future directions of study need to be outlined to reduce the existing gaps in understanding and make model integration more feasible. Third, evaluating which ecosystem functions are most integratable in each strategy based on strong empirical data, available models, and direct connections to important ecosystem services is critical. Finally, selection of most effective conventional models for each strategy is essential considering their structure, scalability, suitability and compatibility under different scenarios. Figure 3 presents the outline of the strategies used in integration.

6.1 Strategy A: Empirical Biodiversity–Ecosystem Function Relationships as a Bridge

A pragmatic way to link biodiversity and the ecosystem services is to utilize established, empirical relationships between species diversity and ecosystem processes. By matching the outputs and inputs of biodiversity projection models with existing empirical relationships, this approach enables the coupling of biodiversity models with ecosystem service models. Human activities on the environment have inadvertently provided large, natural experiments that show the effects of major changes to species richness and composition (Petroni et al., 2022). Human-caused changes to ecosystems are documented in a variety of biodiversity assessments—for example, the adverse effects on marine food webs

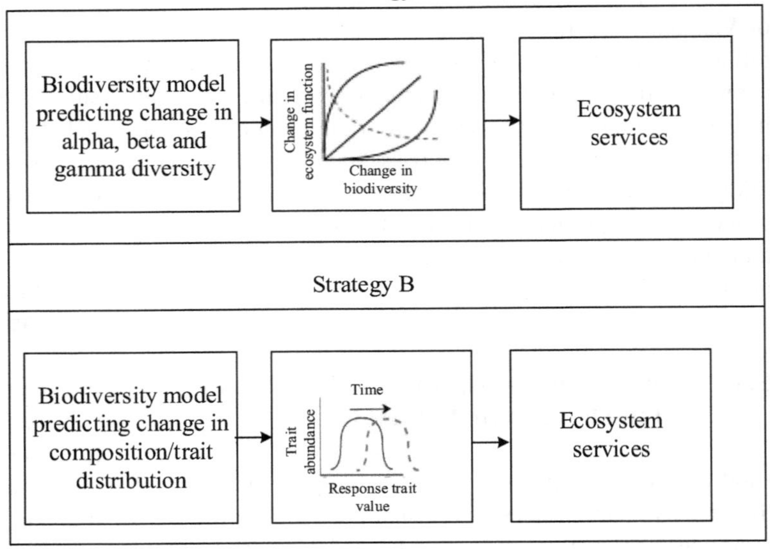

Fig. 3 Outline of integration strategies

due to such activities as overfishing. While observational studies provide critical evidence for large-scale model predictions, they typically do not allow scientists to distinguish how changes in species richness and changes in species composition affect ecosystem function. On the other hand, biodiversity experiments are specifically designed to detect and examine these impacts, which is a robust foundation for their inclusion in modeling frameworks. Biodiversity experiments have explored how plant species richness affects productivity, providing insights for integrating biodiversity models with ecosystem service models. By analyzing these relationships, researchers can predict how environmental changes, such as climate shifts or land-use alterations, impact plant species richness and subsequently influence ecosystem services like biomass production.

Bottom-up species models: By stacking individual species distribution models, these models produce summarized predictions of shifts in species richness.

Top-down community-level models: These models determine variation in beta, alpha, and the gamma diversity without demanding species-level data.

Proportional shifts in species richness can be estimated by transforming model outputs. These values can then be applied to empirical relationships between biodiversity and productivity to predict shifts in plant productivity. When the modified productivity values are obtained, they can be used in ecosystem service models. Some studies have started the efforts to implement this approach. Scientists have equated estimates of species extinction due to changes in land use to biodiversity–ecosystem function relationships based on both theoretical and empirical work. They then linked these relationships to carbon ecological production functions and global land-use and biomass maps are used to estimate the expected slow carbon losses resulting from past land-use changes. Conversely, much of the research to date has been dedicated to the identification of tree and shrub species at a global scale, but relatively little work is done to model their distributions under current and future environmental conditions. Several modeling techniques were combined to establish how tree and shrub species richness may vary between the years 2005 and 2070. The adjustments were converted into proportional changes in forest productivity, represented as elasticity of substitution values, which indicate how different species can compensate for each other to maintain ecosystem function. Looking at a variety of climate projections brings out the fact that this approach enables the assessment of how conserving forest plant diversity contributes to mitigation impacts at regional and global levels. This approach can also be applied to other groups of organisms and ecosystem services, provided that the required empirical knowledge about biodiversity–ecosystem relationships is available. Most biodiversity experiments have been conducted for plant communities in temperate grasslands, with few comparable studies reported for other taxonomic groups or aquatic environments. Thus, in the absence of direct experimental data, it may be crucial to estimate relationships by looking at biodiversity depletion in spatial or temporal patterns—bearing in mind the additional uncertainties such confounding factors may add. By providing a mechanistic view of how biodiversity affects ecosystem services, this framework also allows for the use of more integrated and scientifically based approaches to sustaining ecosystems.

6.1.1 Strategy A: Assumptions and Challenges

When using biodiversity experiment data to support ecosystem service models, some important aspects must be considered. The aim of most biodiversity experiments is to isolate species richness from the effect of species identities. When a species is lost from an ecosystem, two key changes occur: a decline in overall species richness and an alteration in species composition (Gama & José, 2023). Research in biodiversity experiments reduces this by assigning plots with random combinations of species richness and species identity variability. This configuration produces a variety of species richness while maintaining uniform species identity at all levels of diversity. With this experimental setup, researchers are able to distinguish the individual effects of richness and composition on the functioning of ecosystems. The fact that ecological variability in function exists as species richness increases, but not within richness categories, shows that biodiversity affects ecosystems more than composition, and internal changes within richness levels emphasize the importance of composition. Notably, the effect of species richness is sometimes comparable to the effect of species composition. In the real world, increases and decreases in both the diversity and assemblage of species are common. The outcomes can amplify or cancel each other, as in the case of sustained losses in high-productivity populations countered by compensatory effects from low-productivity species. The intricacy of modeling the interactions between richness and composition continues to be a challenge in biodiversity science.

Trait-based frameworks offer a useful solution by predicting which species will be most impacted by future environmental changes. If some traits are associated with environmental change and the delivery of ecosystem services, then such frameworks may explain compositional effects on ecosystem function better. In natural systems, where species composition develops organically, biodiversity is often positively linked to ecosystem function. However, the interconnection of these relations varies according to both the type of ecosystem and the specific ecosystem function at hand. Biodiversity studies frequently encounter the problem of using significantly fewer species than are observed in actual field communities. Therefore, the degree to which species loss affects function in experiments may not reflect how it occurs in more diverse natural settings. The difference in scale between experiments and real-world applications makes it difficult to use local experimental data to make broader predictions. Model comparison enables the study of ecological variation and can enhance the confidence in predicting biodiversity–ecosystem function connections at various scales. However, a major problem is the difference between the spatial scale at which biodiversity and the ecosystem function is measured in experiments and the scale needed for practical applications. Although experiments tend to occur at localized, small-scale sites, real-world scenarios often require knowledge at a larger, regional scale.

Local-scale experiments have examined how interactions among species can influence productivity, typically through reduced competition, increased facilitation, or both. However, larger-scale processes such as species dispersal are generally excluded from these designs. Dispersal can significantly affect both biodiversity and productivity. At low dispersal rates, species may not reach optimal habitats, reducing both productivity and biodiversity. High dispersal rates can also lead to low biodiversity and ecosystem function if a single dominant species out-compete others. Intermediate dispersal rates tend to support higher beta diversity (i.e., species turnover across space), allowing species to thrive where they are most productive and resulting in greater landscape-level biodiversity and ecosystem productivity. These larger-scale dynamics are typically not captured in local biodiversity experiments. Therefore, while these experiments provide valuable insights into biodiversity's influence on ecosystem function, they likely account for only a subset of the relevant processes. Incorporating theoretical scaling relationships and spatial dynamics into ecosystem service models could allow for better projections of how changes in biodiversity, particularly in beta diversity, affect ecosystem function and multi-functionality over time. In future work, combining empirical data with theoretical models could provide a more comprehensive understanding of how biodiversity loss and homogenization of species across landscapes influence ecosystem services. This approach holds promise for more accurately informing sustainable management strategies through improved integration of biodiversity and the ecosystem function across scales.

Research at the local level has revealed that species relationships matter for productivity, which frequently leads to less competition and more mutual benefits. However, these designs usually do not incorporate major processes such as species dispersal that occur at larger scales. Dispersal process has significant effects on biodiversity and overall productivity. If the dispersal rate is low, it is difficult for the species to migrate to optimal habitats, which reduces biodiversity and productivity. If dispersal is very effective, one competitive species can dominate, which will lead to loss of biodiversity and ecosystem function. With intermediate dispersal, beta diversity increases, allowing species to be present in optimal habitats and increasing biodiversity and function at a landscape scale. Local-scale biodiversity experiments often fail to consider important dynamics that take place at larger spatial scales. As a result, although biodiversity experiments provide valuable information about ecosystem function, they generally concentrate on a few underlying processes. The inclusion of theoretical scaling relationships and spatial dynamics in ecosystem service models may increase our capacity to project the impacts of changes in biodiversity, particularly beta diversity, on ecosystem function and the multiple services they offer over long periods. Developing a synthesis of empirical findings and theoretical models could result in enhanced understanding of the impact of biodiversity decline and homogenization of species on ecosystem services. This approach can help to greatly enhance the way in which sustainable management is guided by stronger integration of biodiversity and the ecosystem function at different scales.

6.2 Strategy B: Trait-Based Modeling to Promote the Connection Between Biodiversity and the Ecosystem Function

The application of trait-based models has been shown to be a useful strategy for ecological modeling. The models emphasize critical traits that define the ecological functions of species and their ability to adjust to environmental conditions. Size is one of the many traits that stand out because it has significant impact on processes like metabolic rates, diet, mobility, and development strategies. The use of this trait as a proxy enables researchers to understand how ecosystems are built and to predict their response to a range of stressors. The importance of size traits and others is of equal importance in marine and terrestrial ecosystems. Traits that affect an organism's ecological function and its ability to withstand environmental stressors are important in ecosystem modeling. Most importantly, it is the interactions and balances between traits rather than individual traits that are important in understanding species contributions to ecosystem processes. As a result, the inclusion of trait compositions in models is a reasonable and useful approach to simulating ecosystem functions. Marine ecosystem models have been employing key species or functional groups for more than four decades to decode and forecast ecosystem reactions to altered environmental conditions such as climate change, ocean acidification, or fishing operations. In the same vein, plant functional traits have been systematically incorporated into terrestrial ecosystem models to simulate important ecological processes including physiology, population growth, and ecosystem dynamics. Recently, models of terrestrial ecosystems have developed to incorporate complete descriptions of species traits and the functioning of ecosystems as a whole. The recent advances in modeling ecosystems have allowed for the incorporation of increased consideration of biodiversity into functional ecosystem models based on existing model designs. Although modeling with the complex diversity of species in ecosystems can be technically challenging, there are three viable modeling approaches that provide solutions:

B1: Modeling Trait-Driven Responses to Environmental Change

One of the main strategies involves the simulation of the variable responses of functional groups, which are differentiated by such traits as trophic strategy, body size, and the thermal tolerance to environmental drivers. These particular traits determine the ability of a species to survive changes in an environment. As environmental pressures favor certain traits over others, the frequency of key traits that influence ecosystem properties can shift. The resultant change may affect ecosystem services and the functions they serve. One of the first steps has been to model evolutionary changes in food webs, with an emphasis on the impact of climate change on regional fisheries. However, such approaches in isolation do not appear to provide sufficient representation of the complex relationships between ecosystem processes, species assemblages, and biodiversity. Authentic modeling of these interactions requires the integration of various processes and temporal-spatial frameworks, bringing together forces that determine the properties of the ecosystem with those that determine species or trait distributions.

B2: Using Species-Level Models to Inform Functional Trait Distributions

Another strategy is to combine biodiversity models at the species level with models that aggregate species into functional cohorts. Using species-level models, we can know which species are present and in what quantities in a given ecosystem, and we can develop functional groups and their initial trait distributions. Some classes of models already take into account the distribution of traits within functional groups, as marine size spectrum models demonstrate. However, there are still significant models that do not take into consideration the variability of trait by using mean estimates instead. In some ecosystem models, functional cohorts are constructed to represent similar organisms, but the cohorts can have different traits in general, but homogeneity of traits is assumed within each individual cohort. An improvement can be made by introducing moment-based traits into models to better reflect trait distributions. This would allow ecological processes to sculpt trait distributions dynamically and would support the modeling of biodiversity's insurance effect—the idea that greater biodiversity stabilizes ecosystem functions over time because of redundancy in functions and varied responses to change.

B3: Embedding Biodiversity–Function Relationships into Ecosystem Models

A third option is to incorporate theoretical links between species diversity and ecosystem processes directly into ecosystem simulations. Data support the concept that different species resource use can enhance ecosystem health, and such effects become more apparent over time. In terrestrial ecosystems, sophisticated demographic models of vegetation are now capable of including demographic processes and individual interactions, thus enabling effective simulations of vegetation succession and community assembly. Modeling efforts enable us to determine how changes in biodiversity affect patterns of vegetation, ecological nutrient cycling, and overall ecological productivity. It is possible to better demonstrate the positive effects of biodiversity on the ecosystem functioning by modeling scenarios where intraspecific competition is minimized or functionally distinct species are preferred when biodiversity is high. This strategy is a strong instrument for promoting the incorporation of biodiversity aspects into the sustainable management plans of ecosystems.

6.2.1 Strategy B: Assumptions and Challenges

One of the fundamental assumptions of trait-based modeling is that functional groups are identical to ecosystem diversity. However, the evidence indicates that this assumption underestimates the complexity of natural ecosystems. It is notoriously difficult to represent biodiversity in an authentic manner in models that predict ecosystem function. Many dynamic global vegetation models aggregate biodiversity by grouping plants into about 10 to 15 functional groups on earth. Even within individual sites, numerous vegetation communities consist of a small number of key species. This approach ignores the specific roles and ecological roles of species that constitute only a part of the community. Notably, changes in biodiversity within functional groups can have a major influence on ecological outcomes, thus relatively small changes in diversity can alter ecosystem functions. One of the major barriers to the introduction of a greater degree of complex functional diversity in these models is the high computational burden required. More advanced modeling frameworks require high computational and data needs, thus making them impractical in large areas and long periods of time. Another challenge is the underrepresentation of fine-scale ecological processes that include such factors as density dependence, competition, and localized interactions. The current modeling approaches often lead to a decrease in diversity during the simulations, which covers up the real dynamics of biodiversity in nature. A key premise behind these models is that the relationship between the response traits that govern persistence under environmental shifts and the effect traits that define ecosystem contributions is properly described. Nevertheless, it is often difficult to make clear connections between these factors. Environmental drivers determine the composition of species communities, but there is no direct relationship between drivers and the ecosystem services provided by species. If response and effect traits are correlated, species typically face multiple environmental drivers at the same time. The contradictory impacts of these drivers on species might prevent the clear identification of the best trait combinations for maintaining ecosystem services.

Also, the effect of global change on plant trait distribution is still largely unknown. For instance, climate change supports plant which are less or more productive under varying environmental scenarios. This raises a question regarding the persistence and functionality of several vegetation types. Species interactions such as predation, competition, parasitism, and mutualism greatly enhance this uncertainty by estimating how the numbers and productivity of species will change. Attempting to model these detailed interactions may initially increase the uncertainties of forecasts of ecosystem functioning. However, it is expected that uncertainty will decline as more empirical research and trait information is obtained. Theoretical ecology developments should not stop. For instance, complementarity among species is recognized as a key driver of ecosystem function, but it can be produced by a variety of processes such as resource partitioning, abiotic facilitation, and biotic feedbacks. Many studies have demonstrated complementarity, but they usually do not differentiate among the various processes that produce it. On the contrary, the application of process-based models in ecological research is limited by the need for well-articulated mechanisms, with little practical application without strong theoretical foundations. However, process-based models are of great value as experimental tools for testing the effects of different mechanisms of biodiversity and ecosystem function. By specifying linkages between traits, environmental influences, and ecological processes, these models allow for useful predictions about sustainable ecosystem management, provided that necessary theoretical and empirical research is further developed.

7 Conclusion

This chapter highlights the importance of integrating biodiversity and ecosystem services for sustainable environmental management. While conceptually linked, their practical interdependence requires urgent attention. Biodiversity loss and ecosystem degradation, driven by human activities, threaten environmental stability, emphasizing the need for effective management strategies. Since, the biodiversity and ecosystem services are inter-dependent, integrating them provide pathways for sustainable environmental management. To integrate them effectively, different strategies are explored which provide pathways to make the integration process feasible and optimal. There are two distinct ways of linking biodiversity and ecosystem function models. Strategy A provides a realistic approach by including empirical evidence on biodiversity–ecosystem function interaction, which enables the integration of biodiversity and ecosystem function models globally, especially where data are available. On the other hand, strategy B was analyzed, which involves incorporating biological trait data into ecosystem function models for wider applicability to different systems and taxa. By integrating both approaches, this chapter provides more accurate insights into future biodiversity and ecosystem service trends, informing optimal environmental resource management.

References

Ayompe, L. M., Schaafsma, M., & Egoh, B. N. (2021). Towards sustainable palm oil production: The positive and negative impacts on ecosystem services and human wellbeing. *Journal of Cleaner Production, 278*, 123914.

Beillouin, D., Ben-Ari, T., Malézieux, E., Seufert, V., & Makowski, D. (2021). Positive but variable effects of crop diversification on biodiversity and ecosystem services. *Global Change Biology, 27*(19), 4697–4710.

Berthon, K., Thomas, F., & Bekessy, S. (2021). The role of 'nativeness' in urban greening to support animal biodiversity. *Landscape and Urban Planning, 205*, 103959.

Croci, E., Lucchitta, B., & Penati, T. (2021). Valuing ecosystem services at the urban level: a critical review. *Sustainability, 13*(3), 1129.

Fang, X., Li, J., & Ma, Q. (2023). Integrating green infrastructure, ecosystem services and nature-based solutions for urban sustainability: A comprehensive literature review. *Sustainable Cities and Society, 98*, 104843.

da Gama, T. J. (2023). The role of soils in sustainability, climate change, and ecosystem services: Challenges and opportunities. *Ecologies, 4*(3), 552–567.

Gómez-Baggethun, E., & de Groot, R. (2010). Natural capital and ecosystem services. The ecological foundation of human society. In R. E. Hester & R. M. Harrison (Eds.), *Ecosystem services, issues in environmental science and technology* (Vol. 30, pp. 105–121). Cambridge: Royal Society of Chemistry.

Gómez-Baggethun, E., & de Groot, R. (2007). Natural capital and ecosystem functions: Exploring the ecological grounds of the economy. *Ecosistemas: Revista Cietifica y Tecnica De Ecologia y Medio Ambiente, 16*(3), 4–14.

González-García, A., Palomo, I., González, J. A., García-Díez, V., García-Llorente, M., & Montes, C. (2022). Biodiversity and ecosystem services mapping: Can it reconcile urban and protected area planning? *Science of the Total Environment, 803*, 150048.

Hecht, J. S., Lacombe, G., Arias, M. E., Dang, T. D., & Piman, T. (2019). Hydropower dams of the Mekong River basin: A review of their hydrological impacts. *Journal of Hydrolology, 568*, 285–300.

Heino, J., Alahuhta, J., Bini, L. M., Cai, Y., Heiskanen, A.-S., Hellsten, S., Kortelainen, P., et al. (2021). Lakes in the era of global change: Moving beyond single-lake thinking in maintaining biodiversity and ecosystem services. *Biological Reviews, 96*(1), 89–106.

Jones, G. M., & Tingley, M. W. (2022). Pyrodiversity and biodiversity: A history, synthesis, and outlook. *Diversity and Distributions, 28*(3), 386–403.

Kattel, G. R. (2022). Climate warming in the Himalayas threatens biodiversity, ecosystem functioning and ecosystem services in the 21st century: Is there a better solution? *Biodiversity and Conservation, 31*(8), 2017–2044.

Kleemann, J., Koo, H., Hensen, I., Mendieta-Leiva, G., Kahnt, B., Kurze, C., Inclan, D. J., et al. (2022). Priorities of action and research for the protection of biodiversity and ecosystem services in continental Ecuador. *Biological Conservation, 265*, 109404.

Le Provost, G., Schenk, N. V., Penone, C., Thiele, J., Westphal, C., Allan, E., Ayasse, M., et al. (2023). The supply of multiple ecosystem services requires biodiversity across spatial scales. *Nature Ecology & Evolution, 7*(2), 236–249.

Li, S., Yu, D., Huang, T., & Hao, R. (2022). Identifying priority conservation areas based on comprehensive consideration of biodiversity and ecosystem services in the Three-River Headwaters Region, China. *Journal of Cleaner Production, 359*, 132082.

Liu, M., Wei, H., Dong, X., Wang, X.-C., Zhao, B., & Zhang, Y. (2022). Integrating land use, ecosystem service, and human well-being: A systematic review. *Sustainability, 14*(11), 6926.

Loos, J., Benra, F., Berbés-Blázquez, M., Bremer, L. L., Chan, K. M., Egoh, B., Felipe-Lucia, M., Geneletti, D., Keeler, B., Locatelli, B., & Loft, L. (2023). An environmental justice perspective on ecosystem services. *Ambio, 52*(3), 477–488.

McPhearson, T., Cook, E. M., Berbés-Blázquez, M., Cheng, C., Grimm, N. B., Andersson, E., Barbosa, O., Chandler, D. G., Chang, H., Chester, M. V., & Childers, D. L. (2022). A social-ecological-technological systems framework for urban ecosystem services. *One Earth, 5*(5), 505–518.

Peacock, R., Bently, M., Rees, P., & Blignaut, J. N. (2023). The benefits of ecological restoration exceed its cost in South Africa: An evidence-based approach. *Ecosystem Services, 61*, 101528.

Pereira, H. M., Martins, I. S., Rosa, I. M., Kim, H., Leadley, P., Popp, A., van Vuuren, D. P., Hurtt, G., Quoss, L., Arneth, A., & Baisero, D. (2024). Global trends and scenarios for terrestrial biodiversity and ecosystem services from 1900 to 2050. *Science, 384*(6694), 458–465.

Petroni, M. L., Siqueira-Gay, J., & Gallardo, A. L. C. F. (2022). Understanding land use change impacts on ecosystem services within urban protected areas. *Landscape and Urban Planning, 223*, 104404.

Pisani, D., Pazienza, P., Perrino, E. V., Caporale, D., & De Lucia, C. (2021). The economic valuation of ecosystem services of biodiversity components in protected areas: a review for a framework of analysis for the Gargano National Park. *Sustainability, 13*(21), 11726.

Qiu, J., Yu, D., & Huang, T. (2022). Influential paths of ecosystem services on human well-being in the context of the sustainable development goals. *Science of the Total Environment, 852*, 158443.

Shennan-Farpón, Y., Mills, M., Souza, A., & Homewood, K. (2022). The role of agroforestry in restoring Brazil's Atlantic Forest: opportunities and challenges for smallholder farmers. *People and Nature, 4*(2), 462–480.

Silva-Junior, C. H. L., Silva, F. B., Arisi, B. M., Mataveli, G., Pessôa, A. C. M., Carvalho, N. S., Reis, J. B. C., et al. (2023). Brazilian Amazon indigenous territories under deforestation pressure. *Scientific Reports, 13*(1), 5851.

Terêncio, D. P. S., da Graça Pinto Varandas, S., Fonseca, A. R., Cortes, R. M. V., Fernandes, L. F., Leal Pacheco, F. A., Monteiro, S. M., et al (2021) Integrating ecosystem services into sustainable landscape management: A collaborative approach. *Science of the Total Environment, 794*, 148538.

van Noordwijk, M. (2021). Agroforestry-based ecosystem services: Reconciling values of humans and nature in sustainable development. *Land, 10*(7), 699.

Weiskopf, S. R., Myers, B. J. E., Arce-Plata, M. I., Blanchard, J. L., Ferrier, S., Fulton, E. A., Harfoot, M., et al. (2022). A conceptual framework to integrate biodiversity, ecosystem function, and ecosystem service models. *BioScience, 72*(11), 1062–1073.

Whitehead, J. C., Ropicki, A., Loomis, J., Larkin, S., Haab, T., & Alvarez, S. (2023). Estimating the benefits to Florida households from avoiding another Gulf oil spill using the contingent valuation method: Internal validity tests with probability-based and opt-in samples. *Applied Economic Perspectives and Policy, 45*(2), 705–720.

Wubalem, A., Woldeamanuel, T., & Nigussie, Z. (2023). Economic valuation of Lake Tana: A recreational use value estimation through the travel cost method. *Sustainability, 15*(8), 6468.

Xu, Z., Peng, J., Qiu, S., Liu, Y., Dong, J., & Zhang, H. (2022). Responses of spatial relationships between ecosystem services and the sustainable development goals to urbanization. *Science of the Total Environment, 850*, 157868.

Yin, C., Zhao, W., Cherubini, F., & Pereira, P. (2021). Integrate ecosystem services into socio-economic development to enhance achievement of sustainable development goals in the post-pandemic era. *Geography and Sustainability, 2*(1), 68–73.

Zeng, Y., Koh, L. P., & Wilcove, D. S. (2022). Gains in biodiversity conservation and ecosystem services from the expansion of the planet's protected areas. *Science Advances, 8*(22), eabl9885

A New Era of Environmental Protection: The AI Revolution

P. Devisri and S. Manjula

Abstract

Artificial Intelligence (AI) and modern technologies are revolutionizing forestry and environmental protection. These tools help tackle deforestation, illegal logging, and forest fires by analyzing satellite and drone images, enabling better forest management strategies. AI also plays a crucial role in biodiversity conservation by identifying species, mapping ecosystems, and assessing human impact. In the fight against climate change, AI enhances carbon sink management and land rehabilitation through predictive models and reforestation planning. The integration of Internet of Things (IoT) and data analytics further improves monitoring by providing real-time insights on soil conditions, moisture, and tree growth, leading to adaptive environmental policies. AI-driven predictive modeling ensures sustainable logging while maximizing economic benefits with minimal environmental impact. Additionally, AI-powered data helps policymakers create balanced strategies that support both environmental conservation and human development. However, challenges like ethical concerns, privacy, and global accessibility must be addressed. Moving forward, the fusion of AI, IoT, and big data will transform forest management, promoting a future that prioritizes both ecological preservation and societal progress. AI-powered platforms like Global Forest Watch detect illegal deforestation activities in real-time, while IBM's Green Horizons predict air quality fluctuations with remarkable precision.

Keywords

Artificial intelligence (AI) · Machine learning · Internet of things (IoT) · Environmental impact · Big data · Reforestation planning

1 Introduction

AI emerges as one of the brilliant tools for sustainability in this new era. Its solutions include machine learning, big data analytics, and automation-changing the way in which one can track, analyze, and work to mitigate those environmental threats. From climate patterns to the optimization of energy consumption and attempts toward wildlife protection, the pathway to protect this planet has since been different in the presence of AI (Jige & Launa, 2024). In this chapter, the reader will find out how AI and new modern technologies change the environmental protection landscape. It ranges from having an understanding of AI-based solutions for areas of climate forecast prediction, pollutant control, and resource exploitation that won't be depleted or overused to make better resilience. Figure 1.1 illustrates numerous applications of AI in mitigation of climate change. It provides an overview of how AI affects energy efficiency using carbon capture and storage, along with renewable energy prediction. Besides, it signifies the contribution of AI in transporting systems optimization, precision agriculture, and natural resources management. Further, AI involvement in energy efficient building design as well as retrofits, prediction of weather conditions, and optimization of industry processes. In addition, it also deals with the growth of sustainable and resilient cities, focusing on their future implications.

Figure 1 reveals the AI revolution in environmental protection illustrates how AI enhances environmental sustainability by enabling pollution monitoring, climate change mitigation, wildlife conservation, and sustainable resource management through advanced data analysis and automation.

P. Devisri (✉) · S. Manjula
Department of Computer Science and Engineering, Ramco Institute of Technology, Rajapalayam, Tamil Nadu, India
e-mail: devisri@ritrjpm.ac.in

M. Ashwin et al. (eds.), *Smart Forestry 5.0 for Powered AI Sustainable Environment and Climate*, Sustainable Artificial Intelligence-Powered Applications, https://doi.org/10.1007/978-3-031-99086-1_8

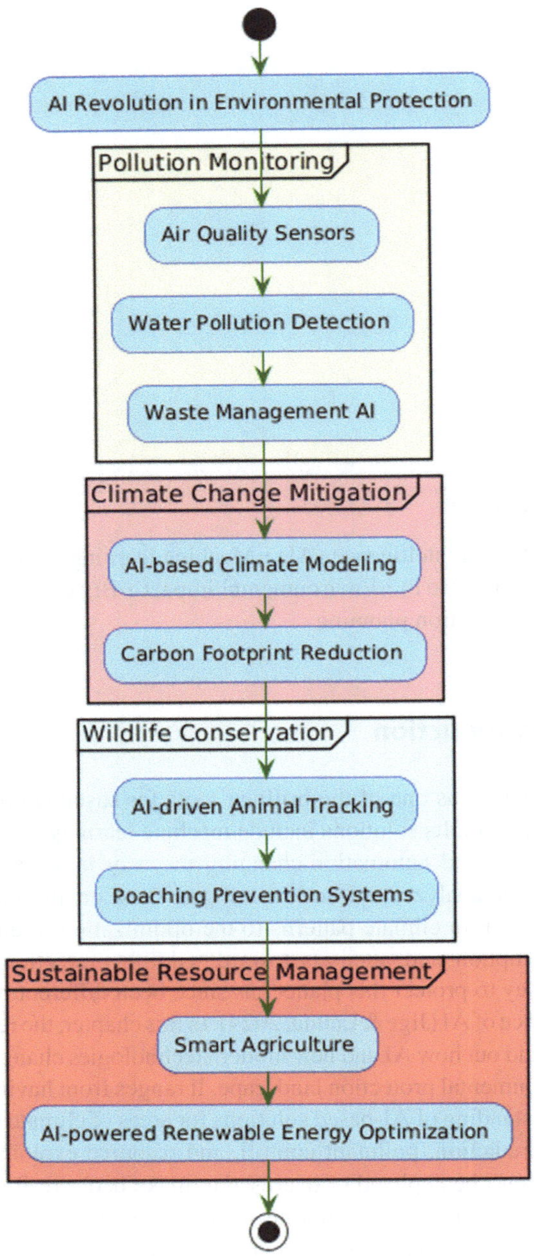

Fig. 1 AI revolution in environmental protection

1.1 Climate Change: A Warming World

Global warming is everywhere, and the signs are there. Summer temperatures are on the rise, storms are becoming more severe, and the ice caps are melting at an alarming rate. More than just the central reason-the persistent effects of inaction on climate change extend much farther than this: greenhouse gas emissions, largely the result of fossil fuel combustion, change the patterns of the weather, cause sea levels to rise, and prompt more intense and frequent natural disasters-all because inaction will ultimately bring food shortages and water scarcity, in addition to the mass displacement of communities (Sandal, 2024).

1.2 Deforestation: Losing Our Green Guardians

Forests are often called the "lungs of the Earth" because they take in carbon dioxide and release oxygen. However, each year, millions of trees cut down for agriculture, for urban expansion and logging. Such vast deforestation does not only accelerate climate change but also removes the habitats, thus pushing the different species on the brink of extinction. These losses of forests also affect local communities that rely on these for their food, medicines, and incomes. Environmental threats are advancing at a speed unprecedented in history. Climate changes, deforestation, and pollution all happen at rates that have dashed forward much quicker than many traditional systems can react to. For example, weather forecasting systems are often limited by the data they can process, resulting in delayed warnings for extreme weather events. Similarly, tracking deforestation in real time through satellite imagery or field surveys can take months or even years, making it difficult to react in time to prevent further damage. Traditional methods rely heavily on manual surveys and observational reports, resulting in delayed responses to deforestation and climate threats (Worku & Ayalew, 2024). It is impossible to make the correct decisions regarding resources allocation and the most urgent interventions needed without fresh data. Most of the more traditional approaches emphasize responses to damage when it is already done. We plant trees after a forest is cut down, or clean up oil spills after they have polluted an area. However, long-term sustain ability will require efforts that prevent such disasters before they occur. Generally, traditional systems do not offer a capacity to predict and, thus, preclude risks in advance; which is why we end up responding to crises and not preventing them.

2 AI and Modern Technology: Game Changers for Sustainability

The world is constantly facing more severe environmental challenges. Climate change, pollution, and biodiversity loss are some of the complex, urgent issues that have made the old ways of addressing them insufficient. But what if we could change the game using cutting-edge technology, such as AI? The truth is that AI and modern technology can revolutionize how we approach sustainability, and that' s exactly what we're seeing today.

2.1 Real-Time Monitoring and Data Insights

Imagine having a view of Earth's health in real time no waiting months for field reports or relying on historical data. AI makes it possible to monitor everything ranging from forest deforestation to air quality right as it happens. Satellites send data from the ground into their respective AI systems, which process them immediately and update the screen in real time. This is the future of saving forests, tracing pollution, and even detecting growing environmental threats even before they start getting out of hand. As an example, AI can, perhaps, even trace illegal deforestation or pollution in remote locations, thus providing the authorities with a chance to act in good time, an event that could not have occurred through traditional approaches. Example: Global Forest Watch for deforestation, Microsoft's AI for Earth for pollution tracking, Google DeepMind energy optimization.

2.2 Predicting and Preparing for Environmental Events

Imagine knowing when a hurricane or flood is going to strike your neighborhood before it happens. AI can do this. Vast reservoirs of historical data make it possible for AI models to predict climate patterns and to foresee extreme weather better than humanly possible. This way, people are prepared beforehand for whatever might cause damage or claim lives in the communities. Not only does AI forecast short-term weather, but also helps us glance at long-term climate trends with insights into all of the matters from changing conditions in agriculture and where we can build infrastructure and avoid disaster altogether. For example, in agriculture, AI decides when to plant crops, how much water is needed, and at what time to harvest all of which are determined with real-time input. As a result, less water is wasted, fertilizers are consumed more judiciously, and plants grow more sustainably. Similarly, AI is enabling decreased wastage in energy by predicting peak times and adjusting the usage of energy, making us solely rely on renewable sources such as wind and solar power.

3 The Role of AI in Environmental Sustainability

As the world grapples with increasingly complex environmental challenges, the need for more accurate predictions and data-driven solutions has never been more urgent. One of the most significant ways in which AI is contributing to environmental sustainability is through machine learning models designed to predict climate patterns, weather events, and long-term changes in our ecosystems. These advanced models are changing the way we understand and respond to climate change, offering new avenues for prevention, adaptation, and mitigation.

3.1 Machine Learning Models for Climate Predictions

Machine learning (ML) has become a game-changes in climate science, providing researchers with tools to predict climate trends, extreme weather events, and the long-term impacts of human activity on the planet (Hamdan et al., 2024).

Machine learning, a subset of AI, uses algorithms to analyze large datasets, identify patterns, and make predictions without being explicitly programmed to do so. In the context of climate predictions, machine learning models are fed with historical climate data, such as temperature records, ocean currents, and atmospheric conditions. The model then "learns" the relationships between these variables and uses this knowledge to make predictions about future climate scenarios. For example, ML models can be trained on decades of weather data to understand how certain patterns (like the El Niño phenomenon) influence weather events such as hurricanes, droughts, or heatwaves. The more data the model is exposed to, the more accurate its predictions become, allowing scientists to project climate changes over the next few decades with greater precision.

3.2 Improving Accuracy and Timeliness of Climate Predictions

One of the primary advantages of machine learning in climate science is its ability to make predictions faster and with higher accuracy. Traditional climate models rely on complex mathematical equations to simulate the Earth's climate system, which can be time-consuming and computationally expensive. Machine learning models, on the other hand, can analyze and process vast amounts of data in real time, making them particularly useful for predicting short-term weather events, such as hurricanes or heatwaves, which require immediate attention (Fente & Dheeraj., 2018). For example, meteorologists use machine learning models to predict the path and intensity of hurricanes by analyzing satellite images, sea surface temperatures, and wind patterns. These predictions, when combined with traditional forecasting methods, provide more accurate and timely warnings, potentially saving lives and reducing property damage. Moreover, ML models can process far more data than traditional methods—incorporating variables like air pressure, ocean temperatures, and even social factors like population density—to make more precise forecasts.

3.3 Predicting Long-Term Climate Trends

While short-term predictions are critical, understanding long-term climate trends is equally important. ML models are increasingly used to analyze the long-term effects of climate change on ecosystems, agriculture, and human populations. For instance, machine learning can help predict how rising temperatures will impact crop yields, the migration patterns of species, or the spread of diseases like malaria. These long-term predictions are essential for developing strategies to mitigate the impacts of climate change, such as planning for sea level rise, devising sustainable agricultural practices, and protecting vulnerable communities. ML models can simulate various scenarios based on different levels of greenhouse gas emissions, giving policymakers valuable insights into the potential outcomes of different climate action plans.

3.4 Enhancing Climate Models with AI

Machine learning is not just improving traditional climate models; it's also helping to create entirely new approaches to understanding climate systems. In the past, scientists relied heavily on Global Circulation Models (GCMs), which simulate the atmosphere, oceans, and land surfaces. While these models are highly complex and accurate, they require enormous amounts of computing power and have limitations in terms of resolution and speed. AI-based models can complement these GCMs by offering more granular insights. For instance, AI can analyze localized climate data to provide highly detailed forecasts for specific regions such as predicting how urban heat islands will affect local weather or how certain agricultural practices could alter the local microclimate. Moreover, AI can integrate data from diverse sources, such as satellite imagery, sensor networks, and climate reports, to offer a more holistic understanding of climate systems (Miller et al., 2025). This ability to combine and analyze diverse datasets allows for more comprehensive models that can be adapted to specific environmental conditions or forecasting needs.

For example, **Climate AI** uses machine learning to forecast agricultural risks by analyzing historical weather, soil data, and satellite imagery. It helps farmers plan crop cycles, manage water use, and prepare for climate impacts, boosting resilience against environmental changes.

Similarly, **IBM's Green Horizons** initiative applies AI to urban climate modeling. It integrates real-time environmental data to predict air pollution, energy demand, and microclimate shifts. Cities like Beijing have used Green Horizons to optimize policies and significantly reduce pollution, showcasing AI's potential for urban environmental management.

4 AI-Driven Carbon Footprint Analysis: Understanding and Reducing Environmental Impact

The world is constantly facing more severe environmental challenges. Climate change, pollution, and biodiversity loss are some of the complex, urgent issues that have made the old ways of addressing them insufficient. But what if we could change the game using cutting-edge technology, such as AI? The truth is that AI and modern technology can revolutionize how we approach sustainability, and that's exactly what we're seeing today.

4.1 How AI Processes Carbon Footprints

In simple terms, AI-processing carbon footprint analysis uses machine learning-based algorithms that run on massive datasets, searching for patterns indicating what an entity is creating in terms of environmental impact. The algorithm works across broad sectors, from energy consumption, transportation logistics, and waste production, to determine resultant carbon emissions. Take an example with an AI system: tracking energy consumption in real time across a company's buildings, machinery, and vehicles. Continuously gathering data from sensors or IoT devices, AI can create a very fine-grained map of energy usage and the corresponding carbon emissions. This exceeds simple reporting and rather provides a dynamic, constantly changing perspective on an organization's carbon footprint—an area that can be scrutinized to locate inefficiencies and improvements.

4.2 Real-Time Carbon Monitoring

One of the most impactful factors of AI-based carbon footprint analysis is its real-time monitoring feature. Unlike reporting at periodic intervals, AI can collect data from various sensors and devices to offer a live feed of carbon emissions. With such real-time data, businesses are able to continuously monitor their carbon emissions and, based on such monitoring, take appropriate adjustments to their activities. For example, smart building technologies use AI to monitor energy consumption and optimize heating, ventilation, and air conditioning (HVAC) systems in real time. By adjusting temperature and lighting based on occupancy patterns, AI can reduce energy waste and minimize a building's carbon footprint. Similarly, AI-based transportation management systems can monitor vehicle fuel consumption, emissions, and routes in real time (Khan et al., 2025). Optimizing delivery routes and predicting fuel consumption reduces the emission related to transportation. The real-time adjustment of energy use, travel, and other factors makes AI-driven carbon monitoring a very effective tool for sustainability. AI models analyze sensor data such as turbine blade vibration, acoustic patterns, and thermal imaging to predict mechanical failures.

4.3 Supply Chain Optimization

A significant portion of a company's carbon footprint lies within its supply chain, from sourcing raw materials to shipping finished products. AI can help organizations analyze their entire supply chain to identify carbon-intensive processes and suggest more sustainable alternatives. AI-powered platforms can report emissions associated with each stage of the supply chain from raw material extraction to transportation to storage (Mazurek, 2025). Historical data can be provided, and runs given to know possible supply chain optimizations through AI, such as recommending more energy-efficient shipping methods, optimum production schedules, or even suggesting alternative materials that are less carbon-intensive. Another benefit of AI tools is that they can track suppliers who meet specific environmental standards so companies can ensure the entire supply chain is responsible for sustainability.

4.4 Customized Carbon Footprint

Reduction Businesses and governments can use AI-driven carbon footprint analysis, but individuals can also use the tools to minimize their personal environmental impact. AI-powered apps and platforms can analyze an individual's lifestyle, including transportation habits, energy consumption, diet, and waste production, to calculate his or her carbon footprint.

4.5 Predicting and Preparing for Environmental

Events Imagine knowing when a hurricane or flood is going to strike your neighborhood before it happens. AI can do this. Vast reservoirs of historical data make it possible for AI models to predict climate patterns and to foresee extreme weather better than humanly possible (Jeremiah et al., 2024). This way, people are prepared beforehand for whatever might cause damage or claim lives in the communities. Not only does AI forecast short-term weather, but also helps us glance at long-term climate trends with insights into all of the matters from changing conditions in agriculture and where we can build infrastructure and avoid disaster altogether.

Figure 2 illustrates an automated workflow where AI collects, processes, and analyzes carbon emissions data. ML algorithms can simulate different scenarios and predict changes in production processes, transportation, or energy sources on overall emissions (Li et al., 2024). For example, AI can predict the amount of carbon saved if there were a switch to renewable energy or electric vehicles. This makes it possible for organizations to better take decisions when formulating strategies based on data. AI-based predictive models would benefit the government by judging its climate

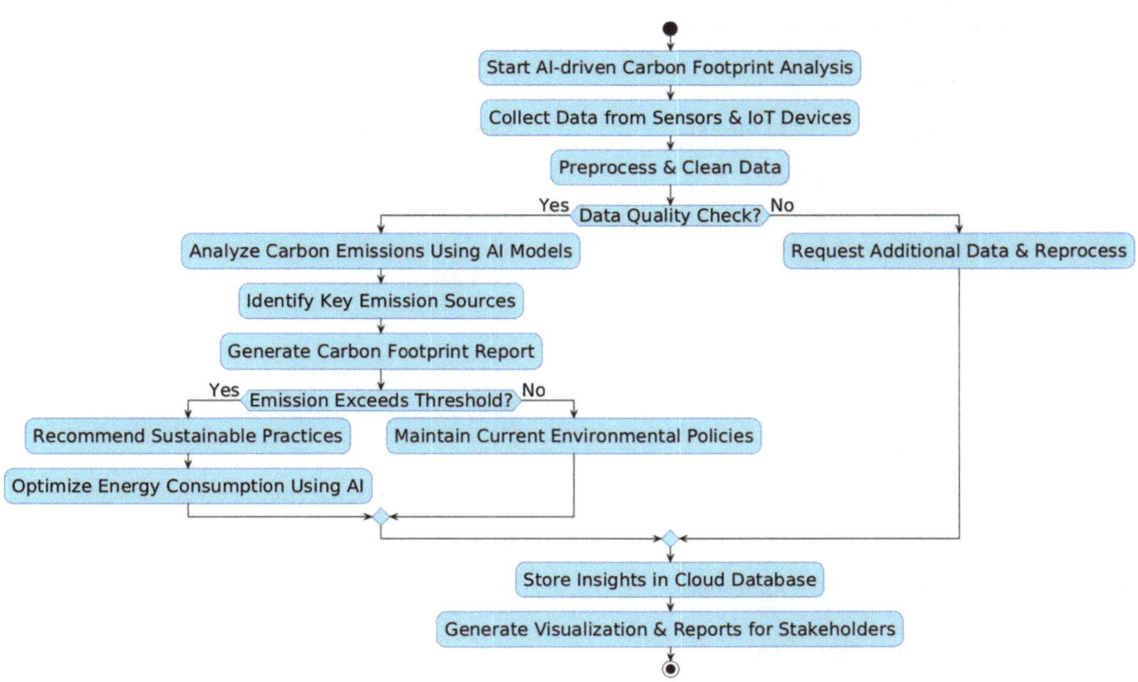

Fig. 2 AI-driven carbon footprint analysis

policies as effective and changing them subsequently. Policymakers can run scenarios on the long-term implications of a whole spectrum of regulations from carbon pricing or emissions trading schemes to enhance the environmental impacts through optimization of strategy.

5 Optimization of Renewable Energy: AI in Smart Grids and Renewable Energy Forecasting

The transition to renewable energy is the largest step taken toward minimizing climatic change and ensuring a future world. In parallel, renewable sources such as solar and wind generation pose challenges in the areas of generation variability and ineffective storage and distribution. Among the most significant contributions of AI to renewable energy optimization, it is one's opinion that smart grids and advanced forecasting models for solar and wind energy hold top priority. Improving the efficiency and reliability in renewable energy generation and distribution causes the world to experience a faster trend toward cleaner energy.

Figure 3 illustrates AI technologies optimize smart grid energy distribution and predict renewable energy outputs, enhancing energy reliability and minimizing carbon emissions AI analyzes the weather and energy consumption data to optimize smart grid energy distribution. Modern renewable energy infrastructure is established by smart grids. Advanced Electricity Grid employs high-tech electronic sensing and managing electrical distribution wherever possible across smart grids to really implement renewable power inputs into that electric grid network. Smart Grid implements AI through its integration at stages for continued improvement in optimum output for every instant in the management of generation and use of power and in supply. Other than energy flow optimization, AI can be deployed toward predictive maintenance and fault detection in smart grids. This is the possibility of predictions regarding the time when equipment failure may occur or when it would need to be sent for maintenance by using historical data in combination with sensor inputs real time; this way, the operators can take pre-emptive action before the breakdown occurs. For example, AI can study the performance of wind turbines, detect patterns that indicate wear and tear or potential faults and therefore allows grid operators to correct such problems in time to ensure that maximum efficiency is maintained in wind farms and that the effect of maintenance on energy generation is reduced.

6 AI for Soil Health and Crop Management

The future of agriculture lies in the adoption of innovative technologies that optimize resource use, improve productivity, and promote sustainability. AI is one such transformative technology that is revolutionizing farming practices (Mishra & Mishra, 2023). Through data-driven insights, AI enables precision farming techniques that enhance soil health, optimize crop management, and ensure more sustainable agricultural practices. The discussion here revolves around AI-driven solutions of irrigation, soil analysis, and pest control; it is an approach that could revolutionize farming and help a farmer increase their yields, impact the environment lightly, and deliver food security to the future generation.

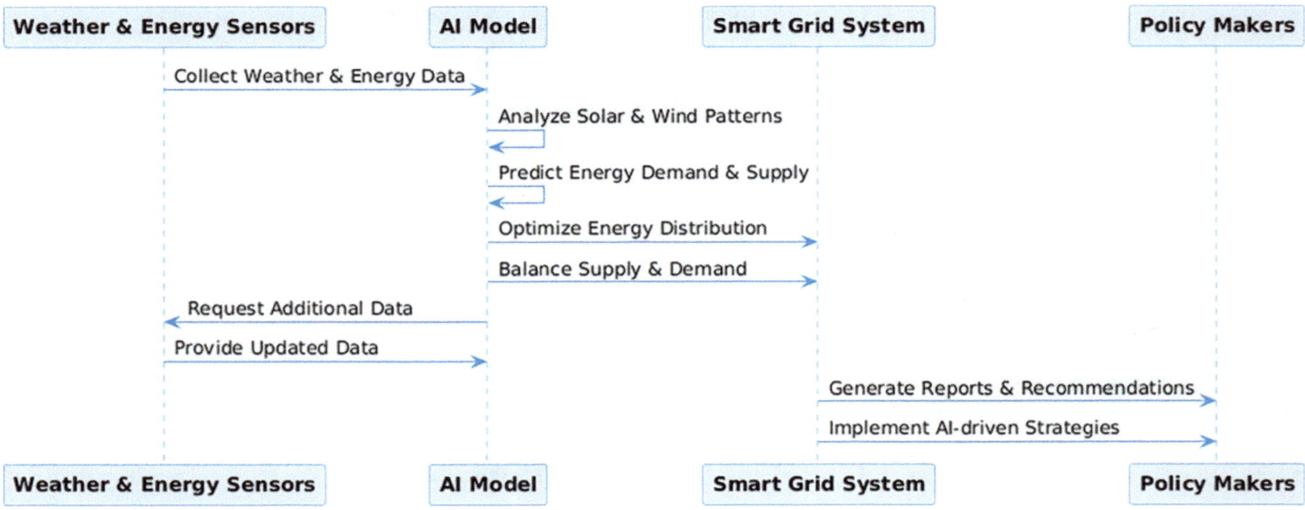

Fig. 3 AI in smart grids and renewable energy forecasting

6.1 Precision Farming: Irrigation, Soil Analysis, Pest Control

Precision farming is the application of advanced technology in monitoring and controlling field variability in crops, soil, and environmental conditions to optimize agricultural productivity and reduce waste. At the core of precision agriculture lies AI, which uses processed vast amounts of information collected from sensors, drones, satellite images, or any other source to take direct real-time decisions for optimizing agriculture farming practices.

6.1.1 AI-Driven Irrigation System

Certainly, water is one of the most important factors in agriculture, and proper management of the element is needed for the productivity of crop yield and concomitantly environmental conservation. The traditional irrigation system mostly results in waste of water due to over-irrigation or absence of monitoring of soil moisture. AI proves to be important to improve irrigation practices as they make better usage of water. AI-powered irrigation systems integrate soil moisture sensors, weather forecasts, and satellite images to track real-time water demands of crops (Dhanke, 2024). Machine learning algorithms process data from these sensors and determine an ideal irrigation schedule for crops to provide the required water at the correct time. In this context, AI can also minimize the usage of water while preventing waterlogging and improve the health of the crops based on the requirement of each field or even plants in the fields. For example, irrigation management can allow AI to automatically detect the timing of rain events so that water can be withheld and not sprayed at the wrong times. Through AI models, soil moisture is monitored, and a farmer receives an alert of when to water, thus conserving water along with energy. Soil health is the foundation of successful farming because it directly relates to crop yield and sustainability. Healthy soils contain organic matter, diverse microorganisms, and nutrients to support plants (Reeve et al., 2016). AI has been used in the management of soil health to analyze and understand in-depth what constitutes the composition, nutrient levels, and overall quality of soil.

6.1.2 AI-Soil Analyzing Instruments Apply Sensors, Data Analytics to Measure Significant Characteristics of Soils

PH, texture, organic content, and nutrients. This software may find what is deficient in the soil and indicate which types of fertilization should be adopted and the patterns of change over time that should be anticipated from the health condition of soil. AI, therefore, through constant monitoring of conditions of the soil, helps the farmers in their decisions on the management of soils, thereby eliminating much chemical fertilizing and improving fertility. For example, AI would help farmers in knowing the quantity and type of fertilizers to be applied on a particular piece of land given real-time information about the land. This means that there will be minimal cases of over-fertilization which would end up in water bodies causing pollution and the crops getting the nutrients for growth.

6.1.3 AI in Pest and Disease Control

Crop management is challenging in the case of pest and disease management. The farmers need to monitor their fields for early infestation signs (Lucas, 2011). The traditional approaches of pest management involved the wide-scale application of pesticides, which are harmful to beneficial insects, cause environmental pollution, and induce pesticide resistance. AI is now being used to minimize the usage of chemical pesticides through more precise and targeted approaches in pest management. AI systems powered by computer vision and machine learning algorithms can analyze images from drones, cameras, and sensors to detect pests and diseases early on. They can be programmed to identify specific types of pests or diseases affecting crops, predict their spread, and recommend targeted treatments. Such precision techniques will enable the farmer to spray chemicals precisely at the right place and time, reducing overall chemical usage while minimizing damage to the environment.

6.1.4 Advantages of AI-Based Precision Farming

AI precision farming has the advantages of serving environmental sustainability along with economic feasibility. AI optimization in resource utilization, enhancement of efficiency, and minimization of wastage is guiding farmers toward the most sustainable way of agriculture and helping them bring about better revenue from their operations.

6.1.5 Increasing Crop Yield and Enhancing Productivity

AI can help farmers increase crop yields by providing real-time insight into soil conditions, crop health, and environmental factors. Precision farming techniques allow farmers to optimize planting schedules, fertilization, irrigation, and pest management, thereby ensuring that crops receive conditions ideal for growth (Monteiro et al., 2021). Hence, even in challenging environments or with limited resources, healthier crops and higher yields result. For instance, AI can assist farmers in determining the ideal planting depth and row spacing for various crops to maximize their growth potential. AI models can also analyze historical crop performance data to suggest the best crop varieties for specific soil and climatic conditions, further enhancing productivity.

6.1.6 Minimized Environmental Impact

The major benefit AI provides in agricultural practice is to minimize the environment's negative influence from farming. Precision farming propelled by AI will reduce overfarming, overdosing of water, fertilizers, and pesticides—all causes for degradation when utilized excessively. AI protects the environment without compromising agricultural productivity. For instance, it saves water from wastage, which would otherwise lead to depletion in areas where water scarcity is a problem. Similarly, AI-based pest control reduces the usage of broad-spectrum pesticides, which can kill non-target species and cause environmental pollution.

6.1.7 Cost Savings and Increased Profitability

Although the initial investment in AI-driven technologies is pricey, long-term cost savings and increased profitability for farmers make these technologies worthwhile. AI optimizes inputs such as water, fertilizers, and pesticides, saving farmers from extra costs (Javaid et al., 2023). Moreover, increased crop yields and lesser environmental damage will also increase the profit, because more can be produced with fewer resources. For example, AI-based soil analysis contributes to reducing the excessive fertilizers' application and saves money and reduces runoff's environmental impact. Similarly, AI-driven pest management reduces the cost of pesticides and labor associated with manual pest control methods.

6.1.8 Integration with IoT Devices

An integration of AI with IoT devices is most likely going to work in favor of precision farming in the future. IoT sensors can provide real-time data regarding soil moisture, temperatures, crop growth, and other environmental conditions, feeding into analysis by AI algorithms. These are already making possible a fully connected, data-driven farming ecosystem where farmers can make informed choices at every stage of the farming process. For example, smart drones will be used to monitor pests, diseases, and nutrient deficiencies in the fields, and AI algorithms would process the information gathered and give specific recommendations. California vineyards use AI-IoT networks to optimize irrigation based on micro-climate data.

6.2 Autonomous Farming and Robotics

Autonomous farming equipment and robotics will also be a part of the future of AI-driven agriculture (Onteddu et al., 2025). This may include self-driving tractors, harvesters, and drones into which AI algorithms have been embedded, and it may perform such functions like planting, fertilization, and harvesting without man's direct intervention. These autonomously functioning systems will further add more precision agriculture by ensuring such functions are performed with maximum efficiency and minimum wastage. For example, autonomous AI-driven robot weeders and fertilizer applicators can perform precise operations by identifying and removing weeds in a timely manner, thereby supporting better crop growth and improving yields at harvest.

Figure 4 illustrates how AI enhances precision farming by optimizing resources, improving crop yield, reducing costs, and minimizing environmental impact.

7 Role of AI in Reducing Food Waste and Optimizing Supply Chains

Food waste today is the most galling issue facing humankind as a whole, ranging across ecological connotations to instances of resource waste. According to the FAO, it estimates that one-third of all the food produced today is lost as waste, further inefficiencies such as the handling, trans-

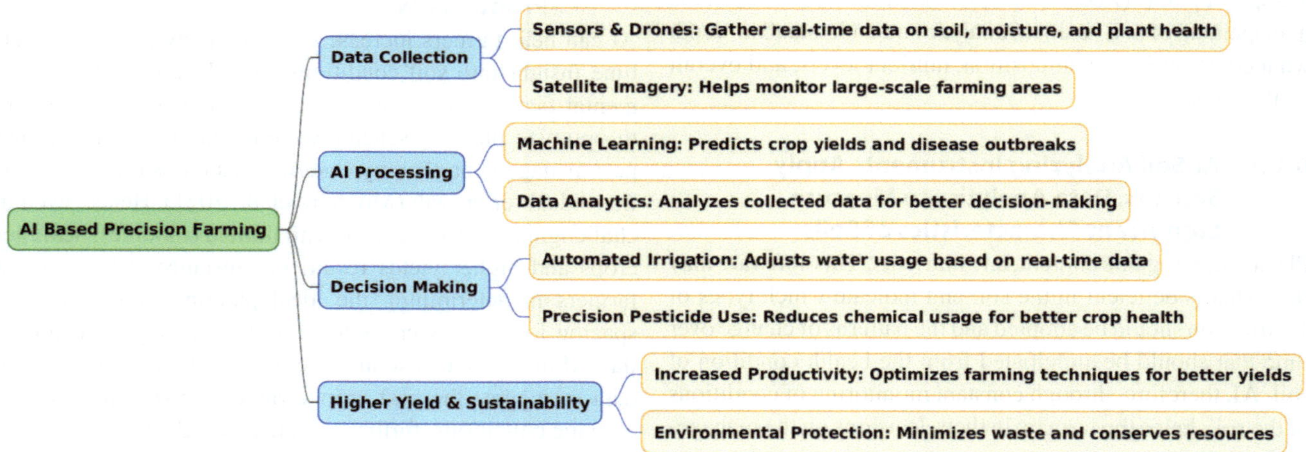

Fig. 4 Cost savings and increased profitability

portation to retail, and into consumption add up to this. AI rapidly turns into a food waste fighting mechanism since this technology has alternative novel methods bottom down, which eradicate much of it during lines of distribution. This results from its integration with data science insights and predicative analytics, accompanied by self-acting procedures that encourage efficiency in the generation, distribution, and consumption of food.

7.1 AI at Consumer Level Loss

Most of the wastages occur at the consumer level-a lot through houses and restaurants. AI technologies have gained a trend in food waste management as they offer ideas that enable better food management, meal planning, and inventory control (Martin-Rios et al., 2020).

7.1.1 Smart Fridges and Food Inventory Management

Smart fridges with AI sensors and cameras allow consumers to keep track of food they have at home, such as expiration dates, storage conditions, and amounts. These appliances can scan food usage patterns and advise on the best ways to use the ingredients before they get spoiled, hence reducing the probability of food wastage. The AI algorithms will also remind the consumer when their food items are approaching their expiry dates, hence helping people consume them before they go bad. Other potential applications of AI would be a presentation of recipes based on what is available inside the refrigerator so that the customer uses it with more creativity to consume. In such a kind of solution, food waste reduces and a more efficient consumption behavior develops, for which the individual benefits in sustaining a culture that is sustainable enough.

7.1.2 AI in Restaurant and Foodservice Operation

Leftover food disposal presents major challenges in restaurants and food service outlets with respect to over-preparation, spoilage, and unsold stock (Rehman, 2020). The potential of AI will help streamline preparation, portions, and stockholding to eliminate most waste. Historical data and current trends coupled with customer preference by machine learning algorithms will assist the right demand assessment for adequate food preparation. Another area where AI-based systems can be useful is dynamic menu planning, considering the real-time data of inventory so that restaurants don't waste their ingredients. Besides, AI-driven demand forecasting models will be able to adjust the quantity of food prepared, which will reduce the chances of overproduction and subsequent waste.

7.2 AI in Food Production: Increasing Efficiency and Reducing Waste

This not only changes the food production from farm to market but saves more useful resources that are used most efficiently so that waste can be minimized while making it sustainable in general.

7.2.1 Precision Agriculture: AI-Fueled Farming Practices

AI supports precision agriculture. This is because AI gathers information from sensor data, drone images, and satellite data that help in optimizing farming practices and waste. Real-time AI monitoring of soil health usage of water and crops growth will help farmers apply water, fertilizers, or pesticides in targeted manners which again reduces wastage and increases yields (Talaviya et al., 2020). AI also provides early identification of plant diseases and infestation of pest; hence farmers can take some preventive measure before it would cause widespread destruction. Thus, this reduces the consumption of pesticides as well as crops which are entirely destroyed, lessen wastes at the same time ensure the sustainability of farming business.

7.2.2 AI in Food Processing and Manufacturing

Through AI implementation, food processing industries cut efficiency and waste by means of automation and optimization. Algorithms help optimize a production line by detecting bottlenecks in order to maximize machine performance and predict maintenance requirements. Food with certain flaws may be detected using AI applied through vision systems before it hits the consumers' hands. Besides that, space in AI is for optimizing food ingredients where it will scan several recipes and offer some alternatives depending on what is found; this leads to reducing waste because it would consume more raw materials available before they waste.

Case Study 1: IBM's Watson Decision Platform for Agriculture

Background: IBM Watson Decision Platform for Agriculture is an advanced AI-based solution, which equips the farmer with an integrated package of tools to improve sustainable farm operations. It thereby enables the combining strength of AI, IoT, weather data, and predictive analytics to provide real-time personalized insights and recommendations to farmers. This case study addresses using such a platform to enhance sustainability in crop production farming practices. AI Implementation: The Watson platform aggregates data from sources such as IoT devices, weather sensors, soil analysis tools, and satellite images. Then, the AI models analyze the data and give actionable insights that help farmers to optimize their activities. For example, it advises in real time which irrigation to use, pest control, crop rotation, and fertilization depending on the general health of the soil and the weather.

8 Predictive Analytics in Waste Reduction and Resource Recovery

Predictive analytics, driven by AI, fundamentally changes the way waste management is done and, in general, offers a better reduction strategy and more efficient resource recovery. Predictive analytics predicts waste generation patterns based on historical data, real-time inputs, and machine learning models; identifies opportunities for reducing waste; and optimizes the process of resource recovery. It is both environmentally and economically beneficial because it reduces the operational cost and enhances the sustainability of the waste management system.

8.1 Forecasting of Generation and Diversion of Waste

The ability of patterns of waste generation to make provision for efficient collection and recycling would be critical, as it requires the application of predictive analytics to forecast the volume and types of wastes that could be generated at specified locations and time intervals. With these, waste management companies can well predict future volume and nature of waste through appropriate machine learning algorithm that considers prior records of the generation of wastes, increase of population, and seasonal changes coupled with the conduct of consumers.

8.1.1 Estimation of the Waste Streams

The predictive models would predict the amount of organic, recyclable, hazardous, and normal waste streams (Johnson et al., 2017). For instance, although food wastes are high during celebrations or festivals at local levels, the recycling rates will be affected by a change in consumer pattern. AI systems can look into all these trends to enable predictions of waste diversion that will allow the companies to adjust collection schedules and allocate available resources properly with a streamlined procedure for sorting.

8.1.2 Waste Collection and Sorting Improvement

Predictive analytics would help companies in waste management predict volumes and types of wastes that would be generated so they can optimize routes, schedules, and strategies to recycle. Predictive analytics prevent the bins being emptied while filled to the brim, or non-recyclable materials going into recyclable bins, meaning contamination rates would be minimized along with recycling efficiencies. More of the recyclable waste will leave the landfill site, and all the valuable material will go toward recovery and subsequent reuse.

8.2 Blockchain-Based Waste Tracking for Accountability

- What will come to the forefront as technology is blockchain-based, which makes it a rather powerful tool that is very securely and transparently designed. Application of blockchain finds its most promise in tracking wastes throughout their lifetime, thereby resulting in greater accountability and transparency at all levels involved (Zhang et al., 2020). Blockchain also addresses several serious challenges in handling waste by virtue of a distributed, immutable log of every part of the processing of waste-management from collection or sorting to recycling or disposal. This technology has excellent potential to upgrade the waste monitoring process, avoid fraud, and enhance the sustainability activities of any organization. For Example:
  - *Plastic Bank* using blockchain for waste collection.
  - *Everledger* for material recycling tracking.

8.2.1 Waste Management Accountability

Proper waste management is very important for environmental protection and public health. However, lack of transparency and accountability often hampers effective waste handling, which leads to illegal dumping, improper disposal of hazardous waste, and low recycling rates. It is also feared that waste materials intended for recycling or reuse end up in landfills, especially when waste is not tracked properly. This problem can be best answered using blockchain by giving an open record and immutable processes concerning waste management. All movement and transactions or even generation and transportation of the waste, together with final treatment, can be included in the blockchain, giving responsibility to both sides.

8.2.2 How Does Blockchain Work to Track Waste

Blockchain is a decentralized ledger that records transactions across multiple computers in such a way that the registered transactions cannot be altered retroactively. For waste management, blockchain can track the waste on its journey from its source to its destination, leading to the generation of a live, transparent, and immutable history for every item of waste.

Case Study: Smart Cities Implementing AI for Zero Waste Initiatives

In a bid to promote sustainability, more and more smart cities are leveraging the power of cutting-edge technologies such as AI. They look for zero waste through less generation, maximum recycling, and forming circular economies with smart usage through data, machine learning, and automated systems. As this AI in waste management comes into place, the smart city sets a new benchmark for sustainability, which in turn raises the waste diversion rates, reduces landfill dependency, and the overall resource efficiency.

8.2.3 AI-Based Waste Collection and Sorting

Songdo employs an automated waste collection system called the "pneumatic waste system." This system would allow residents to deposit waste in chute openings that lead underground pipes. Air pressure then sends the waste through the pipes to central collection stations, which significantly minimizes the reliance on traditional waste trucks and the pollution created by the operation of the said trucks. The system is powered with AI-driven sensors that assess the composition and analyze the waste material in real time. The AI algorithms based on the composition decide which material can be recycled and which not even before reaching a waste sorting plant. With this advanced method of sorting, the recycling rate is higher with fewer contaminants within the recyclables and proper diversion of wastes.

8.2.4 Data-Driven Waste Management

AI also aids Songdo to optimize waste management activities by collecting information on the real-time behavior of waste generation. Monitoring the volumes, types, and frequency of collection at different times in different areas will help Songdo's AI system to generate more valuable insights on waste behavior at different times in different areas. This will afford city officials with the opportunity to retime their schedules for waste collections, resource re-allocation, and efficient operating of waste management systems.

9 The Future of Clean Air: AI's Role in Reducing Global Pollution

This is one of the most advanced innovations in environmental monitoring, with AI combined with IoT sensors. Convergence of the technologies enhances the tracking and response capacity to air quality issues in an unprecedented level of precision and timeliness that has never been seen before (Durlik et al., 2023). Today AI-enabled networks continue to provide continuous and real-time results across an entire geography, whereas the classical systems may provide only a fragmented view of the air quality data. Edge computing capabilities have been quite transformative in deployment. Such a system processes data directly at the sensor level; hence, they could identify and even respond to pollution events in seconds rather than hours. Figure 9.1 illustrates how AI processes environmental data from various sources to generate real-time alerts, forecasts, and policy recommendations. It feeds in data from various sources, including IoT Sensors, Satellite Data, Weather Data, and Traffic Data. Each of the above data sources is critical in supplementing air quality monitoring. The IoT sensors provide actual time information on air quality, while satellite data offers atmospheric insight at a global level. The weather data shall correlate with the environmental conditions and the traffic data will help to understand the magnitude of pollution occurring through vehicular emissions. All these sources fall within the first "package" titled "Data Sources." These data feed into the AI Processing Layer. Then, this layer processes, aggregates, and analyzes the data that it receives. Important components in this lay feeds data from several sources, including IoT Sensors, Satellite Data, Weather Data, and Traffic Data.

All the above data sources are crucial in supplementing air quality monitoring is shown in Fig. 5. The data produced by the IoT sensors is based on actual-time information regarding air quality, and satellite data provides atmospheric insight at a global level (Fadhil et al., 2023). Correlation with environmental conditions is available from weather data; the traffic data will ascertain the extent of pollution that results from vehicular emissions. These sources are available inside the first "package" entitled "Data Sources." Data Sources feed their information into the AI Processing Layer. This layer processes, aggregates, and analyzes the data received. Important parts of this layer include the following:

- Edge Computing: The raw data processing begins at the sensor level for instant responses.
- Data Integration Hub: This interface ensures all ingested data gets integrated into one format and later on used for analysis.
- AI Analysis Engine: This uses complicated algorithms toward integrating the results for analysis to draw useful patterns.
- Predictive Models: AI models predict events concerning pollution and their trends, levels ahead of time.

Recent developments in neural network architectures further improve the ability to make such predictions (Tedjopurnomo et al., 2020). Modern systems are enabled to account for seasonal changes, special events, and even long-term climate trends. In summary, the diagram visually outlines how AI and IoT technologies combine to track, analyze,

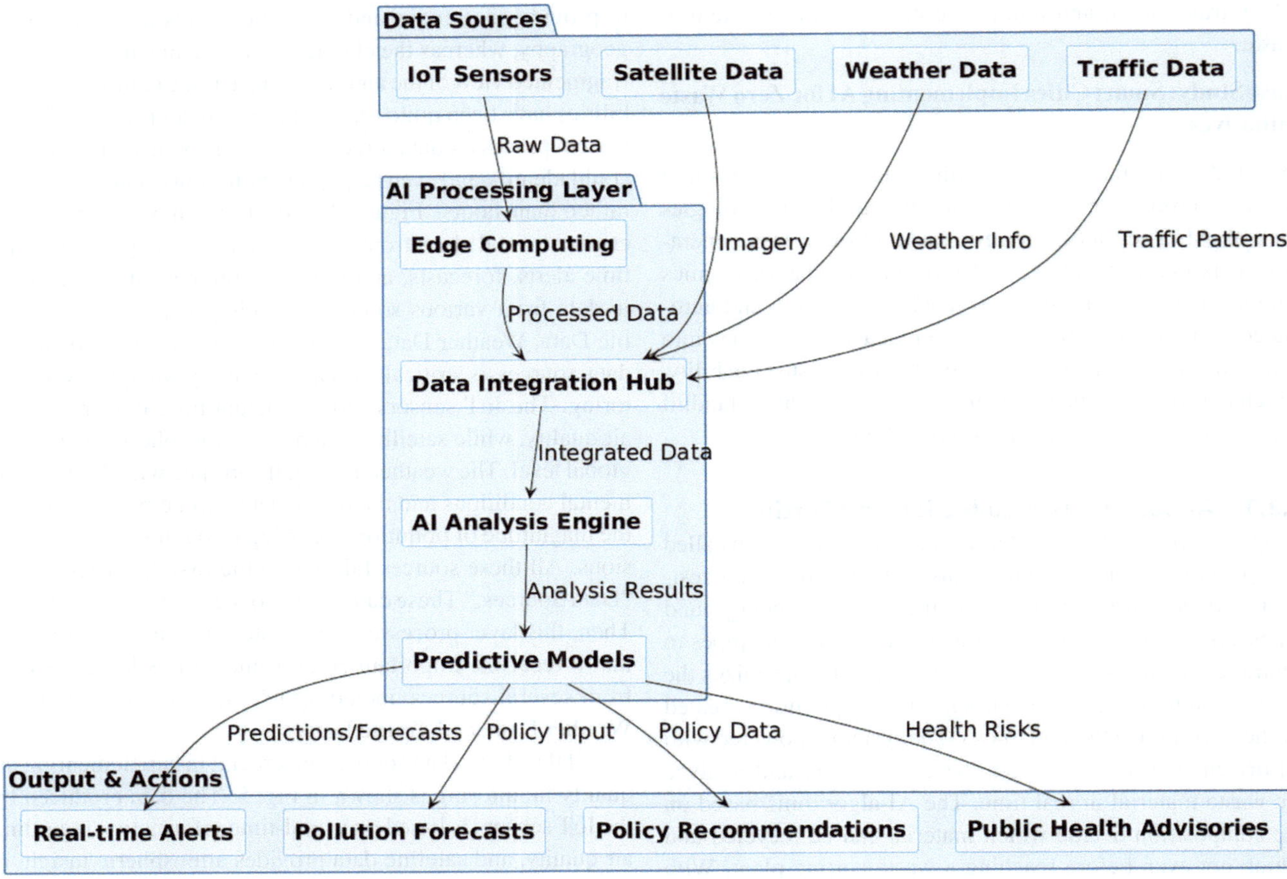

Fig. 5 AI-driven environmental monitoring system

and predict air quality events. It emphasizes the importance of real-time data processing, integration, and forecasting in combating air pollution and its impacts on public health and urban environments. The visual separation of the components and their roles, enhanced with colors, ensures the diagram is both informative and easy to understand. Recent advances in neural network architectures further enhance the capabilities to make these predictions. Modern systems are now enabled to account for seasonal changes, special events, and even long-term climate trends. AI systems extract useful patterns from these complex data streams excellently. They will detect slight, subtle changes in air quality to indicate emerging new pollution sources, or they'll track the effects of long-term environmental policies, which is also valuable for observing remote areas for which it may be impractical or impossible to deploy ground sensors.

Figure 6 illustrates Real-Time Air Quality Detection and Alerts Using AI First, it aggregates data through IoT sensors that continuously monitor air pollutants and meteorological conditions. It runs in real-time analysis to spot the patterns and anomalies; these are then inputted into generation of predictions wherein the ML models will analyze the risk and the trend of the pollution. In case the alert system has identified a risky situation, it alerts the concerned authorities and stake- holders. Relevant intervention steps are taken, like controlling traffic or imposing an emission restriction. Then, their consequence is analyzed at the learning phase of the system to rectify the errors in the prediction made by the AI model. Finally, it will be observed in optimization in that data collection methodology and model parameterization is readjusted in order to be ready for repetition, hence ensuring continued and proactive management of pollution. Good proof of its efficiency is offered by the installation of AI-driven air quality management systems in large urban centers (Bainomugisha et al., 2024). Comprehensive monitoring networks, involving ground-based sensors and satellite data and predictive modeling capabilities, have been set up in urban centers around the world. Integrated systems offer unique insight into city air quality dynamics and enable immediate response to issues that arise. Probably one of the most dramatic examples has been in a large Asian metropolis where the management of air quality was enhanced. Within two years of installation of dense networks of AI-enabled sensors and predictive modeling systems, the city had reduced severe air quality events by 40%. Predictions of pollution hotspots enabled targeted intervention with maximum impact and minimal

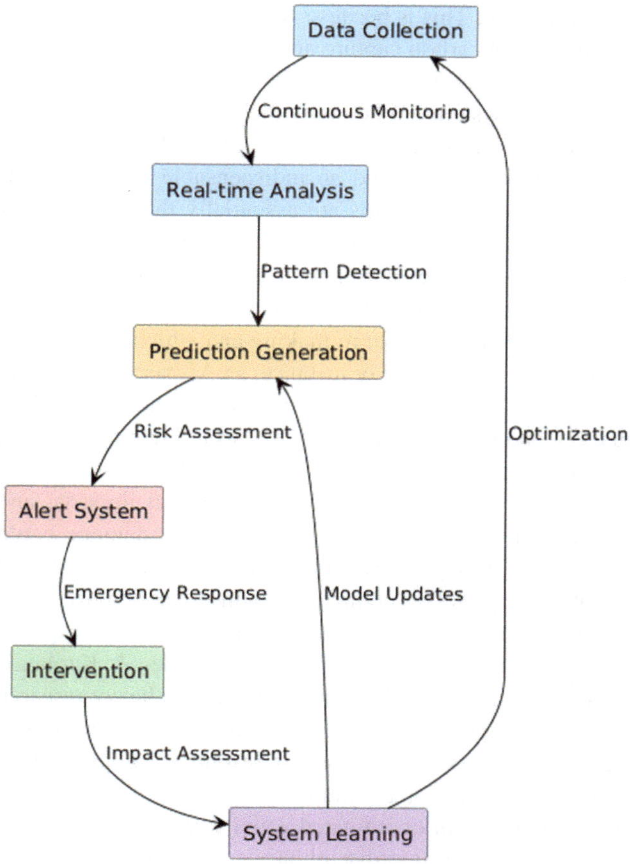

Fig. 6 AI-driven decision-making and optimization cycle

disruption to daily operations within the city. Real-time monitoring systems generated immediate feedback as to the situations prevailing in conditions of air quality, and predictability models also aided in utilizing proactive management methodologies. Its adaptation over time into historical data patterns for prediction updates made the entire system improve from one time-step to another for the interventions becoming effective over that time. This should be a great example of AI-driven systems in changing the air quality within cities. It will serve lessons to other cities that might plan such an undertaking. Success will, however, depend on effective integration of numerous technologies and a response protocol system that harnesses the full benefits of AI. The future seems brighter for AI in air quality management. But as more sophisticated technologies are developed, we should be able to expect even better solutions for air pollution monitoring and control. It shall require considerably committed investment in technology and infrastructure, and willingness in international cooperation, for this potential to come to fruition. With all the current implementations gaining experiences that indicate success in the context, it is about not just technology but the development of relevant governance frameworks and public engagement strategies. Thus, for the way forward, there must be an intention to develop an integrated solution by fusing together technological innovation and effective policy measures with public participation. The main challenge lies in rolling out these techno- logical capabilities in ways that maximize their impact while offering cost-effectiveness and public acceptance. Real-time air quality monitoring using AI raises privacy issues, as sensors may collect personal data inadvertently. Ensuring data anonymization and compliance with regulations is critical.

10 AI and Environmental Justice: Bias and Fairness in Decision-Making

AI is revolutionizing every industry from healthcare to finance, and it is not going to be different for the environment (Scott & Oelofse, 2005). There's no denying that AI will contribute to enhanced sustainability, climate resilience, and environmental protection. The reliance on AI for environmental decision-making raises significant questions regarding fairness, bias, and perpetuating historical inequities. The advocates of environmental justice argue that, if not properly designed and monitored, AI systems could disproportionately affect marginalized communities. The challenge therefore lies in the use of AI to promote fairness and justice instead of exacerbating existing disparities. This essay explores how AI can be used to address environmental challenges while tackling bias and fairness in decision-making. AI is currently changing the face of environmental decisions by scanning extremely large volumes of environmental data to predict trends, optimize resource usage, and evaluate risks. AI models can help governments, organizations, and communities in more informed decisions over air and water quality, land use, biodiversity, and disaster preparation. For instance, machine learning algorithms currently exist that can predict climate change impacts on crop yields, urban flooding, and energy consumption patterns. These technologies can promise more effective and efficient mitigations of these environmental challenges. However, only as good as the data trained on are the AI systems; environmental datasets frequently include information such as climate patterns, pollution levels, and socio-economic conditions—all of which are prone to biases. The reflection of past injustices in terms of more exposure of environment hazards to the deprived communities is contained in historical data. Therefore, usage of this may further aggravate the issue at the time of training AI models. So, these biases need to be addressed first so that it is used as a tool in environmental decision-making to bring fair justice. Different regions face challenges like limited data availability, differing privacy laws, and algorithmic biases, complicating global ethical AI deployment.

Case Study 1: Environmental Justice and the Flint Water Crisis

One of the most famous environmental injustices was the Flint water crisis in Michigan. In 2014, the city decided to switch from its previous source of water from Lake Huron to the Flint River as a cost-cutting measure, and this led to lead contamination in the drinking water. The poor African American community was the hardest hit by the crisis, having already been subject to social and economic disadvantages. Although the crisis was not designed in AI, it reveals an even more gigantic concern about how environmental decision-making may forget to consider vulnerable communities. In this case, equitable representation in decision-making processes both human and algorithmic was missing, which led to the disastrous outcome. Figure 7 outlines a structured approach to ethical AI policy implementation, ensuring data quality, fairness, regulatory compliance, community impact assessment, and continuous evaluation for just and sustainable outcomes. In this process, policy implementation starts by following international standards and national regulations which then implement policies correspondingly.

The process then evaluates the community impact, focusing on environmental justice, achieving Sustainable Development Goals (SDG), and gathering community feedback. After the initial evaluation, the system analyzes the fairness of the outcomes (Siddiqui, 2023). If the outcomes are fair, the policy proceeds to execution; otherwise, the system loops back to reassess data and policy implementation, ensuring improvements are made. The diagram itself utilizes color coding for the distinct phases: gathering and analysis data is pink while policy implementation is going to be light blue, community impact will be light green with the yellow color being utilized for fairness review as it makes proper flow to an understandable decision. Perhaps the most important issue for addressing climate change is to reduce greenhouse gases in the atmosphere. Behind all these efforts at fighting climate change are carbon capture and storage technologies. AI serves as an enabler to optimize and advance those technologies, such as capturing carbon dioxide from power plants or even directly from the atmosphere for safe underground storage. AI will also play a crucial role in the formulation of geoengineering options, especially methods of solar radiation management and Carbon Dioxide Removal (CDR). Geoengineering, briefly, represents massive interventions into Earth's climate system with an intent to undo harmful effects caused by climate change. AI could predict and simulate what geoengineering does and would thereby be deployable in a manner that ensures that no deleterious, unexpected effects emerge during deployment. These changes in the atmosphere would be predicted by machine learning models of their effects on weather, ocean health, and ecosystems to inform decisions about how these interventions might best be applied. This is because AI would be a tool of major significance in carbon capture and geoengineering due to its facility for handling vast volumes of data and simulating complicated systems. When such technologies are scaled up toward attaining the targets of reduction at the global scale, then it would have played an indispensable role. Environmental impact assessments are essentially the critical tools used to assess what probable effects proposed projects or policies will have on the environment. AI can also predict from a historical record made for cumulative impacts that two or more projects are capable of making and which areas would need to bear more care. For instance, an AI-based system can follow up on deforestation patterns with the help of satellite imagery and pinpoint which regions would degrade further. Similarly, industrial activities like mining or agriculture could be analyzed in terms of their environmental impacts, and such insights could provide advice on how damage could be curtailed through measures to be taken. Another aspect is that AI is perfectly compatible with a Geographic Information System (GIS), which allows more advanced capabilities of spatial analysis and modeling (Pandey & Karnatak, 2014). The future of AI and sustainability is heavy research and interdisciplinary collaboration. Issues such as climate change, biodiversity loss, loss of resources, among others are collective challenges; multifaceted and reflective of the sheer magnitude of coordinated activity required toward the resolution of each. Interdisciplinary collaborations will lead to innovation through AI for sustainability. In that regard, the consensus from computer science, environmental science, economics, and social science experts shall have to complement each other in order to achieve the goal of responsible, ethical deployment of AI technologies. Secondly, AI can also be better aligned toward ideas of sustainability work in policy development through the development, community engagement and public awareness of more holistic inclusive measures toward holistic solutions that could eventually have that impact. Figure 8 illustrates AI enables resource management, smart infrastructure, and environmental protection for a sustainable future. This system includes these three major interacting pillars, showing the whole eco-system through which next-generation AI is driving these sustainability initiatives. There is flow out from that central "advanced AI" at the top through three critical areas: Resource management—to the left-energy, water, and waste optimization; and smart infrastructure centered with smart grids, cities and IoT networks as the middle item, and right-side Environmental Protection for climate, biodiversity, conservation efforts.

Each of the three main streams further breaks into much more Next Gen AI in Sustainability. All these three major interlinking pillars represent the whole ecosystem through which next-generation AI drives sustainability initiatives, as in Fig. 8. Starting from the central hub of "Advanced AI"

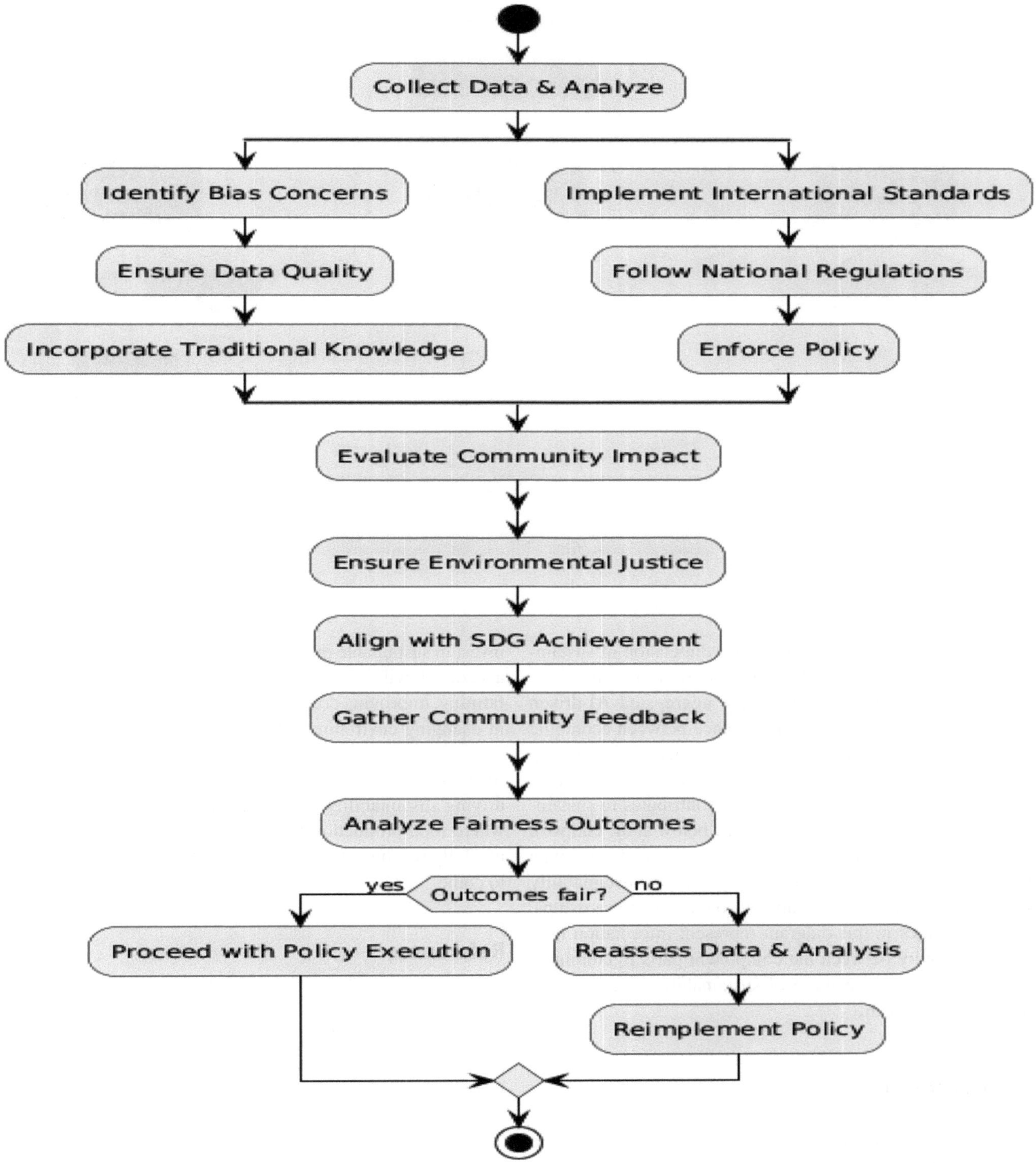

Fig. 7 Ethical AI policy implementation framework

in the middle part, the flows split into the three important areas: to the left is the arena of Resource Management that addresses and optimizes energy, water and waste, and to the center, the Smart Infrastructure arena such as smart grids and cities and IoT network; and on the right-hand side is Environmental Protection focused on climate action and biodiversity, besides conservation efforts. From the three main strands, each then further branches out into even more specific sub-

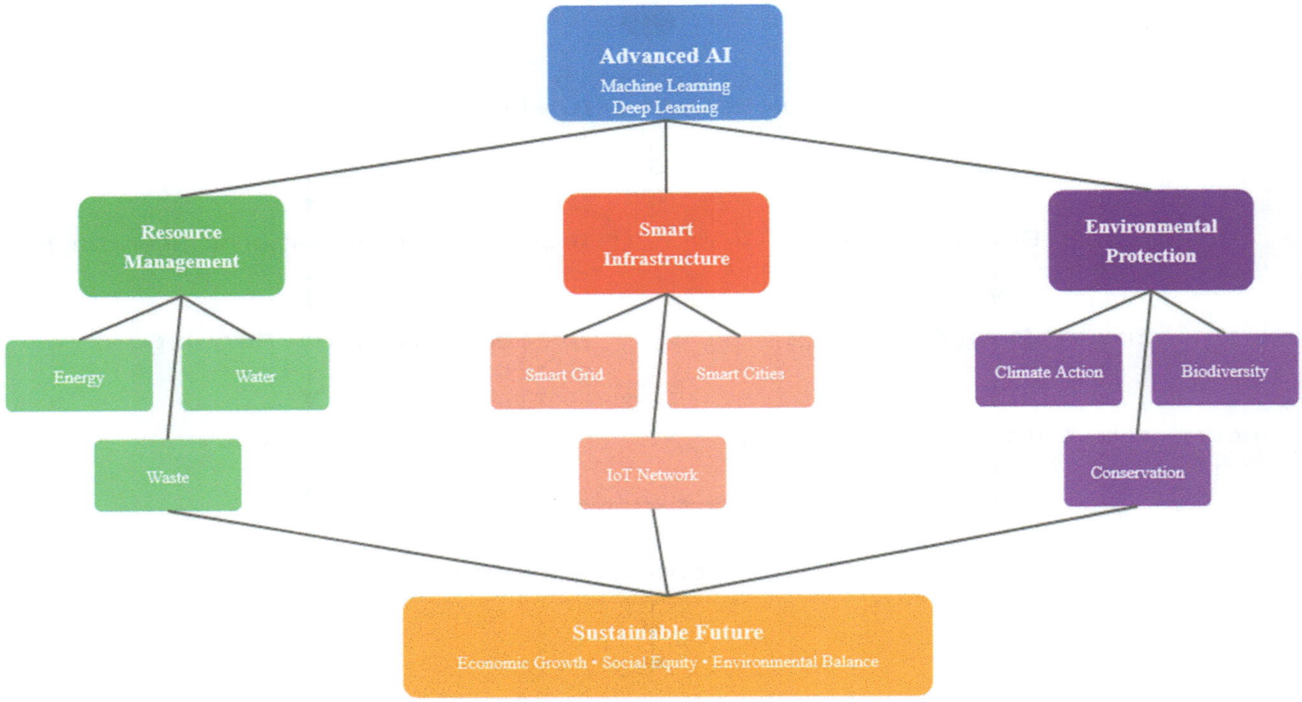

Fig. 8 AI-driven sustainable development framework

components that focus on a particular dimension of sustainability. All of them then merge at the bottom into the "Sustainable Future," meaning that these aggregated AI-driven forces aim at meeting a point of balance between economic development, social equality, and environmental stewardship (Bibri, 2024). The interconnecting lines illustrate data flow, showing how each AI subcomponent contributes to sustainability. All the threads finally converge below at the end with the "Sustainable Future," so what these aggregated, AI-driven powers target is attainment at a tipping point of sustainability between economic, social, and environmental stewardships. The linked lines in this diagram represent information flow and connectivity between the component parts to culminate in the harmonious execution of sustainability within an AI integration approach.

11 Conclusion

AI is transforming agriculture and paving the way for a more sustainable future. AI helps farmers use resources efficiently, reduce waste and minimize the impact of farming on the environment. Precision agriculture improves soil fertility, monitors crop health and reduces the overuse of water, fertilizers and pesticides. AI also enhances supply chains by preventing food loss and ensuring better distribution. This transformation depends on a collective effort. Farmers, businesses, policymakers, and consumers must adopt AI-driven innovations, invest in sustainable practices and share knowledge. Looking ahead, AI will play an even greater role in climate-smart farming, improving crop resilience and ensuring food security in changing environmental conditions. Google DeepMind's energy-efficient data centers and Conservation AI's wildlife tracking projects illustrate real-world AI implementations driving sustainability. The AI revolution is shaping a world where food production is not just efficient but also environmentally responsible, securing a better future for generations to come.

References

Bainomugisha, E., et al. (2024). AI-driven environmental sensor networks and digital platforms for urban air pollution monitoring and modelling. *Societal Impacts, 3*, Article 100044.

Bibri, S. E. (2024). *Artificial Intelligence of Things for smarter eco-cities: Pioneering the environmental synergies of urban brain, digital twin, metabolic circularity, and platform*. CRC Press.

Dhanke, J. A., et al. (2024). Climate-based AI-powered precision irrigation: Sustainably smart agriculture frameworks for maximum crop yields. *Remote Sensing in Earth Systems Sciences*, 1–12.

Durlik, I., et al. (2023). Navigating the sea of data: A comprehensive review on data analysis in maritime IoT applications. *Applied Sciences, 13*(17), 9742.

Fadhil, M. J., Gharghan, S. K., & Saeed, T. R. (2023). Air pollution forecasting based on wireless communications. *Environmental Monitoring and Assessment, 195*(10), 1152.

Fente, D., & Singh, D. (2018). Weather Forecasting using artificial neural network. 1757–1761. https://doi.org/10.1109/ICICCT.2018.8473167

Hamdan, A., et al. (2024). AI and machine learning in climate change research: A review of predictive models and environmental impact. *World Journal of Advanced Research and Reviews, 21*(1), 1999–2008.

Javaid, M., et al. (2023). Understanding the potential applications of artificial intelligence in the agriculture sector. *Advanced Agrochem, 2*(1), 15–30.

Jeremiah, S. R., et al. (2024). A comprehensive survey of digital twins: Applications, technologies, and security challenges. *Journal of Systems Architecture*. https://doi.org/10.1016/j.sysarc.2024.103120

Jige, S., & Launa, J. J. (2024). Role of artificial intelligence in sustainable development. *Practice, Progress, and Proficiency in Sustainability*, 311–326. https://doi.org/10.4018/979-8-3693-3410-2.ch014

Johnson, N. E., et al. (2017). Patterns of waste generation: A gradient boosting model for short-term waste prediction in New York City. *Waste Management, 62*, 3–11.

Khan, M. I., et al. (2025). Integrating industry 4.0 for enhanced sustainability: Pathways and prospects. *Sustainable Production and Consumption, 54*, 149–189.

Li, S., Tong, Z., & Haroon, M. (2024). Estimation of transport CO_2 emissions using machine learning algorithm. *Transportation Research Part D: Transport and Environment, 133*, Article 104276.

Lucas, J. A. (2011). Advances in plant disease and pest management. *The Journal of Agricultural Science, 149*(S1), 91–114.

Martin-Rios, C., Hofmann, A., & Mackenzie, N. (2020). Sustainability-oriented innovations in food waste management technology. *Sustainability, 13*(1), 210.

Mazurek, M. (2025). Smart cities practices applied in the selected city of Slovak Republic. In *5G/6G advancements in communication technologies for agile management* (pp. 333–344). IGI Global Scientific Publishing.

Miller, T., et al. (2025). Integrating artificial intelligence agents with the Internet of Things for enhanced environmental monitoring: Applications in water quality and climate data. *Electronics, 14*(4), 696.

Mishra, H., & Mishra, D. (2023). Artificial intelligence and machine learning in agriculture: Transforming farming systems. *Research Trends in Agricultural Science, 1*, 1–16.

Monteiro, A., Santos, S., & Gonçalves, P. (2021). Precision agriculture for crop and livestock farming—Brief review. *Animals, 11*(8), 2345.

Onteddu, A. R., et al. (2025). Enhancing agricultural efficiency with robotics and AI-powered autonomous farming systems. *Malaysian Journal of Medical and Biological Research, 12*(1), 7–22.

Pandey, K., & Karnatak, H. (2014). Geographic information systems. In *Advances in geospatial technologies for natural resource management* (pp. 41–63). CRC Press.

Reeve, J. R., et al. (2016). Organic farming, soil health, and food quality: Considering possible links. *Advances in Agronomy, 137*, 319–367.

Rehman, H. U. (2020). Food waste reduction in Oslo restaurants: An approach toward sustainability (Master's thesis).

Sandal, N. (2024). Global warming: A comprehensive analysis of causes, impacts, and solutions. *International Journal of Global Research Innovations & Technology (IJGRIT)*, 2(4), 70–78. https://doi.org/10.62823/ijgrit/2.4.7010

Scott, D., & Oelofse, C. (2005). Social and environmental justice in South African cities: Including 'invisible stakeholders' in environmental assessment procedures. *Journal of Environmental Planning and Management, 48*(3), 445–467.

Siddiqui, M. N. (2023). AI revolution: Empowering the future with artificial intelligence. *Pakistan Journal of International Affairs*, *6*(3).

Talaviya, T., et al. (2020). Implementation of artificial intelligence in agriculture for optimization of irrigation and application of pesticides and herbicides. *Artificial Intelligence in Agriculture, 4*, 58–73.

Tedjopurnomo, D. A., et al. (2020). A survey on modern deep neural networks for traffic prediction: Trends, methods, and challenges. *IEEE Transactions on Knowledge and Data Engineering, 34*(4), 1544–1561.

Worku, A., & Ayalew, S. (2024). Conflict of interest: The authors declare that there is no conflict of interest regarding the publication. *Vegetable Crops of Russia, 3*, 112–119. https://doi.org/10.18619/2072-9146-2024-3-112-119

Zhang, A., et al. (2020). Blockchain-based life cycle assessment: An implementation framework and system architecture. *Resources, Conservation and Recycling, 152*, Article 104512.

Forest Fire Prediction Using Machine Learning with Effective Feature Scaling and Selection

Soubraylu Sivakumar, Tamanna Dash, Sukhumjeet Singh, P. Saravanan, U. M. Prakash, and R. I. Minu

Abstract

The wildfires in fire-sensitive ecosystems and regions have ruined fertile land, causing consequences which could well be dealt with by vegetative and environmental regimes. Global climate change is altering fire regimes, most especially with regard to fire frequency and intensity and timing. Some ecosystems are shifting favorably toward fire-adapted species. Fire remains a dominant factor promoting deforestation and extinction on a global scale they frequently spread uncontrollable before suppression. Hence, a serious concern arises in the context of heavy destruction possible due to lack of wise prediction and immediate response to fire events. Forest fires interrupt vegetation and wildlife and usually seem to endanger human habitations and pose a considerable environmental threat. While certain ecosystems such as the grasslands and temperate forests have adapted to regular fires for their regeneration, uncontrolled wildfires devastate beyond recovery. The understanding of where fires will start or how intense they become contributes to minimizing their effect. The aim of this work is to create a machine learning (ML)-based model able to forecast the probability of fires from environmental variables and weather. The model takes into account elements including temperature, humidity, how fast and in what direction the wind blows, rain, Fire Weather Index (FWI), Duff Moisture Code (DMC), Drought Code (DC), Initial Spread Index (ISI), Buildup Index (BUI), Fine Fuel Moisture Code (FFMC), and past fire events. The methods used to pick the most important factors (Extra Trees, Forward Selection, Chi-Square, Pearson Correlation) showed that FFMC, ISI, DC, BUI, and temperature play the biggest role in predicting fires. These factors kept showing up in all the methods, which shows they have a big effect on fire risk. The team chose these factors because they're known to affect how fires behave and they're available in the Algerian Forest Fires Dataset. This means the model looks at both weather and environmental things that can lead to wildfires. The aim of this study, therefore, is to build a ML-based model for predicting the risk of potential fires through meteorological and environmental parameters. By assessing the condition of the weather and its different factors, the model aims to predict the probability of fire outbreak and its intensity in particular regions. Accurate prediction of fires will enable prevention, monitoring, and more effective management strategies, which will help to conserve forests by minimizing fire-induced losses. The k-nearest neighbor model achieved the highest prediction of wildfire risk with an accuracy of 90% on the Algerian forest fire dataset. Its high predictability indicates a significant contribution to managing wildfire hazards while supporting proactive strategies for disaster management. Real-time decisions in case of wildfires can also be aided by the model alone for faster and proper operational responses.

Keywords

Forest fire prediction · Machine learning · Feature selection · K-nearest neighbors (KNN) · Fire Weather Index (FWI) · Wildfire risk assessment · Disaster management · Root mean squared error (RMSE)

S. Sivakumar (✉) · S. Singh · P. Saravanan · U. M. Prakash · R. I. Minu
Computing Technologies, School of Computing, SRM Institute of Science and Technology, Kattankulathur, Chennai, India
e-mail: sivas.postbox@gmail.com

S. Singh
e-mail: ss9084@srmist.edu.in

P. Saravanan
e-mail: saravanp7@srmist.edu.in

U. M. Prakash
e-mail: prakashm3@srmist.edu.in

R. I. Minu
e-mail: minur@srmist.edu.in

T. Dash
Networking and Communication, School of Computing, SRM Institute of Science and Technology, Kattankulathur, Chennai, India
e-mail: td5647@srmist.edu.in

M. Ashwin et al. (eds.), *Smart Forestry 5.0 for Powered AI Sustainable Environment and Climate*, Sustainable Artificial Intelligence-Powered Applications, https://doi.org/10.1007/978-3-031-99086-1_9

1 Introduction

Forest fires are marked as among the most ravaging natural calamities, with immense destruction to ecosystems, biodiversity, and human settlements. Wildfires greatly contribute to extensive deforestation and worsen climate change because they emit massive amounts of carbon dioxide and other greenhouse gases into the atmosphere. Because of extended dry seasons and human activities like deforestation and unmanaged land clearance, climate change appears to be making forest fires more frequent and severe. What is most dangerous about fires is that they achieve rapid spread in areas before one can jump to protect it and, hence, early detection and preventive action are important to reduce their damage. Forecasting bushfires depends on traditional methods based chiefly on meteorological status and satellite imaging that lacks accuracy because of the frequently changing parameters of the environment. This calls for urgent development of sophisticated predictive models that will effectively deal with multidisciplinary and highly complex datasets and give timely notices whenever conditions are favorable.

Recent advancements in artificial intelligence have opened new avenues of opportunities for improving forest fire prediction (FFP). However, perhaps the biggest worry about new (ML) machine learning forms—such as (NN) Neural Network—is that they can be fed and troublingly ridden with enormous multivariate datasets, which may otherwise be overlooked by classical statistical methods in dealing with problems of the size and complexity that they pose. In essence, fire prediction can only produce valid and reliable results if a balanced integration of multiple computational techniques is adopted. Furthermore, the feature selection discards irrelevant and redundant data, thereby improving computational efficiency and the accuracy of prediction. With these newer frameworks, it is possible to establish a more robust and scalable fire prediction system, hence timely notifications to the authorities for proactive actions before wide destruction.

The motivation but not in any way limited has been rooted in the need for a precise and effective fire prediction system which would aim to do away with shortcomings that are seen in modern methods. Current techniques quite commonly fail to take into account the complicated interrelations between various environmental factors, thereby giving very poor predictions. Furthermore, very cumbersome data preprocessing techniques have led to a very high computational cost and a lessening of the efficiency of the model itself. The study aims to create a ML model that could help with fire prediction based on feature scaling and feature selection in order to realize more accuracy and maintain scalability.

Traditional methods like meteorological analysis and satellite imaging are not reliable because they are limited in handling rapidly changing environmental parameters and nonlinear multi-factor interactions. They cannot typically handle nonlinear interactions among parameters like temperature, humidity, and fire indices. Satellite imaging is limited by resolution, cloud cover, and delayed updates, which restrict real-time predictive capability. These methods are also limited in scalability when handling large, multivariate data sets in a timely manner. Machine learning bridges these limitations by using dynamic data and optimizing feature selection for real time, accurate prediction.

Implementing a reliable and accurate prediction system will aid disaster management authorities to act timely, thus minimizing environmental and economic losses and saving human lives. This research not only pushes forward the eventual enhancement of the fire prediction models but also features how advanced data-driven techniques can be used toward the prevention of environmental disasters. The models developed for fire prediction, can aid policymakers in defining better land management plans, the effective allocation of resources, and emergency response planning. With forest fires causing great concern across the globe, this platform provides the groundwork for setting up advanced prediction systems that limit the damage caused. The results of this research will not only be instrumental in bringing about technology-based approaches to enhanced fire prevention but also will play a role in achieving climate change mitigation and sustainable environmental management.

2 Related Work

Recent developments in the prediction and detection of wildfires have witnessed enormous contributions from different machine learning and AI approaches. A thorough analysis of pertinent research studies in this field is given in this section. Deep Belief Network (DBN) wherein Jang Hyun Sunga et al. tackled the shortcomings of the standard regression model for FFP. They noted a nonlinear relationship between weather and consequent forest fires and used the inputs of meteorological parameters and drought indices: SPI and SPEI determined over 3, 6, and 9 months. The study found strong associations between fire occurrences and the climate variables, especially with respect to relative humidity and wind speed (Seong et al., 2021). Rakhit et al. compared various classifiers to assess the level of fire threat. Their comparative study classified the areas into different susceptibility levels and found that the Decision Tree algorithm had the highest rate of classification and accuracy for their dataset (Rakshit et al., 2021).

Extensive research has been performed by Priti et al. to analyze various regression techniques, including but limited to Random Forest, Decision Trees, and SVR for fire prediction. Their methodology for the research consisted of exploratory analysis and data pre-processing. Some outcomes of their work were validation of the fact that such predictors as temperature, relative humidity, and wind speed have a statistically proven influence on fire frequency, for example (Beelagi et al., 2021). Sinha et al. offered energy-efficient nodes localization algorithms based on SVMs for nodes that lack technical capability to use GPS to, for instance, for every Forest to be divided into distinct zones of its activity. The results include increasing the importance in zone prediction that, in turn, improves the efficiency of forest node localization, shortens the time of fire occurrence detection, and boosts energy-saving (Vikram et al., 2020).

Ansori et al. used Extreme Learning Machines architecture in prediction, focusing on atmospheric parameters. They used a single-hidden-layer structure that performed better in shorter time compared to standard feed-forward networks (Mahmudy et al., 2020). Tripathi et al. used a semi-supervised clustering method which had a prediction accuracy of 96%. The model was able to prolong the network lifetime using selective data transmission from high- and medium-activity regions (Sinha et al., 2019).

For mapping the danger of forest fires in the Lao Cai region of Vietnam, Jones et al. suggested a decision tree methodology based on the Logit-Boost ensemble. The model was tested against a Geographic Information System (GIS) database containing 256 fire areas and achieved 92% predictive capability (Tehrany et al., 2019). Singh et al. created a parallel SVM model that claimed to use parallel processing to cut down on the amount of technical time and data effort needed. For Portuguese data, the RMSE was 63.45, with the model incorporating the FWI (Singh et al., 2021).

In 2021, Vikram et al. introduced the Fog-Fire Model, a fog-enabled IoT-based forest fire management system. Their three-layer classification model achieved 94% prediction accuracy, which outperformed the existing methods (Vikram & Sinha, 2023a). In their work, Peng et al. achieved 72.3% overall prediction accuracy through ensemble learning techniques on the UCI Repository dataset, and the Ensemble Gradient Boosting model was especially strong in predicting large-scale fires (Xie & Peng, 2019).

Rajesh Kumar Singh et al. built the eForest-Fire system, an integrated GIS-based mapping solution. The system successfully identified 560 out of 5258 high-risk settlements through the integration of various factors, which included a mobile and web portal for citizen reporting (Qayum et al., 2020). Sinha et al. mode of operation consisted of numerous detectors used to monitor the environment. Their work achieved 83% accuracy in predicting zone fire status through integration of direction of classification based on the Neuro-fuzzy model with a convolutional neural network (CNN) model of images (Vikram & Sinha, 2023b).

These studies in total have shown how FFP and detection have developed, moving from single-algorithm treatments toward multi-systems of integrated algorithms. More and more, energy efficiency, real-time observation, and field implementation issues have taken precedence while maintaining the important aspect of prediction accuracy. The numerous varieties of techniques and methodologies each researcher has contributed to forest-fire prediction and management have indeed advanced our understanding and capacity in this domain.

3 Materials and Methods

3.1 System Architecture and Modules

The FFP system's machine learning architecture, which uses feature scaling and feature selection, is depicted in Fig. 1. First, the dataset with the required features is chosen. After that, the dataset is pre-processed to remove outliers and noise. The optimal scaling technique is selected for additional pre-processing. The second step is feature selection, which

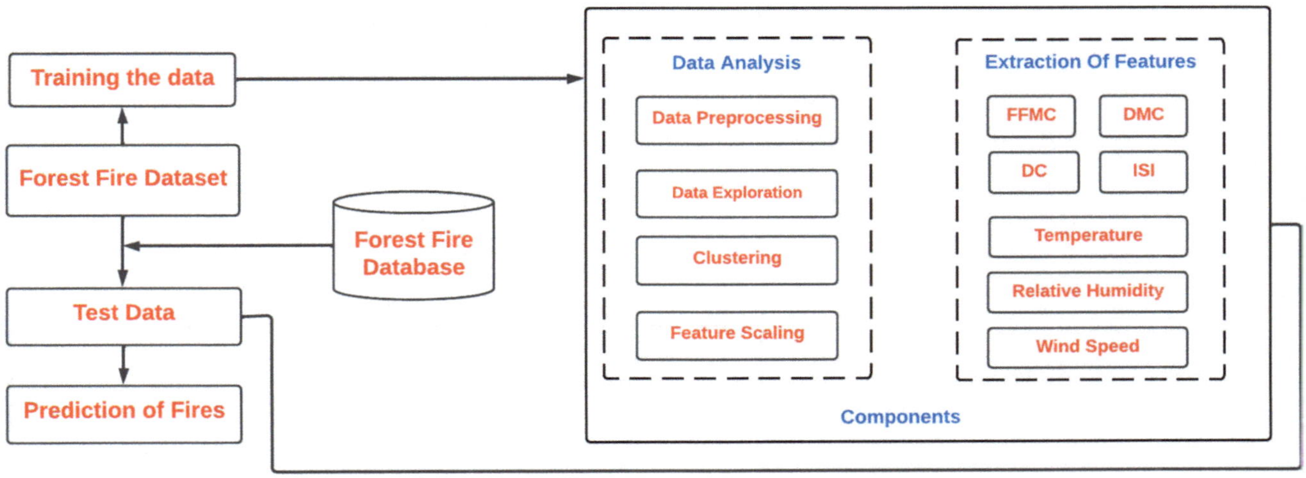

Fig. 1 System architecture diagram

discards unnecessary features. The most efficient method is selected from the methods provided: Pearson Correlation, Chi-Square, Forward Selection, and Extra Tree Classifier. Upon pre-processing, the model is trained with SVM, KNN, and logistic regression algorithms and then tested.

Emphasizing the need of feature selection and scaling in enhancing model accuracy and lowering computing expenses, the data flow diagram (Fig. 2) shows the workflow of the FFP system. The system integrates data from multiple sources, preprocesses it, performs feature engineering operations, and utilizes machine learning algorithms to efficiently predict fire risks. It begins with Input Data Sources, from which inputs are received from diverse sources such as weather stations, ground sensors, and historical fire histories. They provide important variables such as temperature, wind speed and direction, relative humidity, precipitation, FWI, DMC, DC, ISI, and historical fire occurrence. These factors make up a comprehensive data set encompassing meteorological and environmental causes of forest fires.

The process moves on to Data Preprocessing and cleans the data of any noise and outlying data. Since rain can't be negative or temperatures need to be within reason, we get rid of all invalid entries in the data. Methodologies like Winsorization and interquartile range filtering are preferred to secure the integrity of data. Feature Scaling is applied next to make sure the value ranges of the different features are standardized. This allows every feature to affect the model's training the same, so wind speed or temperature cannot overwhelm

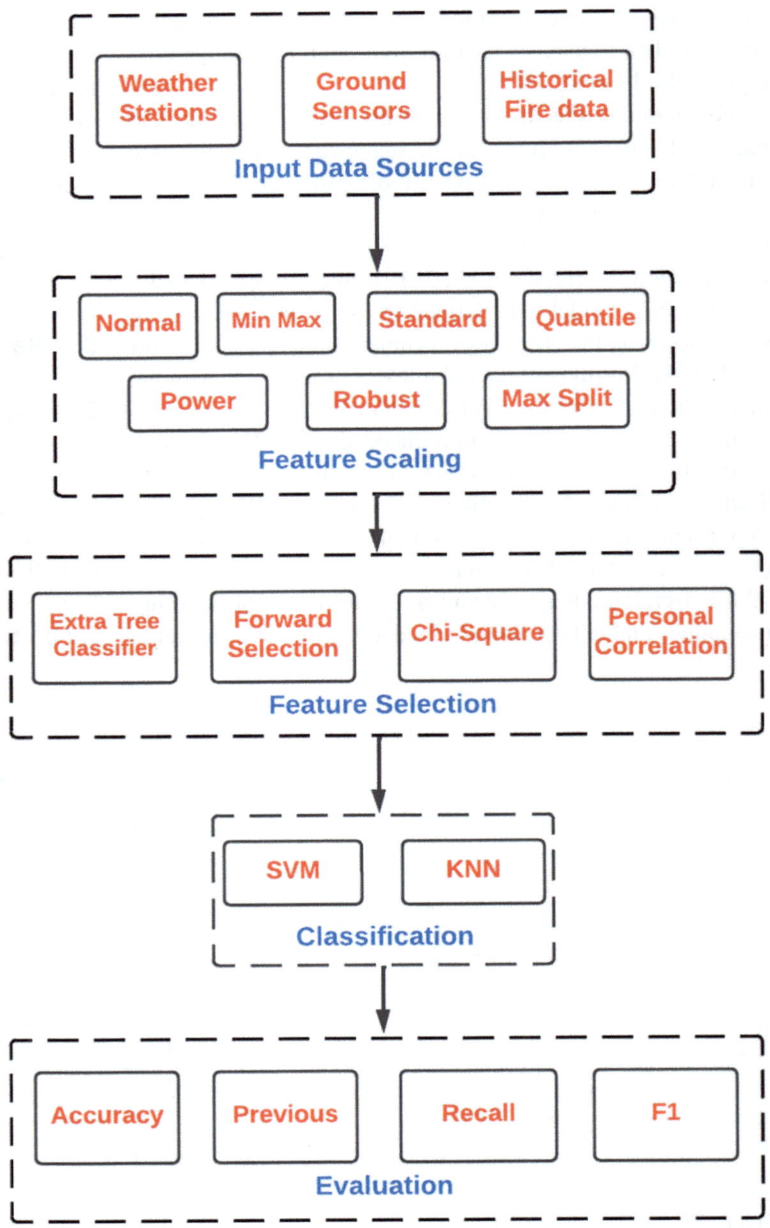

Fig. 2 Data flow diagram

the learning process. The scaling techniques used are z-score normalization, scaling based on the median and interquartile range, Min–Max Scaling, Quantile Transformer Scaling (uniform distribution), and Power Transformations (Box-Cox or Yeo-Johnson). Several scaling procedures are chosen and measured to verify whether they can help reduce RMSE and increase model stability.

Feature selection is carried out at the end to select the key factors that can predict forest fire risk and remove unnecessary or unimportant information. Filter ranking based on Extra Tree Classifier, analyzing category connections with Chi-Square Test, evaluating features incrementally by Forward Selection and studying linear connections with Pearson Correlation are all applied to shrink the dataset. This reduces the efficiency of calculations by focusing only on main variables such as temperature, relative humidity, wind speed, Fire Weather Index, Daily Meteorological Condition, Daily Comfort number, Index of Summer Discomfort and Biologically Equivalent Temperature.

After that, the processed data is put through a Neural Network Model made up of linked layers to spot complex, irregular associations between input variables and fire danger. In typical neural network architecture, dense layers use ReLU activation functions to ensure that the model becomes nonlinear. Training is prevented from overfitting by using dropout layers in the network. The model is made better using Adam optimization along with hyperparameter adjustment by grid search.

Ultimately, the system delivers Fire Risk Predictions (Devi et al., 2024), helping guide activities to reduce fire risk in geographic areas. We check the confidence level of these predictions with accuracy, precision, recall, $F1$-score, AUC-ROC curves, and confusion matrices. The findings are tested using k-fold validation. Using this process, raw environmental information becomes intelligence that forest fire managers can use. With the help of advanced technique and ML, the system supports choosing preventive measures ahead of various disasters.

3.2 Feature Scaling

The data flow diagram highlights feature selection and scaling, showing the data flow within the FFP system. The system begins with data input sources where data is accumulated from a combination of sources such as weather stations and ground sensors. Data may originate from various (Shreya et al., 2023) independent variables such as temperature, relative humidity, past incidents, FWI, precipitation, wind direction and speed, etc. Feature scaling is the first step in standardizing or normalizing the range of values of different elements (Sharma & Khanal, 2024) or features in a dataset. Feature selection makes the module a more accurate one by retaining the most essential and dropping the trivial features and dictating new configurations of the predictive and classification capabilities of the algorithms. The following subsections detail several widely used scaling techniques and their respective advantages.

Data preprocessing is also critical in ensuring data reliability and quality before the modeling process. It comprises dealing with missing values, eliminating outliers, and, if required, converting categorical data into numerical representations. When the data is preprocessed, selecting and scaling features work more effectively, so the dataset is easier to look over and compare. After applying pre-processing, Min–Max scaling, Z-score normalization and Robust Scaler are performed to make all features fall into the same range. Particularly, this setup plays a crucial role for key features used in algorithms, like SVMs and k-NN. Alternative methods include Recursive Feature Elimination (RFE), SelectKBest and methods included in tree-based estimators, including Random Forests. With them, we can prevent the model from fitting the data too well, be more efficient and better understand the model.

3.2.1 Standardization

By dividing a feature by its standard deviation after taking away the feature mean, Z-scores give both a mean of zero and a standard deviation of one. It is helpful if data is normally distributed and enhances model stability at training time (1).

$$X' = \frac{X - \mu}{\sigma} \tag{1}$$

where μ = mean of the feature and σ = standard deviation of the feature.

3.2.2 Normalization

Usually, normalization scales would map values from a feature to an interval of [0,1]. This is useful for non-Gaussian data and in cases where there are outliers present.

Min–Max Scaling

Min–Max scaling rescales data by taking the minimum value away from each value and dividing by the range of the feature (max–min). This makes all the feature values fall within a fixed interval, often [0, 1] (2).

$$X' = \frac{X - X_{\min}}{X_{\max} - X_{\min}} \tag{2}$$

Max-Absolute Scaling

To use Max-Absolute Scaling, every value is divided by the biggest absolute value in the feature to keep relative sizes while keeping them within the interval [− 1, 1]. Much like

Min–Max Scaling, it is easily affected by outliers (3).

$$X_i' = \frac{X_i}{\text{abs}(X_{\max})} \tag{3}$$

where $\text{abs}(X_{\max})$ is the maximum absolute value of the feature.

Robust Scaling

With Robust Scaling, data points far removed from the typical results have less impact because scaling is based on median and IQR instead of mean and standard deviation. Each feature is made stable to outliers by being centered around its median and multiplied by IQR (4).

$$X' = \frac{X - \text{median}(X)}{\text{IQR}(X)} \tag{4}$$

where $\text{IQR}(X) = Q3 - Q1$ (Interquartile Range).

Quantile Transformer Scaling

With this method, the data is converted so that it is uniformly or normally distributed using the inverse of the CDF. Skewed data and outliers aren't a challenge for robust regression. Here is Eq. (5) as I have shown:

$$X' = \Phi^{-1}(F(X)) \tag{5}$$

where $F(X)$ = CDF of X and Φ^{-1} = Inverse CDF (Quantile Function) of a normal distribution.

Power Transformer Scaling

It fixes the skewed shape of the data so that it more resembles a normal distribution. It includes:

(1) **Box-Cox Transformation**: By using a power parameter (lambda) adjusted to produce a normal score, we obtain larger values for positive outcomes (6).

$$X' = \begin{cases} \frac{X^{\lambda}-1}{\lambda}, & \lambda \neq 0 \\ \log(X), & \lambda = 0 \end{cases} \tag{6}$$

where λ is a parameter chosen to make data more normal.

(2) **Yeo-Johnson Transformation**: Now allows transformations when the input values can be positive or negative (7).

$$X' = \begin{cases} \frac{(X+1)^2-1}{\lambda}, & \lambda \neq 0, X \geq 0 \\ \log(X+1), & \lambda = 0, X \geq 0 \\ \frac{-(|X|+1)^{(2-\lambda)}-1}{2-\lambda}, & \lambda \neq 2, X < 0 \\ \log(|X|+1), & \lambda \neq 2, X < 0 \end{cases} \tag{7}$$

where λ is a parameter chosen to make data more normal.

3.2.3 Root Mean Squared Error (RMSE)

RMSE shows the effectiveness of a FFP model that uses ML Feature Scaling and feature selection (Sun et al. 2024). The first step is to separate the data into a training set and a test set, normalize your features for accuracy and remove unimportant features. While you train, apply (8) to RMSE to assess the result. The test set having smaller RMSE represents better accuracy, such that the model predicts forest fires reliably.

$$RMSE = \sqrt{\frac{1}{n}\sum_{i=1}^{n}(y_i - \hat{y}_i)^2} \tag{8}$$

where n = Number of data points, y_i = Actual value, and $\widehat{y_i}$ = Predicted value.

3.3 Feature Selection

The dataset's variables that are either unnecessary or redundant can't be used to build a ML model (Latifah et al., 2019). The overall effectiveness and accuracy of the model may suffer if the dataset contains all of these unnecessary and redundant pieces of information. Finding and choosing the best features from the data is therefore crucial, as is eliminating any irrelevant or superfluous features. The process of selecting features for a model is known as feature selection. An attribute that influences or facilitates the resolution of a problem is called a feature. Techniques for feature selection and extraction differ greatly from one another, even though they have the same objective. The primary distinction between the two is that feature selection only selects a subset of the original feature set, while feature extraction adds new features. Feature selection can reduce overfitting and limit the model's input variable by utilizing only relevant data. It aids in improving the accuracy of the procedure. It improves the algorithms' capacity for prediction (Hidayanto et al., 2021). It chooses the most important components and removes the unnecessary or irrelevant ones. Using this has the advantages of reducing overfitting, increasing accuracy, and reducing training time. The many methods employed in this specific study include:

3.3.1 Extra Tree Classifier

The Extra Tree Classifier, a tree-based supervised model, selects features using a filter-based approach by randomly building multiple tree models from the training dataset. Instead of using Gini or entropy (9), it ranks features based on votes, with random splitting values reducing variance and overfitting risk.

$$H = -\sum_{i=1}^{c} p_i \log_2 p_i \tag{9}$$

where c = Number of classes and pi = Probability of class i.

3.3.2 Forward Selection

Forward Selection is a wrapper method that assesses the combined predictive ability of features by iteratively training models with individual features, selecting the best-performing one, and adding new variables until no significant improvement is observed. Incrementally adds features based on their contribution to model performance, typically measured (10) using:

$$R^2 = 1 - \frac{\sum (y_i - \hat{y}_i)^2}{\sum (y_i - \overline{y}_i)^2} \quad (10)$$

where y_i = actual values, $\widehat{y_i}$ = model's predicted values, and $\overline{y}_i$ = mean of actual values.

3.3.3 Chi-Square

Meanwhile, the Chi-Square test assesses feature dependency by comparing expected and actual values, where a higher Chi-Square value indicates stronger correlation. The null hypothesis assumes feature independence, and calculations follow (11).

$$\chi^2 = \sum \frac{(O_i - E_i)^2}{Ei} \quad (11)$$

where O_i = observed frequency and E_i = expected frequency.

3.3.4 Pearson Correlation

Pearson Correlation assesses the linear relationship between two variables using a correlation matrix, with values ranging from − 1 to 1 (12). A value of 1 or − 1 indicates a strong correlation, while 0 signifies no relationship. It is also known as the Product Moment Correlation Coefficient (PMCC) and should be used only for metric variables.

$$r = \frac{\sum (X_i - \overline{X}) \sum (Y_i - \overline{Y})}{\sqrt{\sum (X_i - \overline{X})^2 \sum (Y_i - \overline{Y})^2}} \quad (12)$$

where X_i, Y_i = data points and $\overline{X}$, $\overline{Y}$ = mean values of X and Y.

3.4 Machine Learning Techniques

Machine learning approaches are critical for data preprocessing and model optimization because they ensure consistency, increase accuracy, and improve computer efficiency.

3.4.1 Logistic Regression

If one is faced with a problem of binary classification, logistic regression, which is a type of directed learning, is generally the ideal choice. It applies a logistic function to give probabilities (between 0 and 1) of a binary outcome. For ensuring the most accurate coefficient predictions for every input attribute, the highest likelihood estimate is used. This method is better applied to classification rather than regression. For multi-class issues, methods like one-vs-rest work very well, and they can be used on linearly separable data. Regularization decreases overfitting (L1/L2), computation is accelerated, and logistic regression is easier to interpret. However, it still assumes a linear correspondence between the qualities of the result and log-odds. Its ease of use and probability-based interpretation have given rise to its extensive application across many disciplines, such as economics, medicine, the social sciences, and business.

3.4.2 Support Vector Machine (SVM)

SVM is a practical way to solve both regression and classification tasks through setting commands. If there are various dimensions, it identifies the hyperplane that keeps classes far from each other. With SVM kernel methods such as RBF and polynomial, LinuxKernal can work with data that is dingle as well as more complex. When all the samples are in well-separated groups, overfitting does not become an issue in it. For that reason, adjusting hyperparameters for large datasets must be done by hand which takes a lot of time. For unclear decision limits, SVMs are designed, but they tend to be less effective on classes where examples are mixed or loaded with noise. Because it adapts well and is backed by theory, more people use it now for image recognition, bioinformatics, and text classification.

3.4.3 K-Nearest Neighbors

The K-Nearest Neighbors method is a clear example of supervised learning and can be used for classification as well as regression. We sort new data points by either averaging the expected results of the closest training samples or by collecting the most votes of those the closest to it. It works for anyone, no matter the data analysis skills and makes no presuppositions about data distribution. Tackling large amounts of data requires more processing power, since every task, like using distances in computing predictions, becomes more computationally intensive. Both the way distance is measured and the value you assign to "K" play a big role in how your cluster works. One problem with 3D is that it does not sense subtle details very well. You can use KNN for pattern recognition, making recommendations and locating unusual patterns. For brief and low-dimensional data, it does remarkably well because it is easy to implement.

4 Results and Discussion

4.1 Experimental Setup

Macs and Windows PCs with CPUs from both the Apple M1 Pro line and Intel Core i5-12450H were used in our experiment. All models had 16 GB RAM and a 512 GB SSD, plus both an Nvidia RTX 3050 and Apple's 16 Core GPU to handle extra computation tasks. The Anaconda Distribution in Jupyter Notebook was used to work with Python 3.12 for this study. The work involved scikit-learn 0.24 for many machine learning techniques and pandas 1.2 and NumPy 1.19 for processing data. The plots in this project were made using Matplotlib 3.3 and Seaborn 0.11. This arrangement clearly highlighted the team's devotion to using new technologies for accurate predictions regarding diseases.

4.2 Dataset and Preprocessing

Data in the Algerian Forest Fires Dataset cover fires that occurred in Bejaia and Sidi Bel-Abbes, Algeria, from June to September 2012. There are 244 examples in the dataset, with 122 from each zone and 138 events of fire versus 106 days that did not involve any fire. There were about 56.56% fire events in the dataset and 43.44% of the days were not registered as fires. The data includes 11 environmental features as well as a class label to study fire occurrences. Using preprocessing correctly leaves the data accurate, consistent, and correct for analysis using machine learning. Preprocessing requires cleaning the data, bringing it together, creating new features as needed and keeping the scales right for the prediction task. It is very important to have consistent and accurate data before starting to analyze it with a machine learning model (Kani & Saudia, 2023).

4.3 Model Training and Evaluation

4.3.1 Performance Evaluation

The performance of the trained machine learning was analyzed by using measures such as Acc, Prec, Rec, and $F1$. They allowed the model to better understand the Acc of its forest fire predictions (Pawar et al. 2022). Although Prec and Rec focused on how well the model separated fire from other cases, accuracy looked at the general accuracy all predictions. $F1$ which is the harmonic mean of Prec and Rec, was given as a balanced measure of model performance when dealing with class imbalances. To judge how reliable the model was, a table with true positives, true negatives, false positives, and false negatives was prepared. Since this technique was followed, it allowed us to recognize possible misclassifications and modify the hyperparameters for better performance.

4.3.2 Feature Importance Analysis

A feature significance analysis was also performed to understand which environmental factors had the biggest impact on fire prediction. It was noted from the results that wind speed, temperature, and the level of humidity are primary factors that determine how a fire will spread. The characteristics of the land and the moisture level of the fuel had a major impact, so it is important to use varied environmental information when making fire prediction models.

4.3.3 Model Validation and Comparison

We applied k-fold cross-validation to ensure our results could be applied to different data. To avoid any data distribution bias, we split the data into several smaller groups and used the model (Bhatt and Kaur et al. 2021) repeatedly on each one. Using cross-validation, it was confirmed that the model works equally well on different sets of data. Also, these methods were compared to traditional ML models including DT, RF, and SVM. Although these models had decent predictive abilities, the ANN performed better in regards to precision and sensitivity to intricate patterns of data.

4.4 Analysis of Forest Fire Prediction Models

Globally, forest fires pose a serious and growing threat to both human populations and ecosystems. Accurate prediction and early detection are therefore critical for developing effective mitigation strategies. This section explores the application of ML techniques for FFP, using the Algerian Forest Fires Dataset. This dataset offers a valuable source of information regarding environmental factors (Thakkar et al., 2022) and their influence on fire occurrence.

A first examination of this dataset shows that it includes data on weather, fire weather indices, and class labels, as shown in Fig. 3. 244 observations are included in 244 entries. The summary lists the names and kinds of the dataset's thirteen columns. Additionally, the dataset is easily manageable for analysis due to its modest memory footprint (24.9 KB) and lack of missing values.

Data distribution can be understood by data visualization using histograms, as shown in Fig. 4. Relative humidity and temperature histograms in FFP are frequently bell-shaped, indicating that they are regularly distributed. This implies that a central value will be the focus of most temperature and relative humidity measurements, with fewer readings taking

	day	month	year	Temperature	RH	Ws	Rain	FFMC	DMC	DC	ISI	BUI	FWI	Classes
0	1	6	2012	29	57	18	0.0	65.7	3.4	7.6	1.3	3.4	0.5	0
1	2	6	2012	29	61	13	1.3	64.4	4.1	7.6	1.0	3.9	0.4	0
2	3	6	2012	26	82	22	13.1	47.1	2.5	7.1	0.3	2.7	0.1	0
3	4	6	2012	25	89	13	2.5	28.6	1.3	6.9	0.0	1.7	0.0	0
4	5	6	2012	27	77	16	0.0	64.8	3.0	14.2	1.2	3.9	0.5	0
5	6	6	2012	31	67	14	0.0	82.6	5.8	22.2	3.1	7.0	2.5	1
6	7	6	2012	33	54	13	0.0	88.2	9.9	30.5	6.4	10.9	7.2	1

Fig. 3 Algerian forest fires dataset sample

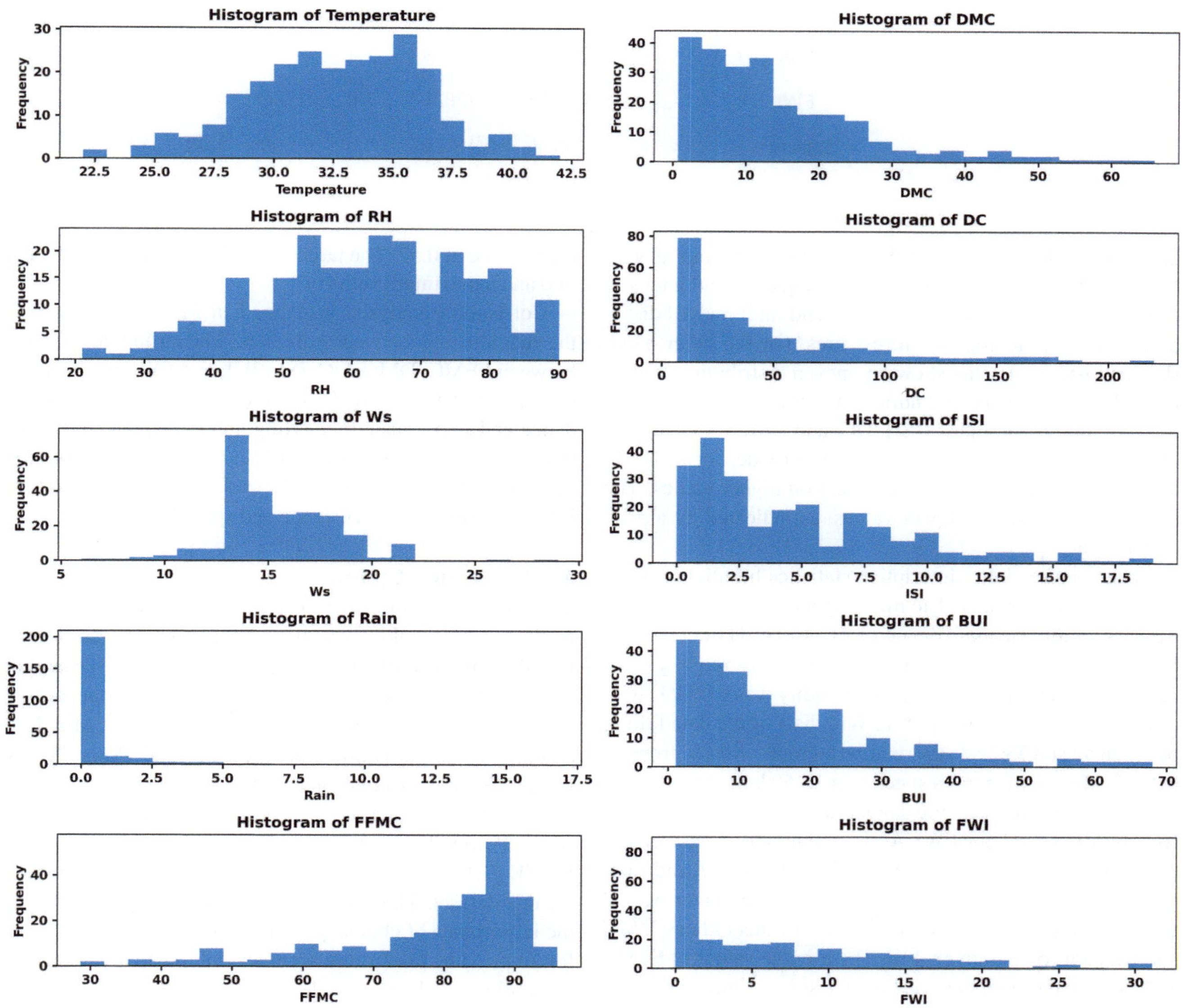

Fig. 4 Histogram of Algerian forest fires dataset

Fig. 5 Descriptive statistics of Algerian forest fires dataset

	count	mean	std	min	25%	50%	75%	max
day	244.0	15.754098	8.825059	1.0	8.000	16.00	23.000	31.0
month	244.0	7.500000	1.112961	6.0	7.000	7.50	8.000	9.0
Temperature	244.0	32.172131	3.633843	22.0	30.000	32.00	35.000	42.0
RH	244.0	61.938525	14.884200	21.0	52.000	63.00	73.250	90.0
Ws	244.0	15.504098	2.810178	6.0	14.000	15.00	17.000	29.0
Rain	244.0	0.760656	1.999406	0.0	0.000	0.00	0.500	16.8
FFMC	244.0	77.887705	14.337571	28.6	72.075	83.50	88.300	96.0
DMC	244.0	14.673361	12.368039	0.7	5.800	11.30	20.750	65.9
DC	244.0	49.288115	47.619662	6.9	13.275	33.10	68.150	220.4
ISI	244.0	4.759836	4.154628	0.0	1.400	3.50	7.300	19.0
BUI	244.0	16.673361	14.201648	1.1	6.000	12.45	22.525	68.0
FWI	244.0	7.049180	7.428366	0.0	0.700	4.45	11.375	31.1
Classes	244.0	0.565574	0.496700	0.0	0.000	1.00	1.000	1.0

place farther from this value. Since temperature and relative humidity have a major impact on the frequency and intensity of forest fires, it is crucial to comprehend their normal distribution in order to forecast forest fires. On the other hand, skewed histograms may show an uneven distribution of data for additional elements including Rain, Buildup Index (BUI), FWI, Initial Spread Index (ISI), DC, and DMC. A histogram that is skewed and has a peak on the left side, for instance, indicates that the data is concentrated on higher values. The data is concentrated at lower values, as indicated by a peak on the right.

When summed up, descriptive statistics highlight important aspects of the data. The mean temperature in Fig. 5 was 32.17 °C, with a standard deviation of 3.63 °C. Half of the data fell between 30 and 35 °C, with temperatures ranging from 22 to 42 °C. The average relative humidity was 61.94%, with a standard deviation of 14.88%. With half of the data falling between 52 and 73.25%, the relative humidity ranged from 21 to 90%. The average wind speed was 15.50 kph, with a range of 6 to 29 kph. The forest fire danger rating indices are DC, ISI, BUI, FFMC, DMC, and FWI. Average values for the aforementioned indices were 49.29, 4.76, 16.67, 77.89, 14.67, and 7.05, in that order. The standard deviations of these indicators varied from 4.15 to 47.62. In comparison with the other indices, DMC and BUI are comparatively low. The average values of FFMC and FWI, on the other hand, are higher, suggesting a higher fire danger rating. 57% of the days in the dataset had a forest fire danger rating of 1, according to the target variable's (classes) mean value of 0.57. The target variable had a minimum value of 0 and a maximum value of 1.

Correlation analysis, visualized in Fig. 6, demonstrates the relationships between variables. The strong connections between FFMC, DMC, DC, ISI, BUI, and FWI are displayed on the correlation map. This demonstrates their interdependence and importance in averting and managing forest fires (Pandey et al., 2023). These indices, along with factors like fuel moisture content and wind speed, are interconnected and influence forest fire occurrence and spread.

4.4.1 Feature Scaling

To enhance model performance, different feature scaling methods were applied, and their impact was assessed using RMSE. The bar chart presents RMSE values (Alauddin et al., 2018) across six scaling techniques: Standardization, Max-Absolute Scaling, Robust Scaling, Min–Max Scaling, Quantile Transformer Scaling, and Power Transformer Scaling. Among them, Min–Max Scaling produced the lowest RMSE (1.19e−16), indicating better performance, while Power Transformer Scaling resulted in the highest RMSE (7.42e−16), suggesting a suboptimal transformation for this dataset. This comparison shown in Fig. 7 highlights the importance of choosing an appropriate scaling technique for improving prediction accuracy in forest fire modeling (Bharathi & PeddaReddy, 2023).

The use of feature scaling methods, Standardization, Normalization, and Robust Scaling, in this study is because

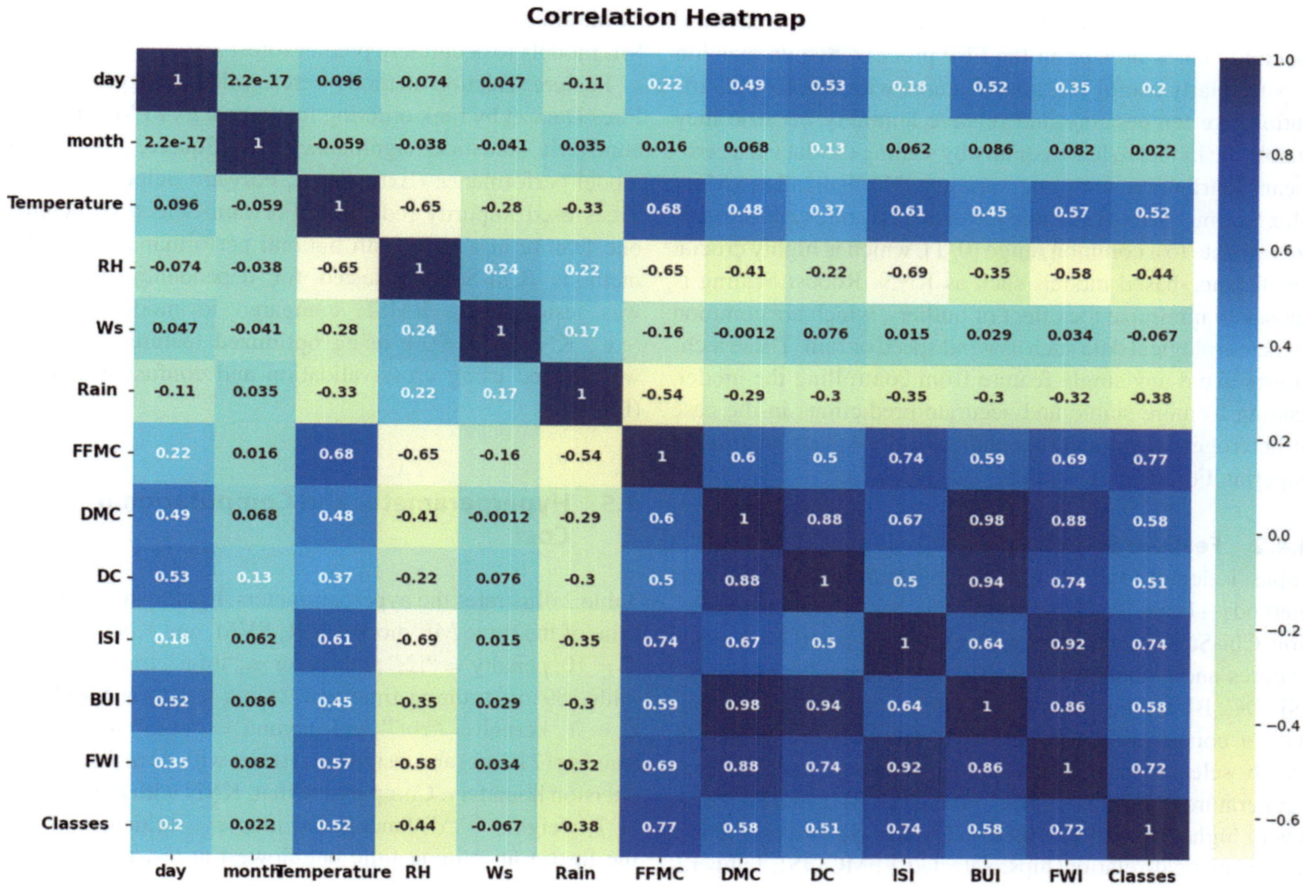

Fig. 6 Correlation analysis of Algerian forest fire dataset

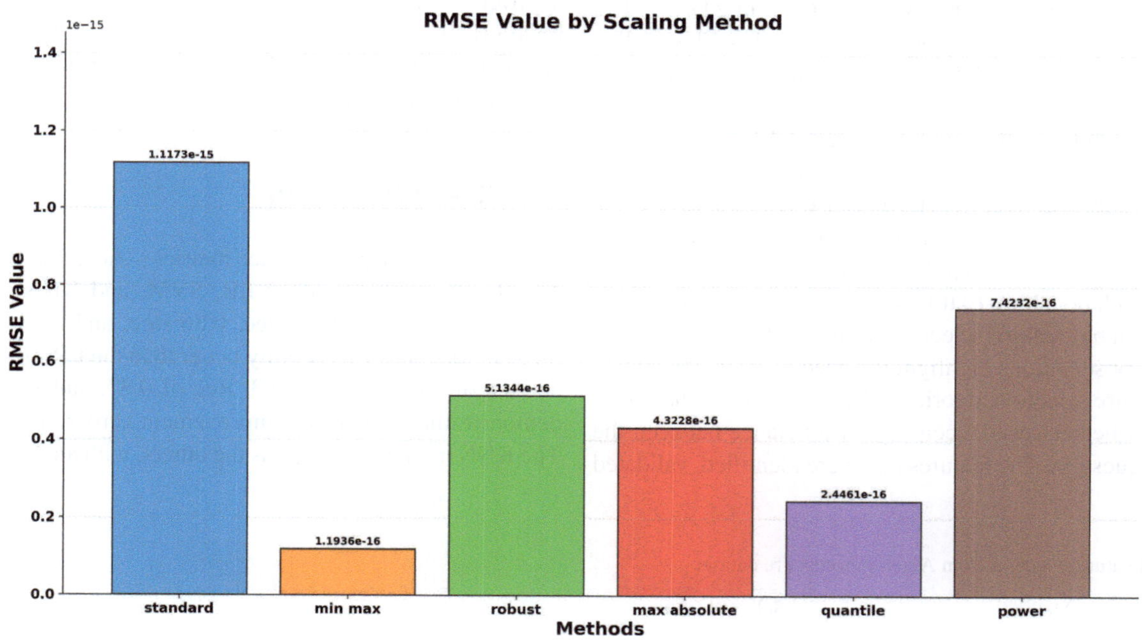

Fig. 7 Comparison table showing the values for the models

they are well applicable in the heterogeneity of the environmental features utilized in the FFP process. Standardization is particularly useful for those features with a normal distribution like temperature and relative humidity because they enable the model to learn equally by scaling the data to a zero-mean distribution with one variance. Normalization (Min–Max Scaling) is most suitable for skewed features and normalizes all data to a common range [0, 1], which is highly crucial for distance-based models such as KNN. Robust Scaling is chosen to minimize the effect of outliers, which are common in meteorological data such as wind speed or rain. These techniques avoid any single feature from controlling the model, leading to more stable and accurate predictions in the case of heterogeneous environmental conditions portrayed in the Algerian Forest Fires Dataset.

4.4.2 Feature Selection

Table 1 depicts the comparison of four feature selection methods—Extra Trees (Ghosh et al., 2024), Forward Selection, Chi-Square, and Pearson Correlation—and their selected features and computation time. Extra Trees selected FFMC, ISI, DC, BUI, temperature, and GWI as significant features with a computation time of 4528 ms. Forward Selection, which selects predictive variables step by step, selected temperature, Rain, RH, FFMC, DMC, ISI, BUI, and DC but in a very higher time of 13,569 ms. The Chi-Square test, which tests statistical relationships, selected FFMC, ISI, temperature, Rain, DMC, DC, and BUI in 397 ms, while Pearson Correlation, which finds linear relationships, selected FFMC, DMC, temperature, ISI, DC, BUI, and GWI in 247 ms. The most common features among all the methods—FFMC, ISI, DC, temperature, and BUI—indicate their importance in the prediction of forest fires.

The computational cost is a trade-off between exhaustiveness and efficiency. Forward Selection is highly accurate but computationally costly because it is iterative. Chi-Square and Pearson Correlation are computationally much cheaper, thus good for rapid analysis, but may not capture nonlinear relationships. Extra Trees takes a middle ground and provides stable feature selection at moderate computational cost. The observations highlight the need to select the appropriate feature selection algorithm depending on the nature of the task, whether speed, accuracy, or a balance between the two is in question. The features that were identified, validated by multiple means, form a good basis for the construction of ML models to accurately predict forest fires.

Feature selection methods were applied to Algerian Forest Fires Dataset by rank ordering features (e.g., FFMC, temperature) via statistical significance (Chi-Square, Pearson) or model performance (Extra Trees, Forward Selection). Extra Trees used impurity reduction, Forward Selection included one feature at a time with optimal performance, and filter methods (Chi-Square) tested for dependency. Efficiency was tested using RMSE compared to model precision (e.g., KNN with 90% using optimized features). Findings were tested using cross-validation and confusion matrices (Fig. 8).

4.5 Hyperparameter and Computational Cost

Table 2 illustrates the hyperparameters, training time, and test time of the three ML models: LR, SVM, and KNN. LR with C = 10, penalty = "l2," and solver = "lbfgs" trained quickly with 189 ms training time and 21 ms test time SVM with C = 1, kernel = "rbf," and gamma = "auto" took longer training (212 ms) and test (34 ms) times with its more complex decision boundary. Compared to that, KNN with n_neighbors = 7, weights = "distance," and metric = "Euclidean" was the fastest at training time but slowest at test time with 10 ms training time but 137 ms test time due to its lazy learning feature where it calculates distances at test time. These results uncover the model complexity, training time, and testing time trade-off, which reflects the choice of models toward some application needs. A k-fold cross-validation with value 5 is used on the three ML models.

4.6 Result Discussion

Table 3 shows the performance metrics—Acc, Prec, Rec, and *F*1—for three ML models: LR, SVM, and KNN. The LR achieves 0.86 Acc, 0.89 Prec, 0.86 Rec, and 0.88 *F*1. The SVM model performs slightly better than the LR model, with an Acc of 0.88, Prec of 0.93, Rec of 0.87, and *F*1 of 0.90, demonstrating a balanced improvement across all measures. The KNN model outperforms the others, with an Acc of 0.90,

Table 1 Feature selection from Algerian forest fire dataset

Sl. No	Method	Selected features	Computation cost (ms)
1	Extra Trees	FFMC, ISI, DC, BUI, temperature, GWI	4528
2	Forward Selection	Temperature, Rain, RH, FFMC, DMC, ISI, BUI, DC	13,569
3	Chi-Square	FFMC, ISI, temperature, Rain, DMC, DC, BUI	397
4	Pearson Correlation0	FFMC, DMC, temperature, ISI, DC, BUI, GWI	247

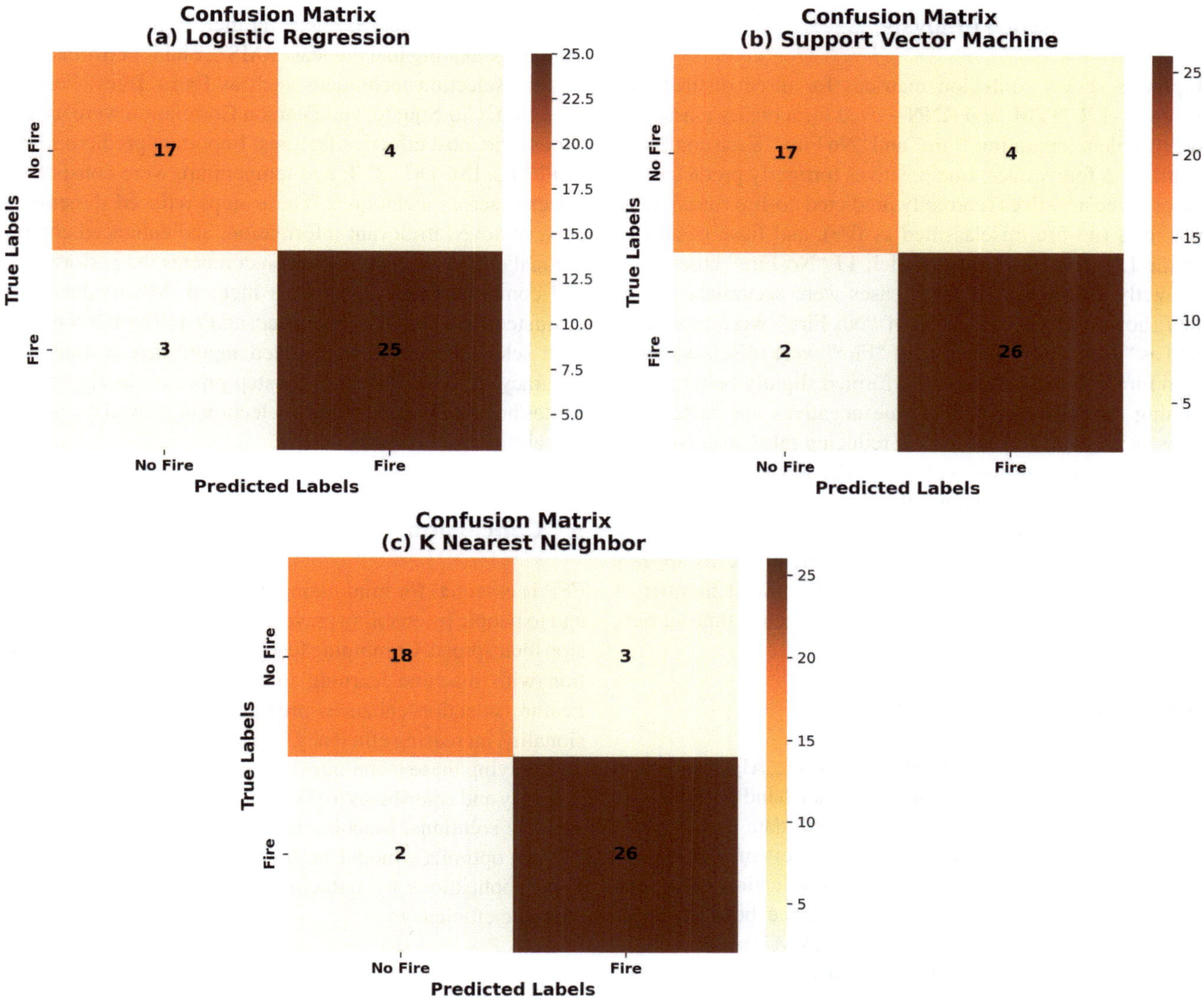

Fig. 9.8 Confusion matrix comparison of three models

Table 2 Hyperparameter, training, and testing time of machine learning models

Machine learning models	Hyperparameter	Training time (ms)	Testing time (ms)
LR	C: 10; Penalty: l2; solver: lbfgs	189	21
SVM	C: 1; kernel: rbf; gamma: auto	212	34
KNN	n_neighbors: 7; weights: distance; metric: euclidean	10	137

Table 3 Performance metrics of three ML models

Machine learning models	Accuracy	Precision	Recall	$F1$ score
LR	0.86	0.89	0.86	0.88
SVM	0.88	0.93	0.87	0.90
KNN	0.90	0.93	0.90	0.91

Prec of 0.93, Rec of 0.90, and $F1$ of 0.91. These findings indicate that KNN is the most effective model, balancing Prec and Rec while getting the maximum accuracy and $F1$. The high Prec values for both SVM and KNN suggest a lower false positive rate; however, the Rec values demonstrate that KNN is superior at detecting true positive situations. The $F1$ results show that KNN has the best overall categorization performance.

4.7 Performance Analysis

Figure 8 shows confusion matrices for three distinct ML models—LR, SVM, and KNN—used for a binary classification problem separating "Fire" and "No Fire." Each confusion matrix has four values: true positives (properly predicted fire cases), true negatives (correctly predicted no-fire cases), false positives (no-fire misclassified as fire), and false negatives. In the Logistic Regression model, 17 "No Fire" cases were correctly classified, 25 "Fire" cases were accurately identified; however, 4 occurrences of "No Fire" were misclassified as "Fire" and 3 instances of "Fire" were misclassified as "No Fire." The SVM model performed slightly better, maintaining the same number of true negatives and false positives as Logistic Regression but reducing false negatives to 2 and increasing true positives to 26. The KNN model showed the best performance among the three, correctly classifying 18 "No Fire" cases and 26 "Fire" cases while reducing false positives to 3 and false negatives to 2. Overall, KNN appears to be the most effective model with the lowest number of misclassifications, making it the best choice for this dataset.

4.8 Results and Challenges

The choice of SVM, KNN, and LR for the Algerian Forest Fires Dataset is suitable since they can handle effectively small, structured, and low-dimensional data unlike deep learning models that are overkill for datasets of such nature. The models are optimal for binary classification tasks such as fire/no-fire prediction and are a balance between accuracy, computation, and interpretability. SVM is stable in high-dimensional space and efficient with clean class margins; KNN offers simplicity and robustness without assuming data distribution; and LR gives probability-based predictions with efficient computation. Because the dataset is small (244 samples and 11 features), the algorithms give reasonable, scalable, and efficient solutions without the overhead of more complex models.

To manage and regulate overfitting or underfitting in feature selection, the study utilized k-fold cross-validation in the way model generalizes to data splits. The study approximated Acc, Prec, Rec, and $F1$ and utilized confusion matrices for the purpose of measuring misclassification. Models were also hyperparameter-tuned and regularization techniques like L2 in LR were utilized in a bid to prevent overfitting. All the above provided the best trade-off between predictive accuracy and model complexity, hence ensuring robustness.

The paper properly explains the process of applying the feature scaling and feature selection methods to enhance the performance of the ML algorithms in the prediction of forest fires. Different feature scaling techniques such as Standardization, Robust Scaling, and Min–Max Scaling are compared and evaluated based on their impact using RMS. Min–Max Scaling had the least RMSE, and it performs well. Feature selection techniques such as Extra Trees, Forward Selection, Chi-Square, and Pearson Correlation were used to choose the most effective features. Important predictors such as FFMC, ISI, DC, BUI, and temperature were consistently selected across techniques. These steps reduced dimensionality, removed irrelevant information, and enhanced computational efficiency. The study also compares the performance and computational cost of each method. ML models were evaluated based on Acc, Prec, Rec, and $F1$. The KNN model, with selected features and scaled input, gave a maximum accuracy of 90%. This step-by-step process clearly demonstrates how scaling and feature selection improved the predictive ability of the model.

5 Conclusion

FFP is essential for minimizing damages to the environment and to people by enabling preventive measures to manage and stop their spread. Combining feature scaling and feature selection with machine learning improves predictive accuracy. Feature selection enhances predictions by reducing dimensionality, increasing efficiency, and isolating relevant features. By applying these techniques, the predictive model increases accuracy and contributes to the provision of scalable and interpretable solutions. Machine learning with adequate preprocessing optimizes model performance and allow for large-scale applications by reducing redundancy and improving compute efficiency.

6 Future Direction

Future research can focus on ensemble learning for resilience, hybrid models that use multiple ML approaches, and DL architectures like CNNs and RNNs for capturing spatial and temporal correlations. By incorporating real-time data sources and geospatial analytics, emergency response planning can become more practically applicable. The use of timely preventative actions, made feasible by the development of real-time prediction frameworks, can reduce the severity of fires. Ongoing research into enhancing FFP models can lead to more effective forest management and disaster mitigation strategies.

References

Alauddin, M. W., Anshori, M., Wicaksono, A. S., & Utaminingrum, F. (2018). Comparative study based on error calculation in multiple linear regression coefficient for forest fires prediction. In *2018 IEEE*

SIET, Malang, Indonesia (pp. 115–120). IEEE. https://doi.org/10.1109/SIET.2018.8693226

Beelagi, A., Preeti, T., Kanakaraddi, S., Malagi, S., & Sudi, A. (2021). Machine learning techniques for predicting forest fires. In *2021 IEEE CONIT*. IEEE. https://doi.org/10.1109/CONIT51480.2021.9498448

Bharathi, V., & PeddaReddy, C. (2023). Comparison of forest fire prediction system using machine learning algorithms. In *2023 3rd international conference on advance computing and innovative technologies in engineering (ICACITE)* (pp. 1539–1542). IEEE. https://doi.org/10.1109/ICACITE57410.2023.10182818

Bhatt, T., & Kaur, A. (2021). Automated forest fire prediction systems: A comprehensive review. In *2021 IEEE ICRITO, Noida, India* (pp. 1–5). IEEE. https://doi.org/10.1109/ICRITO51393.2021.9596528

Devi, K. L. S. R., Kumar, G. N., Narayana, P. A., Ramana, K. V., Amarendra, K., & Gullipalli, T. R. (2024). Forest fire prediction and management using AI (artificial intelligence), ML (machine learning) and deep learning techniques. In *2024 8th international conference on inventive systems and control (ICISC)* (pp. 324–327). IEEE. https://doi.org/10.1109/ICISC62624.2024.00062

Ghosh, P., Shukla, A., B. C. V., & Yakub, F. (2024). Predicting illnesses through symptomatic patterns through ensemble voting regressor. In *2024 second international conference on advances in information technology (ICAIT)* (pp. 1–6). IEEE. https://doi.org/10.1109/ICAIT61638.2024.10690506

Hidayanto, N., Saputro, A. H., & Nuryanto, D. E. (2021). Peatland data fusion for forest fire susceptibility prediction using machine learning. In *2021 IEEE ISRITI, Yogyakarta, Indonesia* (pp. 544–549). IEEE. https://doi.org/10.1109/ISRITI54043.2021.9702762

Kani, D. C. J., & Saudia, S. (2023). Analysis on the performance of machine learning models for forest fire prediction. In *2023 IEEE ICSSIT*. IEEE. https://doi.org/10.1109/ICSSIT55814.2023.10060870

Latifah, A. L., Shabrina, A., Wahyuni, I. N., & Sadikin, R. (2019). Evaluation of random forest model for forest fire prediction based on climatology over Borneo. In *2019 international conference on computer, control, informatics and its applications (IC3INA)* (pp. 4–8). IEEE. https://doi.org/10.1109/IC3INA48034.2019.8949588

Mahmudy, W. F., Anshori, M., Mar'i, F., & Alauddin, M. W. (2020). Extreme learning machines (ELM)-based neural network for forest fire prediction. In *2020 IEEE SIET*. IEEE. https://doi.org/10.1109/SIET48054.2019.8986106

Pandey, S., Singh, R., Kathuria, S., Negi, P., Chhabra, G., & Joshi, K. (2023). Emerging technologies for prevention and monitoring of forest fire. In *2023 international conference on innovative data communication technologies and application (ICIDCA)* (pp. 1115–1121). IEEE. https://doi.org/10.1109/ICIDCA56705.2023.10099572

Pawar, S., Pandit, K., Prabhu, R., Samaga, R., & Geethalaxmi. (2022). A machine learning approach to forest fire prediction through environment parameters. In *2022 IEEE AIDE, Karkala, India* (pp. 1–7). IEEE. https://doi.org/10.1109/AIDE57180.2022.10059873

Qayum, A., Ahmad, F., Arya, R., et al. (2020). Predictive modeling of forest fire using geospatial tools and strategic allocation of resources: eForestFire. *Stochastic Environmental Research and Risk Assessment, 34*, 2259–2275. https://doi.org/10.1007/s00477-020-01872-3

Rakshit, P., Sarkar, S., Khan, S., Saha, P., Bhattacharyya, S., Dey, N., Islam, S. M. N., & Pal, S. (2021). Prediction of forest fire using machine learning algorithms: The search for the better algorithm. In *2021 IEEE CITISIA* (pp. 1–6). IEEE. https://doi.org/10.1109/CITISIA53721.2021

Seong, K. W., Sung, J. H., & Ryu, Y. (2021). Deep learning-based prediction of fire occurrence in South Korea in relation to hydroclimatic conditions and drought phase. *Sustainability, 14*(9), 5494. https://doi.org/10.3390/su14095494

Sharma, S., & Khanal, P. (2024). Forest fire prediction: A spatial machine learning and neural network approach. *Fire, 7*(6), 205. https://doi.org/10.3390/fire7060205

Shreya, M., Rai, R., & Shukla, S. (2023). Forest fire prediction using machine learning and deep learning techniques. In S. Smys, P. Lafata, R. Palanisamy, & K. A. Kamel (Eds.), *Computer networks and inventive communication technologies. Lecture notes on data engineering and communications technologies* (Vol. 141). Springer, Singapore. https://doi.org/10.1007/978-981-19-3035-5_51

Sinha, D., Tripathi, S., & Kumari, R. (2019). Semi-supervised classification based clustering approach in WSN for forest fire detection. *Wireless Personal Communications, 109*, 2561–2605. https://doi.org/10.1007/s11277-019-06697-0

Singh, K. R., Neethu, K. P., Madhurekaa, K., Harita, A., & Mohan, P. (2021). Parallel SVM model for forest fire prediction. *Soft Computing Letters, 3*, 100014. https://doi.org/10.1016/j.socl.2021.100014

Sun, X., et al. (2024). A forest fire prediction model based on cellular automata and machine learning. *IEEE Access, 12*, 55389–55403. https://doi.org/10.1109/ACCESS.2024.3389035

Tehrany, M. S., Jones, S., Shabani, F., et al. (2019). A novel ensemble modeling approach for the spatial prediction of tropical forest fire susceptibility using LogitBoost machine learning classifier and multi-source geospatial data. *Theoretical and Applied Climatology, 137*, 637–653. https://doi.org/10.1007/s00704-018-2628-9

Thakkar, R., Abhyankar, V., Reddy, P. D., & Prakash, S. (2022). Environmental fire hazard detection and prediction using random forest algorithm. In *2022 IEEE ICONAT, Goa, India* (pp. 1–4). IEEE. https://doi.org/10.1109/ICONAT53423.2022.9726029

Vikram, R., Das, A. K., Sinha, D., & De, D. (2020). Energy-efficient data forwarding for forest fire detection utilizing localization algorithm via wireless sensor network. *Wireless Networks*. https://doi.org/10.1007/s11276-020-02393-1

Vikram, R., & Sinha, D. (2023a). FogFire: Fog-assisted IoT enabled forest fire management. *Evolutionary Intelligence, 16*, 329–350. https://doi.org/10.1007/s12065-021-00666-y

Vikram, R., & Sinha, D. (2023b). A multimodal framework for forest fire detection and monitoring. *Multimedia Tools and Applications, 82*, 9819–9842. https://doi.org/10.1007/s11042-022-13043-3

Xie, Y., & Peng, M. (2019). Forest fire forecasting using ensemble learning approaches. *Neural Computing and Applications, 31*, 4541–4550. https://doi.org/10.1007/s00521-018-3515-0

IoT System for Real-Time Monitoring and Management of Artificial Water Resources in Elephant Habitats to Mitigate Human–Elephant Conflict

S. J. Sugumar, V. Ramesh Babu, Sivabalan, J. Kalyan Srinivas, A. Kishore, G. Mohan Raj, and Hemanth Gouda S. Patil

Abstract

Forest ecosystem management by elephant habitat preservation helps in maintaining biodiversity ecosystem services by using Assisted migration by providing man-made water resources that helps the species to stay in the habitat without migrating. Elephants require about 68.4–98.8 L (18 to 26 gal.) of water every day. During summer season, the natural water resources start to dry and gradually decrease in level, as a result the elephants start to find alternate water sources. Most of the farmers in the forest border areas use borewell to fill surface water tanks to feed their catties and watering their agriculture lands. This water sources attract the elephants and leads to HEC. To overcome such situations, artificial water resources are needed in the elephant habitat regions to avoid smaller tenure migration of elephants during search of water and also provides a better ecosystem to improve the population growth rate of the wild elephants. There are 370 big and small waterholes which are almost (< 100 sq. ft.) in Bandipur in Karnataka state that are built for elephants. At Bandipur, majority of waterholes dry up by February and March leading to severe water stress situation for elephants. These waterholes are refilled manually using tanker lorry and solar-powered borewells. Forest department is making efforts in maintaining the required water level in these ponds by regular inspections and maintenance that are carried out to ensure the functionality of waterholes and borewells. This work aims sustain water availability for elephants, preventing their migration to human settlements and reducing the likelihood of HEC by providing a 5G-based IoT system to monitor the water levels in the manmade artificial water resources that are interior inside forest and very difficult to monitor manually. The deployment of the proposed system ensures monitoring the availability of water at certain level through periodic updates and if water levels are consistently low proactive steps are taken via alerts to notify on the schedule to refill the water. This methodology is a real-time monitoring system which measures the water level through sensors and updates the data by IoT-based GUI-based cloud server with visualization of information about the current water level at each pond. The outcome of the work provides the trends in water level fluctuations between seasonal rainfall and the drought impact indicating how often and how much water is being consumed by the elephants. Identifying frequently visited elephants at the ponds, which can help map their movement and habitat preferences and migration pattern. The major benefit is to prevent elephants from venturing into human settlements in search of water to avoid HEC and analyses data, generating insights and predictions for easy understanding by forest officials and researchers. Early warnings can trigger preventive measures, such as supplying water or improving alternate sources for better water management strategies for elephant conservation. Insights can also help in creating artificial ponds in strategic locations.

S. J. Sugumar (✉) · J. Kalyan Srinivas · A. Kishore · G. Mohan Raj · H. G. S. Patil
Department of ECE, Cambridge Institute of Technology North Campus, Bengaluru, Karnataka, India
e-mail: drsugumar.ece.nc@cambridge.edu.in

J. Kalyan Srinivas
e-mail: kalyansrini9845@gmail.com

A. Kishore
e-mail: kishorestarone@gmail.com

G. Mohan Raj
e-mail: mohanrajg684@gmail.com

H. G. S. Patil
e-mail: hemanthgoudapatil9980@gmail.com

V. Ramesh Babu · Sivabalan
Department of CSE, Cambridge Institute of Technology North Campus, Bengaluru, Karnataka, India
e-mail: vrb.cse.nc@cambridge.edu.in

Sivabalan
e-mail: sivabalan.developer@gmail.com

M. Ashwin et al. (eds.), *Smart Forestry 5.0 for Powered AI Sustainable Environment and Climate*, Sustainable Artificial Intelligence-Powered Applications, https://doi.org/10.1007/978-3-031-99086-1_10

Keywords

HEC · Waterholes · Migration · Water level monitoring · Elephants

1 Introduction

Human–elephant conflict (HEC) is a reason for the rising problem among human and elephant for living area, food, and water in the forest border areas. This HEC incidents within Western Ghats of Karnataka and Tamil Nadu, India are considered in this work. Search of food and water due to habitat loss and deforestation are the major life threat of the elephants living in this zone.

1.1 Overview of Human–Elephant Conflict

Elephants and humans are frequently contending for space; human elephant conflict is on the increase and it is a clash that the elephants are decreasing. As elephant habitat reduces, the elephants are pulled out of their living areas and entering human living areas. The shortage of food and water has a damaging impact on rates of reproduction so the number elephants found to decrease. The result of shortage of foods in the forest leading to rise of crop grazing in the human living areas (Lenin & Sukumar, 2011). The shortage of food and water is due to reasons such as forest fragmentation, deforestation, human encroachment and development in forest areas, climate changes, and longer drought seasons. The decrease of forest areas leads to shortage of space and lack of water that makes the elephant to look for food and water in human development areas. This increases the human–elephant conflict as they both come close together in use of space and resources. But, it is the elephants losing its areas, and its life is the focal point of this work which directly and indirectly impacts the forest development. Compared to other animals, elephants are the one of the popular and decreasing wildlife species in the country. Each year, nearly 400 humans and 100 elephants are affected by the conflict related instances in India, and nearly 500,000 farmers are affected by crop damage due to elephant intrusion into the their cultivated land (Kumara et al., 2012). Several reasons including elephant habitat area fragmentation, poor management of physical barriers, and other causes have been cited for the elephant intrusion in to human areas.

Humans when the elephants enter in to their settlements or in their agriculture lands in search of water leads to higher risk of humans being killed by elephants. They also tend to raid the elephants back to forests by hurting with stones and rods for which the elephants react severely and attacks the human in back. Elephants that wait in the dawn near villages to drink water in the tanks which are set up on the upper ground when obstructed by humans the elephants react badly and conflict gets to a higher level and both the elephant and human are affected. The farmer fills the tanks for feeding their catties and watering their vegetation when they see the elephants fully drink, they get angry on the elephants and they react badly. When the elephants go out of forest and look for water sources, they sometime get hit by train which takes place when elephants attempt to cross the railway pathway along the segment of the forest area; this has cause loss of many elephants. The sick and elderly elephants when they could not travel to reach water sources they sometime die or get sickness even more without water as shown in Fig. 1.

Elephants are known to raid over 12 types of crops cultivated in more than 50 settlements along the forest peripheral areas, with the highest incidence of crop raiding occurring during the fruiting season of cereal crops, particularly in December and January. It has also been observed that groups of elephant herds often venture as far as 10 km into human settlements from the forest peripheries in pursuit of water. Physical breakage to water pipes, water tanks, and grain storage facilities, as well as livestock deaths and reduced availability of farm labor due to lower attendance, are significant challenges. Additionally, the increased need for nighttime guarding efforts negatively impacts daytime productivity. Given that elephants possess complex social structures and exhibit high intelligence with advanced cognitive abilities, any management strategies must be carefully designed with strong ethical and welfare considerations in mind.

1.2 Role of Artificial Water Resources in HEC Mitigation

The Western Ghats at the south India region which consists of Tamil Nadu, Karnataka, and Kerala region is a biodiversity zone for the country and the source of major rivers for these states. The forest conservation of this region is important for the biodiversity and ecological operations as it holds about one-fourth of the world-wide elephant population. The Western Ghats (Baskaran, 2013) divided into: (1) Uttara Kannada, (2) Brahmagiri-Nilgiris, (3) Anamalai-Nelliyampathy-High Range, and (4) Periyar-Agasthyamalai, covering a landscape of 30,000 km^2 as shown in Fig. 2, and Table 1 provides information about the split of landscape within the region of Western Ghats providing living space of elephants which move across the Western Ghats and migrate from one forest zone to another in search of food and water. The change in the seasonal rainfall results in increased droughts inside forest where the natural water sources are getting dried that impacts the growth of plants and trees that results in reduced forest density and lack of food and water which is crucial for elephant sustenance. Reduced natural water availability in the elephant habitats during dry

Fig. 1 Elephant death reasons

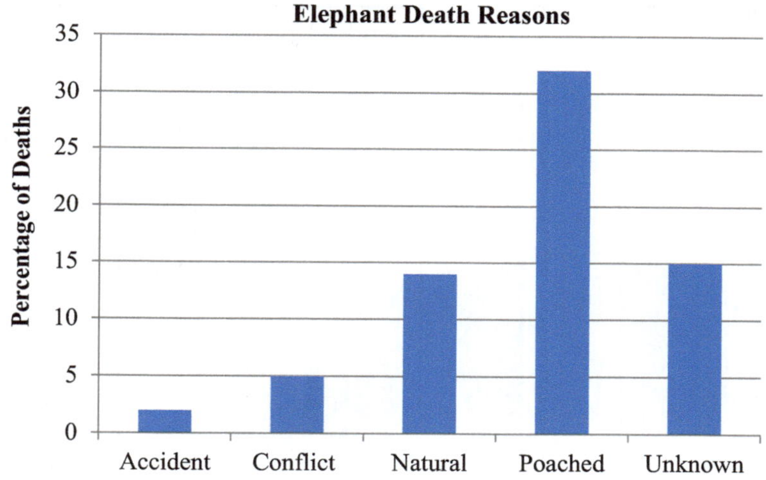

seasons, forcing them to move to areas with water sources which are man-made to source their agricultural and farming land when these elephants trying to drink the man-made water leads to conflict with humans. Increasing temperatures can stress elephants, especially during hot summer months, as they struggle to regulate their body temperature.

As elephants are enforced to move towards human living areas in pursuit of food and water, conflicts with local communities can escalate, leading to crop damage and potential elephant casualties. Reduced access to food and water, combined with increased stress factors, could lead to a decline in elephant populations in the Western Ghats. Elephants may alter their migratory routes to find suitable habitats with adequate resources. Fragmentation of elephant populations can lead to decreased inherited range, making them more susceptible to disease and environmental challenges. Strengthening existing protected areas, establishing wildlife corridors, and implementing effective management strategies to reduce migration of elephants. Supporting initiatives to decrease conservatory gas discharges to mitigate the severity of climate change impacts. Due to rise in the temperature, the natural sources of water get dried up in forests which leads to a hard lifetime to sustain. To keep water access easier for the elephants in forest region, the forest department builds artificial water holes in identified location where shortage of water in the natural sources. These artificial water resources are frequently filled by the forest officials to keep the water at a higher level for the elephants to drink. These artificial water resources (AWR) are made in potential points where multiple herds of elephant can come to drink water so that they do not want to roam around to look for waters around the agricultural fields or human living areas.

As a conclusion, we can say these AWR meets a basic requirement of need of water for elephants in the summer dry seasons so that the stress of search of water is reduced and avoids HEC significantly.

1.3 Artificial Water Resource Monitoring and Management

Forest ecosystem management by elephant habitat preservation helps in maintain biodiversity ecosystem services by using Assisted migration by providing man-made water resources that helps the species to stay in the habitat without migrating. Elephants require about 69 to 99 L (20 to 30 gal.) of water every day. During summer season the natural water resources start to dry and gradually decreases in level, as a result the elephants start to find alternate water sources. Most of the farmers in the forest border areas use borewell to fill surface water tanks to feed their catties and watering their agriculture lands (Gowda et al., 2024). This water sources attract the elephants and lead to HEC. To overcome such situations, artificial water resources are needed in the elephant habitat regions to avoid smaller tenure migration of elephants during search of water and also provides a better ecosystem to improve the population growth rate of the wild elephants.

Artificial waterholes are the man-made water resources that support elephants and other wildlife to survive throughout scarcities. The artificial water resources can be important in the regions where longer drought session and lack of number of natural water resources in the closer regions (Mpalo, 2024). The Artificial waterholes that are created in forests regions should provide a reliable water availability so as to full fill the source water for elephants and other wildlife. The artificial waterholes are designed favors elephants in such way they can drink water with the trunks, where it will be difficult for other small sized animals. These kind of water holes can be exclusively installed for the access of elephants where other waterholes can be designed suited for smaller animals. These waterholes differed in size and shape based on the land terrains, and it can be surrounded with mud shallows, which are important for thermoregulation and parasite control for elephants. Four types of artificial waterhole are presented in

Fig. 2 Western Ghats map

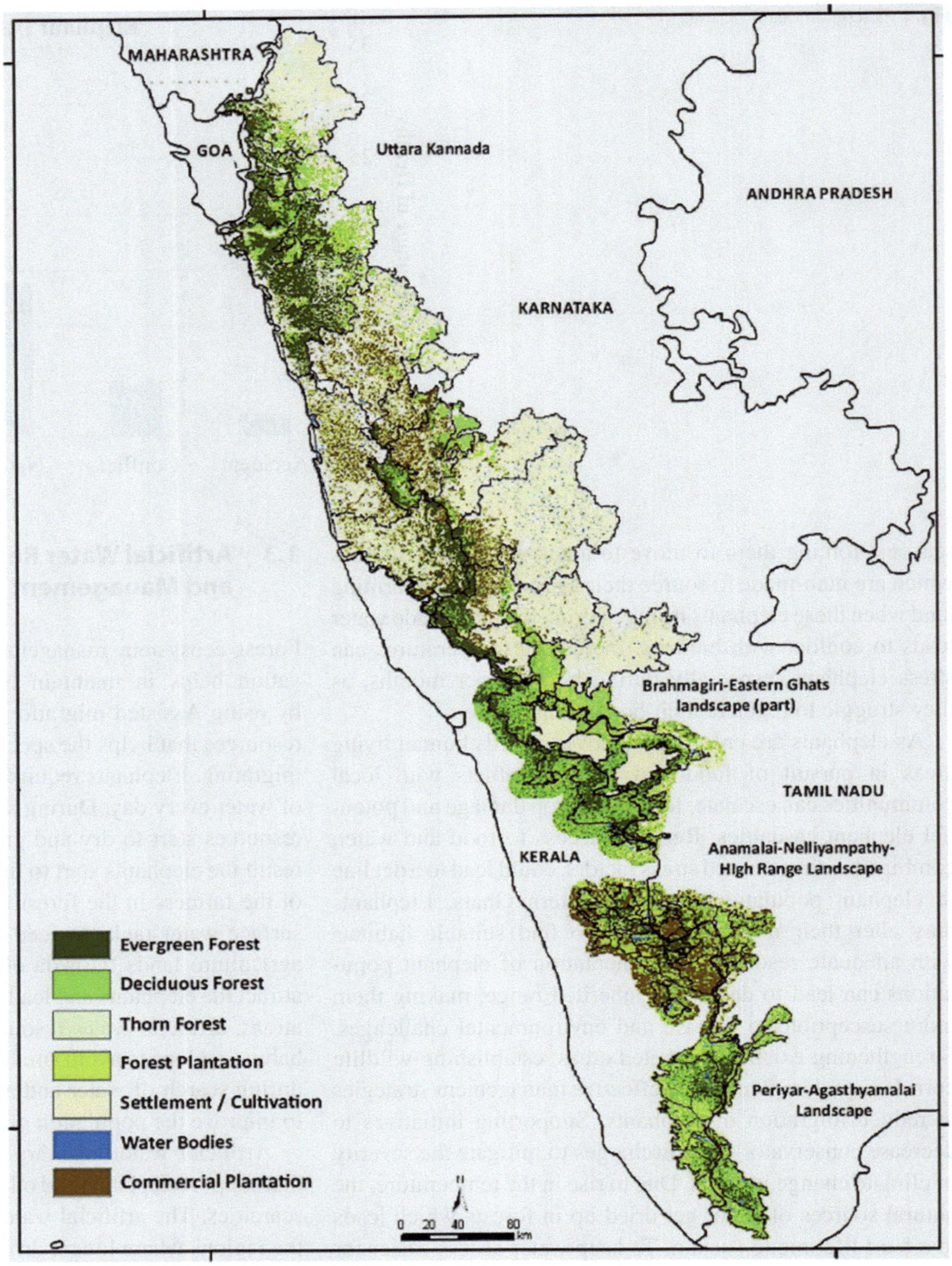

Table 1 Elephant landscape and habitat

S no.	Western Ghats	Elephant landscape	Elephant habitat (km^2)
1	Northern Karnataka	(i) Uttara Kannada to Maharashtra and Goa	5081
2	Brahmagiri and the Nilgiri area Eastern Ghats	(ii) Malnad to Crestline, Mysore to Nilgiris and Wynad to Nilambur plateaus	3058
3	Anamalai to Nelliyampathy high range	(iii) Anamalai to Nelliyampathy and Palani hills (iv) Idukki to Kothamangalam	(i) 5352 (ii) 296
4	Periyar to Agasthyamalai	(v) Periyar Plateau to Varushanad to Mehamalai hills (vi) Agasthyamalai to Mahendragiri hills	(iii) 3860 (iv) 1758
	Total elephant migration area in Western Ghats		29,400

this work are the earth dams, concrete pans, reservoirs and troughs. These artificially made waterhole are constructed based on some nearby water sources which flow or stored artificially. Earth dams are not constructed with a concrete base or sides, and these were formed along natural sources. Reservoirs were built with the concrete base with high sides walls and these were constructed close to boreholes nearer to dense woody vegetation. Concrete pans had a concrete base with low sides and were in relatively open areas dominated by grass. Troughs were often associated with reservoirs, had a concrete base, and were usually placed some distance away from reservoirs in open areas dominated by grass. All waterhole types provided water perennially and were chosen based on their spatial orientation and ease of access from the road network in the reserve. There are 370 larger and smaller waterholes which are almost (< 100 sq. ft.) in Bandipur in Karnataka state that are built for elephants. At Bandipur, most waterholes dry up by February and March, resulting in severe water stress for the elephants. These waterholes are refilled manually using tanker lorry and solar-powered borewells. Forest department are making efforts in maintaining the required water level in these ponds by regular inspections and maintenance that are carried out to ensure the functionality of waterholes and borewells.

Availability of water resources in the forest areas to meet the demands of water usage within a region is a big challenge in the summer season. If these water sources are not managed effectively, it causes a severe drought and the animals in specific elephants will look for alternate water sources and will migrate to other forests or enter in tom human habitat in search of water. Change of climatic conditions are inevitable hence looking for artificial water sources to feed the elephants are utmost importance in securing the elephant habitats (Uloko et al., 2023). The elephants are one of the important animals in protecting the forest ecosystem and being a national heritage. These AWR are to be filled with water on regular basis so as to maintain a required water level to feed the elephants. As the elephant's drinks 100–150 waters a day the requirement is heavier. The refilling of water sources can be done with borewell or pipeline of by filling with tanker lorries when the level of water falls down a required limit.

The monitoring of these artificial water ponds is very essential because when elephants reach out the ponds to drink water, and if it finds empty and not enough water to drink, then it might look for other sources around forest border areas when human built water bodies for agriculture and other purposes. It leads to severe HEC effects. So, it is important to keep the water level at required level of drinking for elephants by filling the AWR ponds whenever it is getting empty. A real-time monitoring system might be helpful in achieving constant water level at AWR ponds. A central monitoring station can be kept to monitor the AWR ponds in a particular forest zone in the forest office. Water level sensors can be kept in the ponds to detect the level and alert can be given for a particular pond for refilling the water by any of the sources. The water level data can feed to a IoT-based module to upload the water level in the cloud. A front end is developed to monitor the levels and once the water level comes down a specific level indication can be given for refilling schedule.

The Internet of Things as a technology has a great potential to solve remote monitoring of artificial water ponds with instant and predictable water level indication. One of the crucial challenges of AWR water management is to determine the amount of water that any particular pond is going to be utilized by the elephants during the next day. This can be determined to accurate with the use of **predictive analytics** using IoT. This is done by monitoring on the history of water consumption in the pond on any given period of time or last three weeks. Based on the previous days of utilization and history of water consumption by herds of elephant are analyzed by predictive analytics and combined with the consideration of climate and weather in that forest region, the predictive tool can determine the amount of water that the elephant population is going to consume in one day. If the requirement of amount of water that is needed for elephants to drink in a pond on any given day is determined, it is convenient for the forest authorities to maintain the level of water in a water reservoir/ tank/pond as required throughout the day.

2 Artificial Water Resources in Elephant Habitats

Artificial water resources, such as man-made ponds, check dams, and solar-powered borewells, have become essential by providing water to elephant in dry seasons so they don't move from their own habitat. This chapter deals the role of the artificial water sources in maintaining ecological balance, reducing human–elephant conflicts, and supporting biodiversity.

2.1 Importance of Water for Elephant Behavior and Movement

Water is critical for elephants' survival and migration, and they use water for three main purposes, viz. (a) drinking, (b) thermoregulation, and (c) social activities. As the size of the elephant is large, these mammals have high water requirements compared to other animals in the forest, their daily routine activities and exodus patterns from one forest to forest are often characterized by the water availability. The water sources determine their habitat range, with herds staying proximity to reliable water sources. Staying within a 10–15 km radius of known rivers and watering holes. The elephants are capable of sensing the aroma of the water as the elephants has about 2000 olfactory receptors in their trunks, while they can

Fig. 3 Elephant migration in the forest

sense a water source up to a distance of 20 km. The elephant nostrils are available at the edge of the trunk, and they utilize this for smelling and breathing. Elephants require huge quantity of water, where as an adult is able to drink about 150–200 L per day. The elephants travel in group of herds to graze and drink water. A herd has number about 6–10 elephants sometime higher value (Uloko et al., 2023). The number of an elephant herd is based upon the family size and the terrain.

In the dry season, the movements of the elephants mostly based upon the water availability. In search of water the herds may move in short distance, or longer distance depends upon temporary or permanent water sources Fig. 3. When they approach the water sources, they move with high-speed bursts to get to water. Elephants uses water sources such as including rivers, reservoirs, lakes, ponds and artificial waterholes. They spend a lot time in the water for drinking, swimming, making them cool etc. Thermoregulation is a process which elephants are used to keep them cool in summer season, for this it uses water-related activities like crawling, rolling, bathing and flushing the water around the body. As the elephants are a bigger animal and have thick skin, the evaporation is higher and heat stays for a longer time; this is the reason the elephants require huge quantity of water for its survival Fig. 4 (Courtesy India Today, Sept 14, 2022). The elephant can only able to survive for a maximum of three days without drinking water. Due to lack of water in the forest in the summer season the elephants are force to travel longer distances to reach water sources and unfortunately those water sources are utilized by human activities majorly agricultural purposes. This increases stress on the elephants and also on humans and leads to human elephant conflicts. In worst situations if water does not find for a longer time, the elephant may end up with giving its life in search of water. Most of the times the causalities are cub, aged, and gestation period elephants as they cannot travel longer and faster.

Fig. 4 Herd of elephants drinking water

2.2 Types of Artificial Water Resources

Artificial water resources for elephants in forests include dams, concrete pans, reservoirs, and troughs (Chakuya et al., 2023). These waterholes are often fed by boreholes using solar-powered pumps. Elephants use artificial waterholes when natural water sources are scarce, such as

(a) During droughts elephants may visit waterholes instead of rivers.
(b) Elephants may use AWR on their way from a feeding area to a river or reservoir.
(c) During migration of one forest to another.
(d) Elderly elephant could not often visit natural water sources.
(e) During sick and gestation period.

Understanding the preferences of elephants for different waterhole types can help with waterhole management. This can help reduce competition and promote the well-being of other species. Artificial waterholes can be placed strategically to prevent conflicts. However, some studies have shown that artificial waterholes can have negative consequences for elephants and their habitats. For example, one study found that elephants stayed near rivers often than artificial waterholes (Smith et al., 2024). Artificial water resources are essential for elephants in forests, especially during dry seasons or in the areas where natural water sources are scarce. Here are some common types:

(i) Man-Made Waterholes

Man-made waterholes are artificial water sources created to provide a reliable supply of water for elephants and other wildlife, particularly in dry regions or during droughts (Smith et al., 2024). These waterholes are designed to mimic natural ponds and can be constructed by digging depressions in the ground and filling them with water from borewells, rainwater harvesting systems, or diverted streams Fig. 5. They are built with clay or concrete to prevent water seepage and ensure long-

Fig. 5 Man-made waterholes

Fig. 6 Check dams and small reservoirs

term availability. They are strategically placed near elephant corridors or within protected forest areas to reduce the need for elephants to wander into human settlements in search of water, thus minimizing human–wildlife conflict.

(ii) Check Dams and Small Reservoirs

Check dams and small reservoirs are essential artificial water resources for elephants in forests, particularly during dry seasons. Check dams are low-height barriers built across seasonal streams or rivers to slow down water flow, allowing water to be stored for extended periods to ensure a continuous water supply for elephants. Constructed using materials like concrete, stones, or earthen embankments, they play a crucial role in forest conservation (Quiñonero-Rubio et al., 2016). Similarly, small reservoirs are artificial water storage structures created by building embankments in valleys or depressions, holding large amounts of water for larger use. These reservoirs not only provide a steady water source for elephants and other wildlife but also support vegetation and reduce human–elephant conflicts by preventing animals from wandering into human living areas in pursuit of water as shown in Fig. 6. Properly designed check dams and reservoirs with gentle slopes and protective vegetation can significantly enhance water availability and sustainability in forest ecosystems (Straškraba, 2004).

(iii) Borewells with Solar Pumps

Borewells with solar pumps are a sustainable solution for providing water to elephants in forests, especially in areas where natural water sources are scarce or dry up seasonally. These borewells are drilled deep into the ground to access underground water, and solar-powered pumps are used to lift the water to the surface Fig. 7 (Courtesy Star of Mysore, Feb 28, 2024) (Courtesy The Indian Express, March 15, 2025). This method ensures a continuous and eco-friendly water supply without relying on electricity or fossil fuels (Straškraba, 2004). The pumped water is directed into artificial waterholes, which are designed to mimic natural ponds, allowing elephants and other wildlife to drink safely. These waterholes are strategically placed along elephant corri-

Fig. 7 Borewells with solar pumps

Fig. 8 Concrete or earthen tanks

Fig. 9 Water tankers to fill waterholes

dors and near feeding areas to reduce stress and prevent human–wildlife conflicts. By utilizing renewable energy and providing a consistent water source, borewells with solar pumps play a crucial role in wildlife conservation and ecosystem sustainability.

(iv) Concrete or Earthen Tanks

Concrete or earthen tanks serve as essential artificial water sources for elephants in forests, particularly in areas where natural water bodies are scarce or dry up during summer. These waterholes are designed to store and supply water throughout the year. Concrete tanks are durable structures built with cement and reinforced materials, ensuring long-term water availability Fig. 8 (Courtesy The Indian Express, March 23, 2016). On the other hand, earthen tanks are made by excavating and compacting soil, allowing water to seep naturally into the ground, aiding in groundwater recharge (Purdon & Aarde, 2017). Both types of tanks collect rainwater or are refilled using borewells, pipelines, or check dams. These earthen tanks are filled with tanker lorries by the forest officials on even interval to make sure water availability in the waterhole as shown in Fig. 9 (Courtesy The Indian Express, March 15, 2025).

2.3 Challenges in Artificial Water Resource Management

Managing water resources in artificial forests presents several challenges, including availability water, monitoring and data collection of water quality, unpredictable weather patterns and soil dynamics (Sutherland et al., 2018),

Key challenges

- Monitoring water availability within a forest can be difficult, requiring specialized sensors and techniques to accurately measure water level and quality of water.
- Higher weather changes like scarcities or higher rainfall can considerably impact water accessibility in artificial water resources in forests, making it challenging to design a sustainable water management strategy.
- The artificial water resource ponds are prone to dry as the requirement of water needs are high as the herds of elephants and other animals use the water to drink. The question of how refilling the water sources is always a challenge if there is no rainfall in the recent days.
- Providing electrical sources for pump and bore well are challenging. Fixing and solar is another challenge.
- Refilling with a tanker lorry at a given schedule or based on the level variation can be a challenge, etc., and the transportation is another risk to reach the animal living areas to fill water.

Potential solutions

- In a forest ecosystem, various wildlife may compete for water, which requires careful consideration when designing a water management system.
- Employing sensors, IoT-based remote sensing, and data analytics to gather real-time information about water status within the AWR at forest.

3 System Architecture Internet of Things-Based AWR

The integration of Internet of Things (IoT) technology with artificial water resources enables water management by real-time monitoring by predicting the water requirement and the probability of water sources to refill. This chapter deals the system architecture of an IoT-based artificial water resource management system with its key components and communication protocols.

3.1 IoT-Enabled Artificial Water Resources Management

The IoT devices and technology are used to monitor, analyze, and optimize water usage, quality monitoring and data communication. It enables real-time data collection and smart water refilling scheduling to enhance overall water availability in the AWR efficiency. The Internet of Things (IoT) is a network comprising physical devices, vehicles, appliances, and other objects embedded with sensors, software, and connectivity, enabling them to collect and exchange data. This technology allows smart devices to communicate with one another as well as with other Internet-connected systems (Ismail et al., 2022). Similar to smartphones and gateways, these create an extensive network of interconnected devices capable of exchanging data and performing various tasks autonomously. This real-time data helps utilities improve utilization of water, reduce waste, and ensure a reliable water supply. An IoT-based smart artificial water source management system utilizes sensors and IoT technology like Wi-Fi and Bluetooth to monitor and manage the water in the ponds, check-up dams, reservoir and other artificial water sources. It provides remote monitoring, predictive maintenance, and data-driven insights to improve the efficiency of water management in these AWR (Bouwer, 2000). It leads to enhanced sustainability and cost-effectiveness in water management for these artificial water resources in forest areas to feed animals during the drought seasons and other water requirements.

3.2 Key Components of IoT for AWR Level Monitoring

The Internet of Things (IoT) refers to a network of physical objects, or "things," that are embedded with electronics, software, sensors, and connectivity, allowing them to collect and share data. IoT enables remote sensing and control of objects through existing network infrastructure, facilitating closer integration between the physical environment and computer-based systems, which leads to enhanced efficiency, accuracy, and economic advantages. In the context of IoT, "things" can encompass a wide range of devices, including heart-monitoring implants, biochip transponders on livestock, electric clams in coastal waters, and sensor-equipped automobiles as shown in Fig. 10. These devices gather valuable data using a range of existing technologies and then autonomously transmit the data between other devices. (Keramidas et al., 2016).

Remote artificial water resource level monitoring helps to provide an early predictions of requirement of water, providing valuable time to save elephant lives, avoid elephant migration into human living areas in search of water and to avoid human elephant conflicts in turn maintain a better eco system of forest with the aid of the above mentioned advantages of IoT (Brooks et al., 2018). Internet of Things along with sensor technology and cloud systems provide a versatile monitoring system of AWR in real time and also uses the AI/ML model for predictions. The major component in the IoT-based AWR is the Rain Water Sensing Unit which can sense whether a conductive liquid like water has affected the PCB. The unit permits to determine moisture via analog output pins, and it generates a digital output when a threshold of moisture is above a level. The module has a PCB to sense

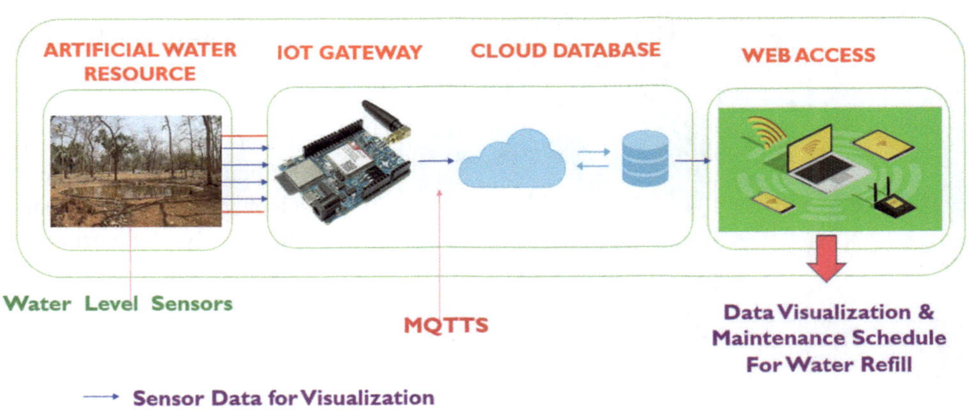

Fig. 10 Functional block diagram of IoT-based AWR monitoring system

the water flow, as the water flows on the PCB the parallel resistance measured via the op amp LM393 is taken in to account. The lower the resistance means water flow on the PCB, the lower the voltage output. Conversely, when no water touches the PCB, the greater the output voltage on the analog pin. The sensor output voltage levels that help to determine if its water level is on a specific range. When water is detected, the comparator sends a signal to the Bharat Pi board, enabling it to trigger actions or receive alerts accordingly. Two water level sensors are used: one at 30% to indicate a critical level requiring a refill and another at 80% to signal a full condition. No specific action is needed for water levels between these two sensor thresholds.

3.3 Bharat Pi 4G/LTE Wi-Fi Module

The Bharat Pi 4G/LTE Wi-Fi Module is a versatile IoT prototyping platform designed to function both as an IoT device and as a gateway. It integrates multiple communication interfaces and processing capabilities, making it suitable for a wide range of applications. Equipped with the ESP32 microcontroller, the Bharat Pi module offers robust processing power and supports various wireless communication protocols: Wi-Fi and Bluetooth Connectivity Fig. 11: Facilitates seamless integration with local networks and peripheral devices. 4G/LTE Cellular Module (SimCom A7672S Chip): Enables reliable internet connectivity in remote locations without Wi-Fi access. These features allow the module to collect data from sensors, process it locally, and transmit it to cloud services or other systems, making it an effective standalone IoT device. When the Bharat Pi LTE Wi-Fi kit runs the code, it collects sensor data, makes AI-based predictions, and publishes the results via MQTT. Below is an example of the output format sent to the MQTT broker and displayed on the console (https://bharatpi.net/product/bharat-pi-4g-lte-module-with-esp32-simcom).

Fig. 11 Bhatth Pi Wif Kit rain water detecting sensor

Table 2 Power budget example

Component	Power consumption	Daily consumption estimate (Wh)
Water level sensor	0.5 W continuous	12 Wh (24 h operation)
Bharat Pi LTE IoT module	3 W active, 0.2 W sleep	36 Wh (assuming 8 h active/day)
Controller (MCU)	0.5 W	12 Wh
Total estimated usage	–	~ 60 Wh/day

There are two major challenges in deployment of 5G-based IoT systems for real-time water level monitoring in elephant habitats Limited 5G Network Coverage in Remote Areas. Forested and rural regions often lack extensive 5G infrastructure. The IoT modules which are used my fail due to lack of internet connections. Therefore, we can go for a 5G with LTE 4G hybrid modules such as Bharath Pi LTE WiFi Kit. Providing a standalone forest controlled 5G Network can also be considered. Power supply issues for the IoT devices in the forest. The interior forest regions do not have electricity connection and getting electricity to the AWR are challenging and cost involved process though we suggest in most cases to go for a renewable energy sources such as solar panel modules to power up the IoT module and the associated sensor modules to collect environmental data. Methods—Solar + Battery Hybrid System: Solar Panels: Solar harvesting is the primary power source. Panels rated at 50W to 100W are generally deployed per IoT unit, depending on load and sunlight availability. Lithium Iron Phosphate ($LiFePO_4$) batteries are preferred due to long cycle life (2000 + cycles) and temperature resilience. A typical setup uses 12 V 20Ah to 40Ah batteries, providing backup for 3–5 days of operation without sunlight as shown in Table 2.

Thus, a 100 W solar panel with 20 Ah 12 V battery ensures safe daily operations for 2–3 rainy days.

4 AI and Machine Learning for Water Level Prediction

Water level prediction plays a crucial role in managing artificial water resources in forests by providing water to elephants in dry seasons thereby protecting the elephant habitat. By implementing the artificial intelligence (AI) and machine learning (ML), predictive models that can analyze the datasets, identify rainfall patterns, and provide a forecast for the requirement of water availability. This chapter describes the application of AI and ML in prediction of water levels and refilling the AWR.

4.1 Importance of Predictive Analytics in Water Management

Predictive analytics is the use of data, statistical models, and machine learning to forecast future events. It analyzes historical and current data patterns to predict what might happen in the future. Uses statistical techniques like data mining, machine learning, and predictive modelling (Krishnan et al., 2022).

Types of predictive analytics

- **Decision trees**: A visual chart that uses data mining algorithms to predict the risks and rewards of different actions
- **Neural networks**: A method that uses complex algorithms to identify relationships in data
- **Regression analysis**: A statistical technique that identifies relationships between variables and predicts future outcomes

Predictive analytics plays a crucial role in water management by allowing water utilities to anticipate future water demand, potential infrastructure failures and enabling proactive measures to optimize water distribution, minimize disruptions, and ensure efficient resource allocation, ultimately leading to improved water security and cost savings. By analyzing data from sensors installed at the AWR, the predictive models can identify potential requirements of water usage by the elephants present at their habitat. Predictive analytics can forecast water requirement by analyzing historical data on the water availability at the AWR ponds and other parameters like rainfall, temperature changes, and climatic change allowing for planned maintenance to prevent unavailability of water. By analyzing historical consumption patterns of elephants and weather data, predictive models can accurately predict future water requirement, enabling better resource allocation and preventing shortages. Predictive analytics can be used to analyze various drought scenarios, helping the forest department water managers to develop proactive strategies for managing water supplies during dry periods. A Predictive analytics model is developed for the artificial water resources within a forest which utilizes machine learning algorithms to analyze data from water level sensors and environmental factors like climate, rainfall, and timestamp allowing for predictions about future water availability, potential drought risks, and optimal management strategies for artificial water sources like ponds or reservoirs within a forested area, enabling proactive decision-making for forest health and ecosystem sustainability (Ayus et al., 2023).

There are four constraints considered in the decision of when the refilling of the AWR should be scheduled. Timestamp, temperature, rainfall, and water level are the constraints which are considered as decision making for the refilling scheduling prediction model. Timestamp is a parameter which are obtained from the on board clock that are used to update the water level and climate data on timely basis. The measured water level is updated in the cloud for every two hours to find whether the water level has reached the lowest level or enough water in the AWR. The temperature around the AWR zone is obtained via a temperature sensor which provides data on the environmental climate, data analytics is performed upon the temperature data obtained in finding out the requirement of water for the AWR. A water requirement prediction is carried out with the aid of the temperature drifts for the period of time. We can also predict the refill requirement schedule for the AWR by determining the predictions of rainfall from the cloud data (Rahbaralam et al., 2007).

An assessment is carried out on the current water level in the AWR and a water availability for the upcoming days for elephants to drink is determined with past water level data and a prediction is provided on the water refilling schedule. There may be water overflow on heavy rainfall or over fill of water is determined and the water is routed to the drainage system to avoid overfilling. This prediction method is a proactive approach that can maintain a specific water level all the season through for the elephants to drink so as to avoid elephants to enter in to the human living areas. The prediction model is explained in the Fig. 12

4.2 AI-Driven Decision Making for Refill Scheduling of AWR

"AI-driven decision making" refers to the process of utilizing artificial intelligence (AI) algorithms and technologies to analyze large datasets and generate insights, enabling more informed and data-driven decisions, often surpassing the capabilities of human decision-making alone, especially in complex scenarios where vast amounts of information need to be processed quickly and accurately; it essentially involves using AI as a tool to assist in making better choices by leveraging its pattern recognition and predictive capabilities. AI systems can quickly analyze large volumes of data from various sources, identifying patterns and trends that might be missed by humans. Various machine learning models like decision trees, neural networks, and regression analysis are employed to make predictions and classifications based on data patterns. To predict water levels using historical data, rainfall, and temperature trends, an LSTM model would generally be the preferred choice due to its ability to effectively capture complex temporal relationships within time series data, making it well-suited for handling the dynamic nature of water levels influenced by changing weather patterns. A recurrent neural network (RNN) architecture is widely used in Deep Learning that can learn and predict

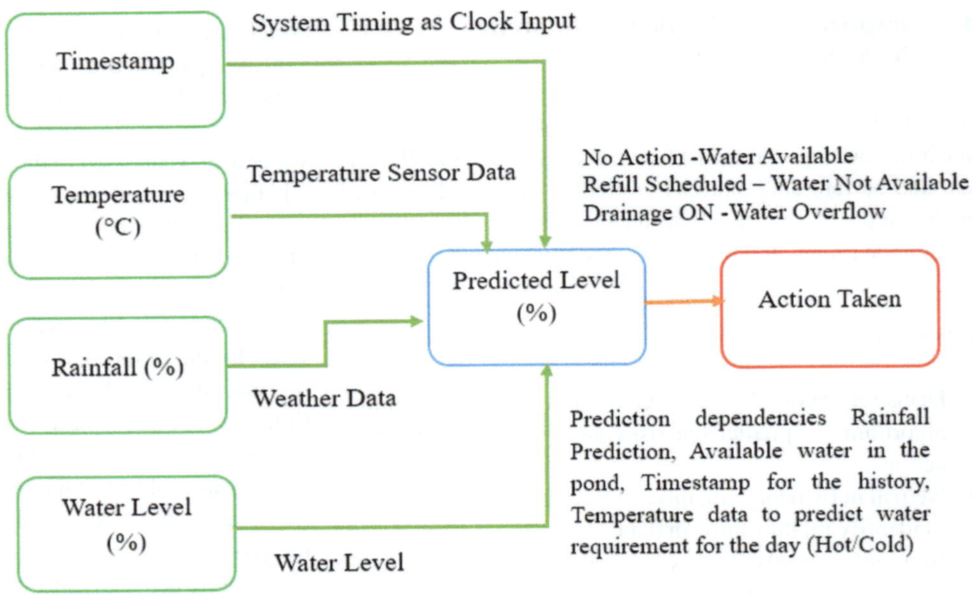

Fig. 12 Proposed prediction model for detecting water level in the AWR

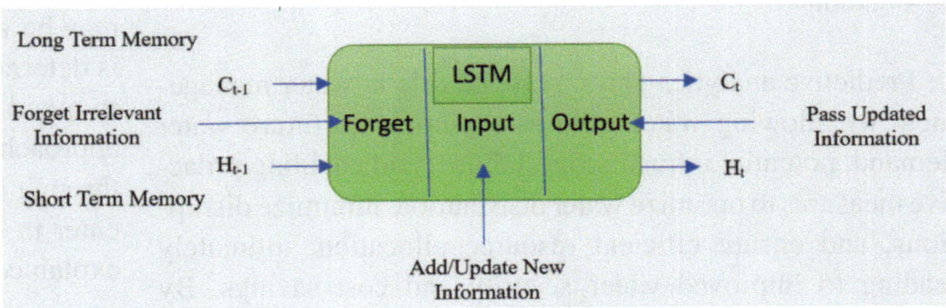

Fig. 13 Long short-term memory (LSTM)

sequences of data (Sherstinsky, 2020). It is used in many applications, including speech recognition, language translation, and handwriting recognition. The Data Collect historical water level data, corresponding rainfall data, and temperature readings for the desired location, ensuring consistent time intervals (e.g., daily, hourly) as shown in Fig. 13.

Preprocessing

Normalization: Scale all features (water level, rainfall, temperature) to a similar range for optimal training. **Time series structuring**: Convert data into sequences where each sequence represents a window of past data points used to predict the next water level.

Model Architecture

Input Layer: Feed the time series data including historical water levels, rainfall, and temperature as input features. **LSTM Layer**: Use one or multiple LSTM layers to learn the long-term dependencies within the time series data. **Output Layer**: A single output neuron to predict the future water level.

Similar to a simple RNN, an LSTM also maintains a hidden state, where $Ht - 1H\_\{t -1\}Ht{-}1$ represents the hidden state from the previous time step and HtH_tHt represents the hidden state at the current time step. Additionally, LSTM includes a cell state, denoted as $C(t - 1)$ for the previous timestamp and $C(t)$ for the current one. In this context, the hidden state represents short-term memory, while the cell state represents long-term memory. The Training Process includes Split data in which to divide the dataset into training, authentication, and testing sets. Next define the loss function (e.g., mean squared error) and optimizer (e.g., Adam) for training. Then train the LSTM model by iterating through the training data, adjusting the model parameters to minimize the loss function. A Hyperparameter tuning is done withexperiment with different LSTM layer configurations, sequence lengths, learning rates, and batch sizes to optimize model performance (Sunny et al., 2020). Using the method of AI/ML model (LSTM) to predict water levels based on historical data, rainfall, and temperature trends. Use the trained model to forecast water level changes for the next week of time or a day. Store predictions in the cloud database for analysis. The automated water management requires actions for filling water ponds. The actions are defined based upon the how the water level is maintained.

If predicted water level < 30%, trigger "Refill Scheduled".
If predicted water level > 80%, activate "Drainage ON".

Table 3 Water level predictions in the AWR and corresponding action for the period Jan 2025

Timestamp	Temperature (°C)	Rainfall (%)	Water level (%)	Predicted level (%)	Action taken
2025-01-01 06:00	28.4	5.2	40.1	38.5	No action needed
2025-01-03 12:00	32.7	0.0	29.5	27.8	Refill scheduled
2025-01-06 18:00	26.3	15.5	65.2	68.3	No action needed
2025-01-09 00:00	30.1	3.7	22.4	25.0	Refill scheduled
2025-01-12 06:00	27.8	19.2	78.3	80.1	Drainage ON
2025-01-15 12:00	31.2	2.5	45.6	42.8	No action needed
2025-01-18 18:00	29.7	6.8	37.4	36.0	No action needed
2025-01-21 00:00	25.9	12.0	55.2	58.1	No action needed
2025-01-24 06:00	34.5	1.1	30.9	28.6	Refill scheduled
2025-01-27 12:00	28.1	10.7	72.8	74.0	No action needed

Table 4 Water level predictions in the AWR and corresponding action for the period Feb 2025

Timestamp	Temperature (°C)	Rainfall (%)	Water level (%)	Predicted level (%)	Action taken
2025-01-01 06:00	28.4	5.2	40.1	38.5	No action needed
2025-01-03 12:00	32.7	0.0	29.5	27.8	Refill scheduled
2025-01-06 18:00	26.3	15.5	65.2	68.3	No action needed
2025-01-09 00:00	30.1	3.7	22.4	25.0	Refill scheduled
2025-01-12 06:00	27.8	19.2	78.3	80.1	Drainage ON
2025-01-15 12:00	31.2	2.5	45.6	42.8	No action needed
2025-01-18 18:00	29.7	6.8	37.4	36.0	No action needed
2025-01-21 00:00	25.9	12.0	55.2	58.1	No action needed
2025-01-24 06:00	34.5	1.1	30.9	28.6	Refill scheduled
2025-01-27 12:00	28.1	10.7	72.8	74.0	No action needed

If predicted water level < 80% and greater than > 30%, No action taken.

For the first condition—If predicted water level < 30%, trigger "Refill Scheduled", an alert message is sent to the water authorities of forest department to indicate them to refill the pond by turning on pump motor or send a water tanker lorry to refill the water pond.

For the second condition, if predicted water level > 80%, activate "Drainage ON", if the pond is filled with higher level and may overflow and cause water spill around the pond, it may harm animals to stand and drink water as the cubs might skid with muddy water on the edges. To avoid the situations, we keep a drainage system when water overflow a certain level the excessive water flow via a pipe to nearby vegetation areas.

In the third case, if the predicted water level < 80% and greater than > 30%, no action taken and to just monitor the water level considering temperature and rainfall (Thaker et al., 2019) as in Tables 3 and 4.

Tables 3 and 4 are a AI-based LSTM model report prediction to analyzes the water level management data for artificial water resources in Coimbatore Forest from January 1, 2025, to February 15, 2025. A total of 186 data points were collected and processed using an AI-based predictive model. The key factors analyzed include temperature, rainfall, actual and predicted water levels, and actions taken to maintain water balance. A dashboard is developed to monitor, where all these data are updated on the webapp created to check the water level and action to be taken on the AWR ponds. An automated GSM message also sent to the officials via Bharath Pi LTE Kit.

4.3 Impact of AI-Based Predictions on Elephant Water Consumption Patterns

Understanding the water consumption patterns of elephants in the **Coimbatore and Bandipur forest divisions** is crucial for effective conservation and mitigation of human–elephant conflicts.

Coimbatore Forest Division

- **Elephant Population Dynamics**: The Coimbatore Reserve Forest, part of the Nilgiris Biosphere Reserve, spans approximately 4662.45 sq.km. It serves as a signifi-

Table 5 Estimated artificial water resource usage in Coimbatore forest division (last 2 years)

Month/year	No. of elephants using water sources	Water source type	Avg. water consumed per day (liters)	Notable observations
Jan 2023	120	Artificial pools	25,000	Dry season, moderate dependency
Jul 2023	180	Check dams	35,000	Rain-fed, high availability
Nov 2023	150	Borewell-fed ponds	30,000	Increased movement near villages
Jan 2024	130	Artificial pools	26,000	Water levels dropping
Jul 2024	190	Check dams	36,500	Frequent herd visits
Nov 2024	160	Borewell-fed ponds	31,000	Conflict incidents reduced

Table 6 Estimated artificial water resource usage in Bandipur forest division (last 2 years)

Month/year	No. of elephants using water sources	Water source type	Avg. water consumed per day (liters)	Notable observations
Jan 2023	200	Perennial waterholes	40,000	High water need
Jul 2023	250	Seasonal streams	50,000	Enough rainfall
Nov 2023	220	Artificial ponds	42,500	Elephant sightings near human areas
Jan 2024	210	Perennial waterholes	41,000	Water deficiency
Jul 2024	260	Seasonal streams	52,000	More elephants movement
Nov 2024	230	Artificial ponds	45,000	Increased usage due to habitat changes

cant habitat and corridor for elephants, facilitating movement between regions like the Silent Valley National Park. The elephant population in this area was estimated at 2761 as of 2017.

- **Water Need and Seasonal Movements**: Elephants in this region exhibit seasonal migration patterns, often influenced by water and forage availability. During dry seasons, the scarcity of natural water sources can lead elephants to search for alternative water supplies, sometimes bringing them into proximity with human settlements. This movement highlights the importance of maintaining adequate water resources within forested areas to reduce human–elephant interactions as in Table 5.

Bandipur Forest Division

- **Elephant Population**: Bandipur National Park, established in 1931, is home to a substantial elephant population. As of 2020, internal surveys indicated that over 1600 elephants wandered in and around the Bandipur Tiger Reserve.
- **Water Consumption Patterns**: While specific daily water consumption data for Bandipur's elephants is limited, it's known that elephants require significant water intake daily, both for drinking and bathing. The availability of perennial and seasonal water sources within the park plays a vital role in supporting this large elephant population. During dry periods, the reduction in natural water sources can lead to increased movement of elephants in search of water, potentially leading them closer to human habitations and increasing the risk of conflict Table 6.

Inferences for Conservation and Management

- **Artificial Water Sources**: The establishment and maintenance of artificial water points within these forest divisions can help in retaining elephants within protected areas, thereby reducing the likelihood of human–elephant conflicts. Ensuring year-round availability of water can influence elephant movement patterns positively Table 7.
- **Monitoring and Research**: Continuous monitoring of water sources and elephant movements is essential. Implementing IoT-based real-time monitoring systems can provide valuable data, enabling proactive measures to manage water resources and mitigate potential conflicts.

From Figs. 14 and 15, we can see the peak consumption is during April-June2025 due to temperature rise during the summer season and expected highest water requirement for the AWR. At the same time, we can see during July to September 2025 there will be less requirement due to monsoon rainfall and the need of AWR will be less. Therefore, during summer March to June due to lack of water in the forest the elephant movement near human settlement can be detected more.

Table 7 Predicted elephant water consumption (Dec 2024–Dec 2025)

Month/ year	Predicted no. of elephants using water sources		Avg. water consumption per day (Liters)		Prediction on requirement of AWR
	Coimbatore	Bandipur	Coimbatore	Bandipur	
Dec 2024	180	250	30,000	4500	Wet condition, water availability
Jan 2025	200	270	35,000	50,000	Constant water availability
Feb 2025	220	290	38,000	52,000	Pre Dry season. moderate water intake
Mar 2025	250	310	42,000,	58,000	Dry season, higher water requirement
Apr 2025	270	340	46,000	65,000	Top dry season, higher water requirement, elephants may move to human areas. AWR needs to filled
May 2025	300	370	50,000	70,000	Temperature increases heat, maximum water usage, chances of HEC. AWR needs to filled
Jun 2025	280	350	48,000	68,000	Monsoon rainfall, AWR starts fill
Jul 2025	240	310	42,000	60,000	Increased natural water availability due to monsoon rains. AWR filled
Aug 2025	200	270	38,000	55,000	Monsoon rainfall continues, lower usage of AWR
Sep 2025	190	260	36,000	53,000	Post-monsoon water available in AWR
Oct 2025	210	280	39,000	57,000	Post winter water available in AWR
Nov 2025	230	300	40,000	60,000	Winter water available in AWR
Dec 2025	250	320	42,000	63,000	Peak winter water available in AWR

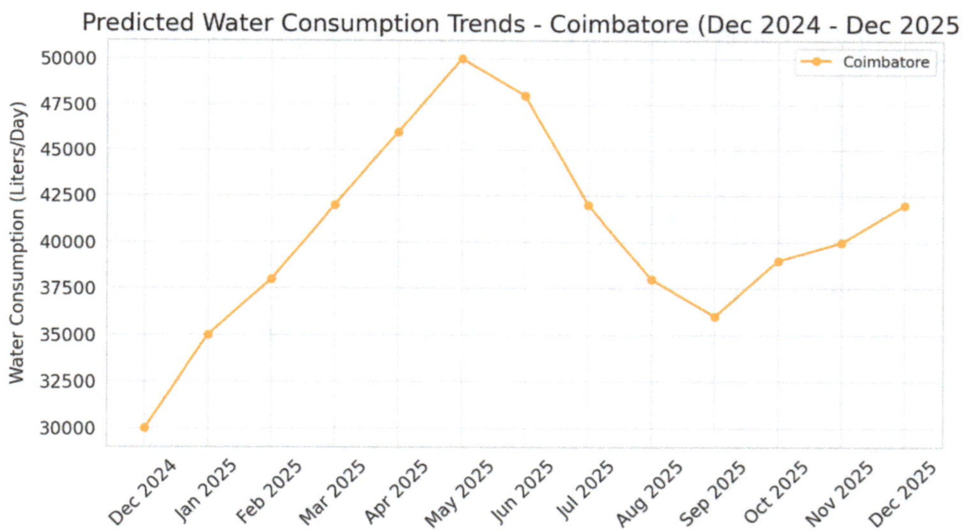

Fig. 14 Predicted elephant water consumption—Coimbatore (Dec 2024–25)

5 Cloud Dashboard and Visualization for Real-Time Monitoring

In this chapter, we discuss a cloud-based dashboard for real-time monitoring of water levels of AWR. This dashboard provides a remote accessibility, allowing the forest personals to monitor water levels and receive alerts by prediction decision-making and visualization techniques, providing a better water resource management.

5.1 Real-Time Graphs: Water Level Trends, Temperature and Rainfall Correlation

Data Collection and Preprocessing

To collect real-time water level trends, temperature changes & rainfall correlation data is fed using rainfall detector temperature sensor and weather update into the Bharath Pi LTE WiFi board. The Bharath Pi board will send water and temperature level data to Firebase real-time

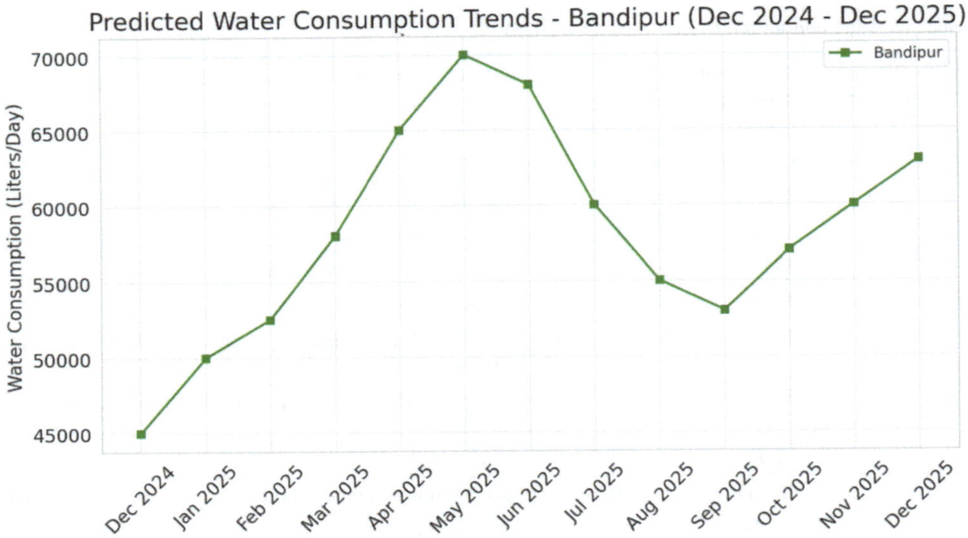

Fig. 15 Predicted elephant water consumption—Bandipur (Dec 2024–25)

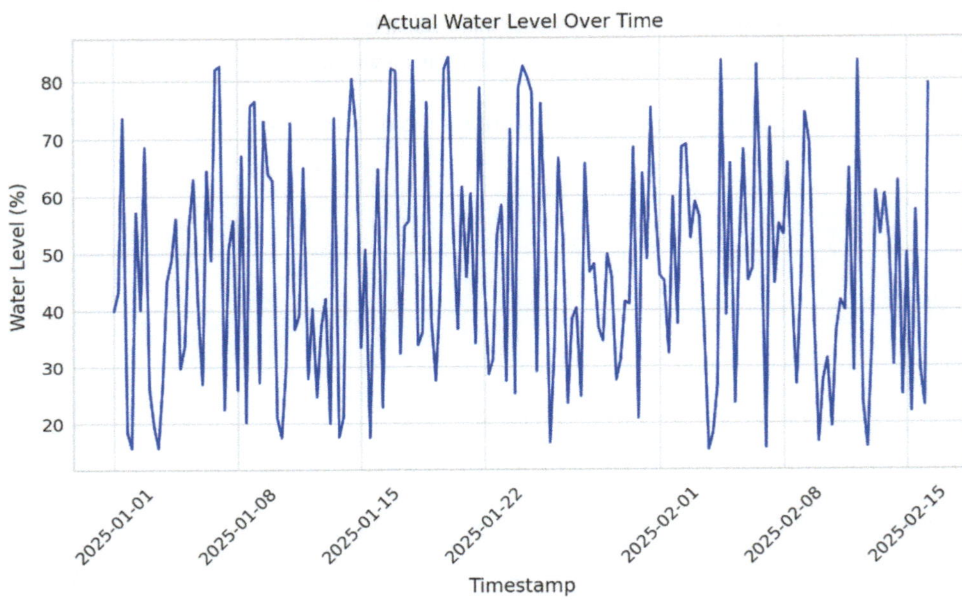

Fig. 16 Actual water level over time in AWR

database using firebase API key. We integrated Smart Artificial Water Resource Monitoring system in to Firebase using API key. It will read data from firebase and display it in website using graph Transmit data to the firebase Cloud via MQTT/HTTP API (Loso et al., 2019). The following are the data display in real-time & predicted water levels using:

From Fig. 16, we can see the water levels fluctuated between 15 and 85% over the period of study time. There are seasonal variations observed with lower levels during dry seasons and higher levels during winter seasons. The AWR Refill Scheduled actions were triggered when water levels dropped below 30%. The Drainage ON was activated when water levels exceeded 80%.

In Fig. 17, we can observe that the AI model showed strong accuracy in predicting water level changes, with deviations mostly within ± 5%. The predictions helped automate refill scheduling and drainage management, preventing critical shortages or overflow.

From the Fig. 18, it is observed there is a negative correlation was observed: higher temperatures (> 35 °C) often resulted in lower water levels due to evaporation. Water levels remained relatively stable when temperatures were between 25 °C and 32 °C.

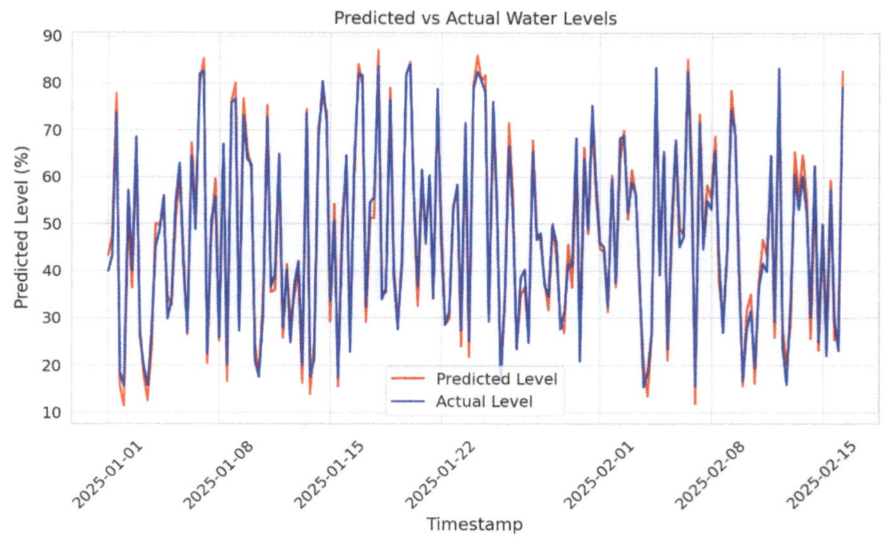

Fig. 17 Predicted water level over time in AWR

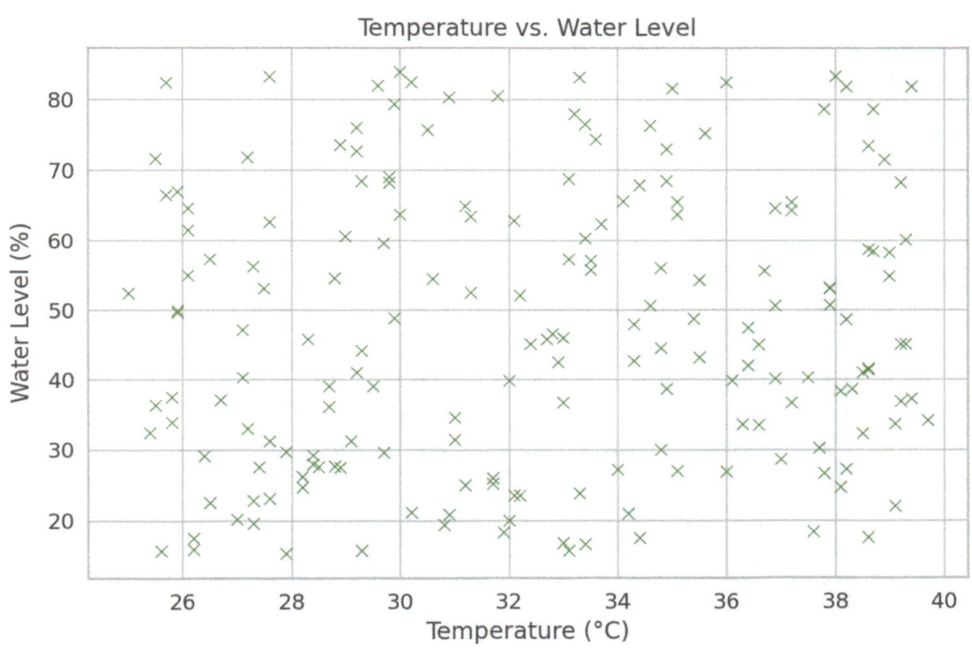

Fig. 18 Temperature versus water level

From Fig. 19, it is observed that a positive correlation was detected: rainfall events above 20% led to significant increases in water levels. In periods of low rainfall (< 5%), the water level steadily reduced due to natural usage and evaporation.

From Fig. 20, we can observe most water levels ranged between 35 and 70%, indicating stable resource availability. Extreme low levels (< 30%) and high levels (> 80%) were infrequent but required automated intervention. From Fig. 21, we can observe Refill Scheduled actions were triggered 22 times, mainly during long dry periods. Drainage ON was activated 18 times, preventing overflows during heavy rainfall. The majority of time, water levels remained within the optimal range (30–80%), requiring no action.

5.2 Firebase Web Server for Real-Time Water Level Monitoring

A Firebase Hosting is developed to deploy a web-based dashboard, providing interactive visualizations of water levels, refill schedules, and alerts. The seamless integration of Firebase Authentication ensures secure access for authorized

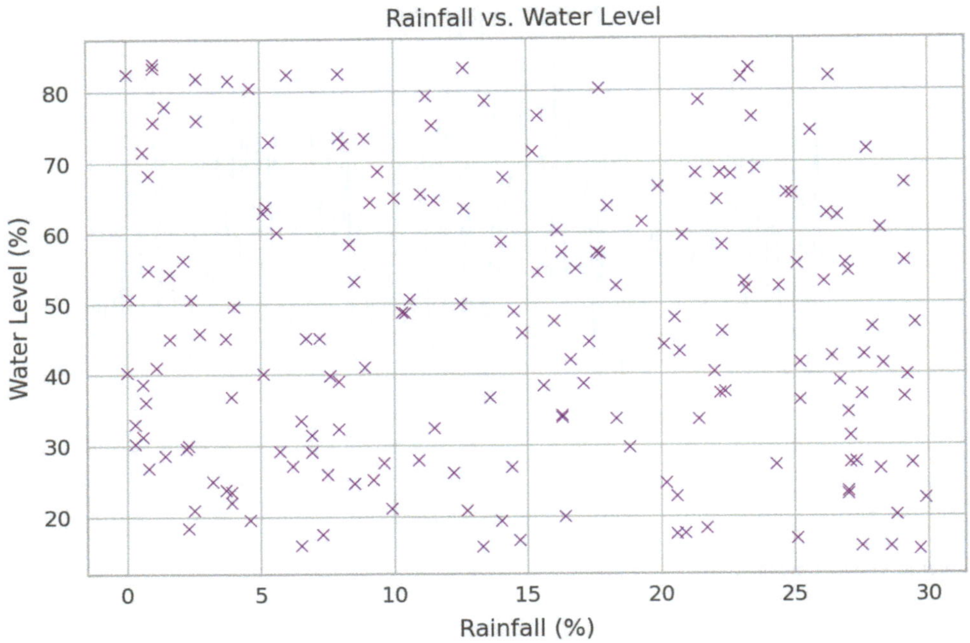

Fig. 19 Actual rain water level and water level over time in AWR

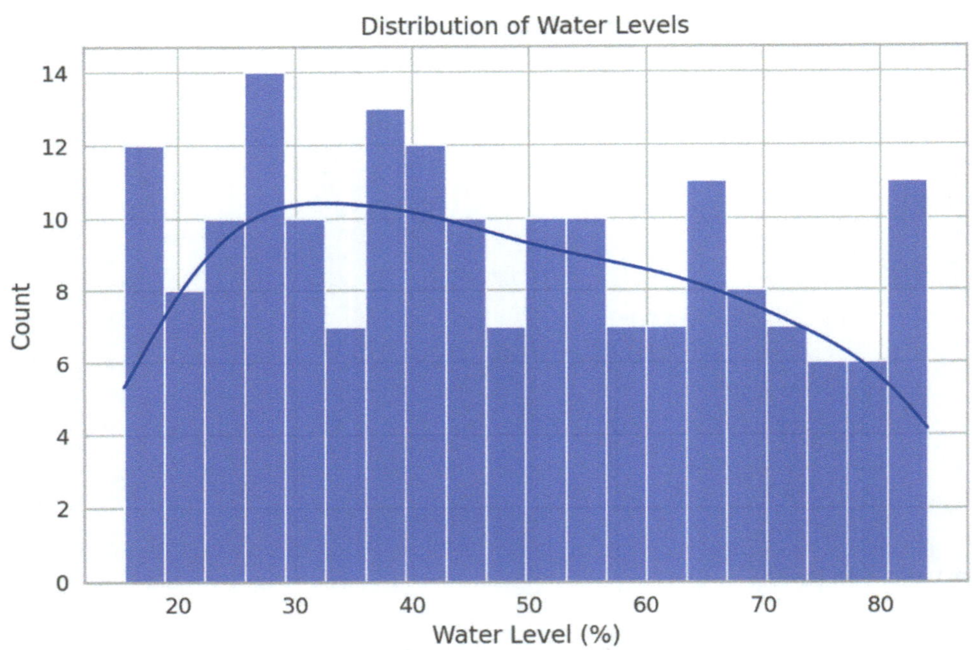

Fig. 20 Distributions of water level

users, while Firebase Firestore efficiently manages structured data for analysis. Overall, Firebase serves as a scalable, low-latency backend solution, enhancing real-time decision-making and proactive water resource management (Marqas et al., 2022).

The Firebase API Key as shown in Fig. 22 is a unique identifier that allows the frontend applications (web or mobile) to interact securely with Firebase services, including the Real-time Database, Firestore, Authentication, Cloud Functions, and Hosting. This key is typically included in the Firebase configuration file and enables authenticated applications to read and write data, trigger cloud functions, and manage user authentication. A Firebase, a cloud-based platform is used as the web server for real-time data storage, processing, and visualization in this work as shown in the landing page of Fig. 23. The Firebase Realtime Database enables instant synchroniza-

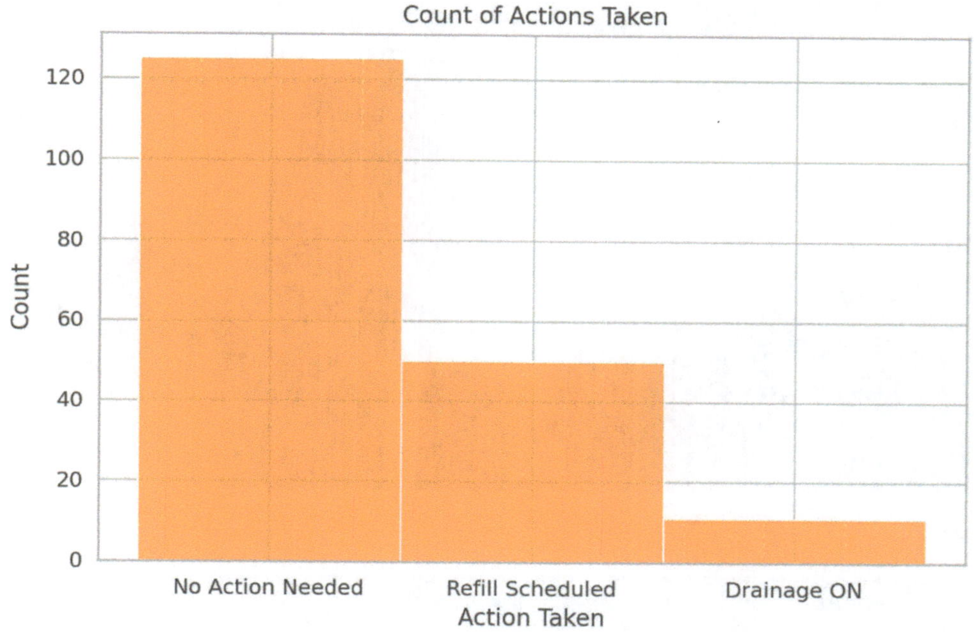

Fig. 21 Count of action taken

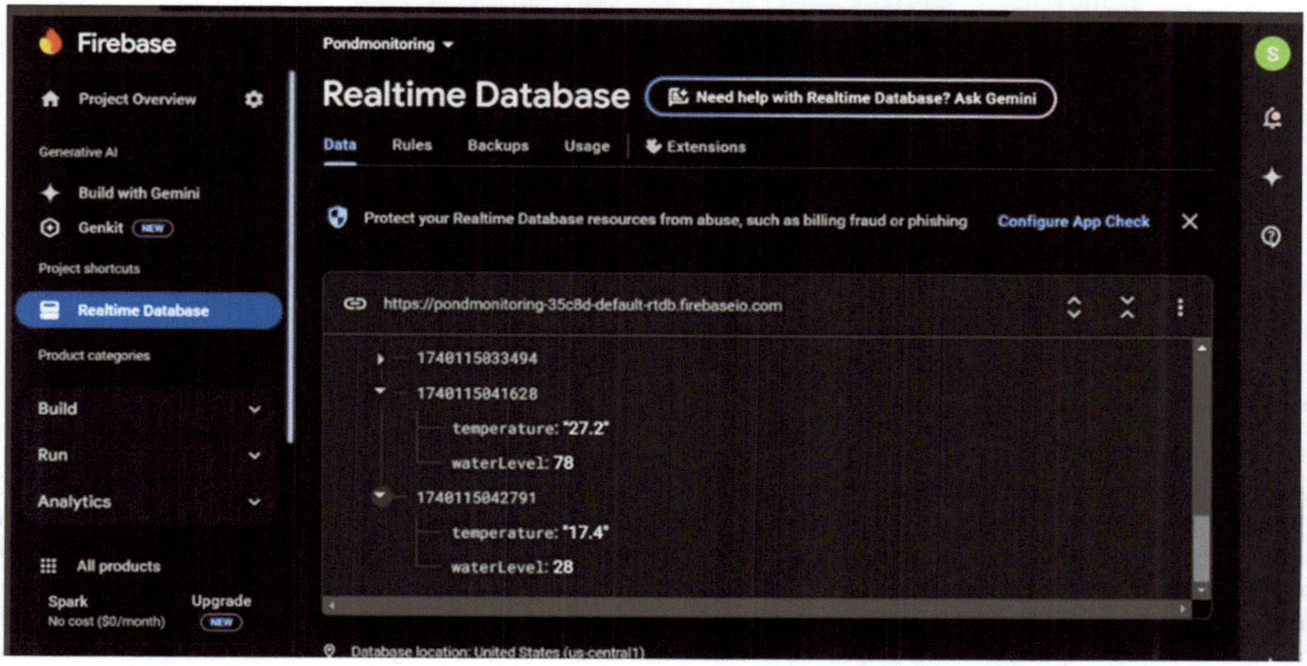

Fig. 22 Firebase API key identifier

tion of water level data collected from IoT sensors deployed in artificial water resources within Coimbatore and Bandipur forests. The Firebase Cloud Functions allow for automated execution of AI/ML models, predicting water consumption patterns based on historical data, rainfall, and temperature variations. Additionally, A webserver cloud is developed as shown in Fig. 24 having options to live monitoring of real-time updates on water level and temperature to ensure maintaining water level. An emergency alerts are notified on water level when they are critically low. AI predictions are also obtained to detect water level and water usage patterns and migration. A GPS tacking system is also provided to identify the location details of the AWR ponds. A Lidar tracking system is also proposed to setup to track number of elephant movement in the AWR zone and to determine the impact of pond usage. The web server created displays information with real-time

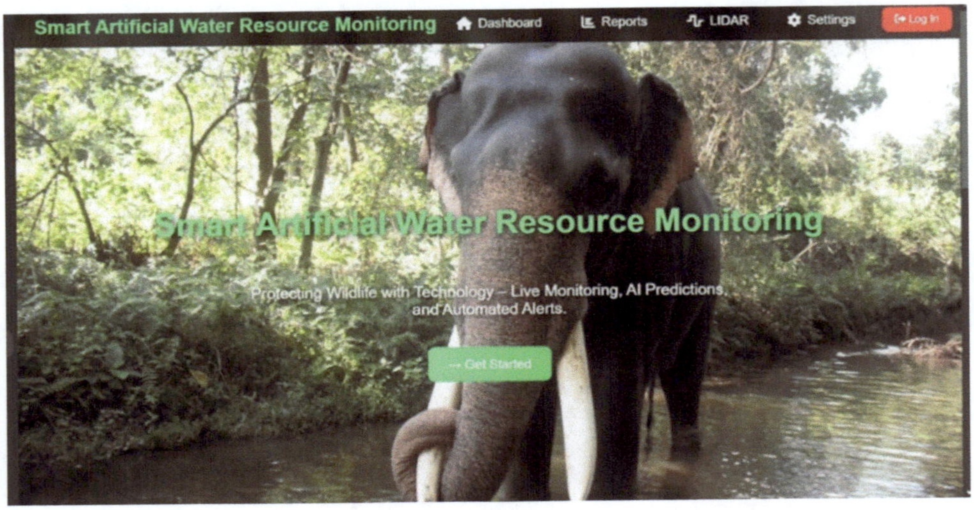

Fig. 23 Landing page

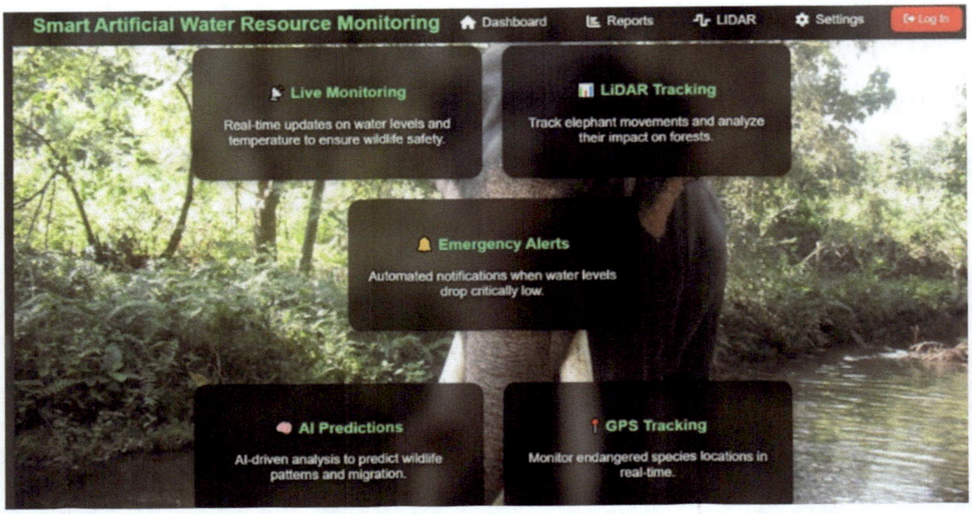

Fig. 24 Smart artificial water resource monitoring

water level and temperature data to check the current condition of the pond as shown in Fig. 25. For refill of water using tanker lorry or other sources we display last filled date and next scheduled date of filling water in to the pond. This data helps in monitoring and tracking of filling of water in the pond. The current temperature and average temperaturreal-time updates on watere around the pond also displayed to understand the climate around the pond. The tools used in this work are shown in Table 8.

5.3 Impact of IoT in Forest Ecosystems for Conflict Mitigation

The forest ecosystem is facing unparalleled challenges due to climate change, animal habitat demolition, deforestation, human settlement, poaching, and other manmade threats. Protecting our forest ecosystem and the animals living and conserving our forest are the highest requirement and it needs to use latest technologies like IoT powered with AI-based data analytics in monitoring the forest climate, rainfall, etc. and providing solution. The objective of the work is to integrate IoT-based system with environmental data to provide real-time solution for conserving the forest. The prediction system integrated with the IoT enables an advanced alert in filling water in AWR which provides water without scarcity. The data communicated to the centralized monitoring system with visualization tools developed helps the forest officials to view the real-time data and take action accordingly. As the IoT is operated with 5G seamless data availability with lower latency helps to monitor the climate and water availability data in real time which helps the

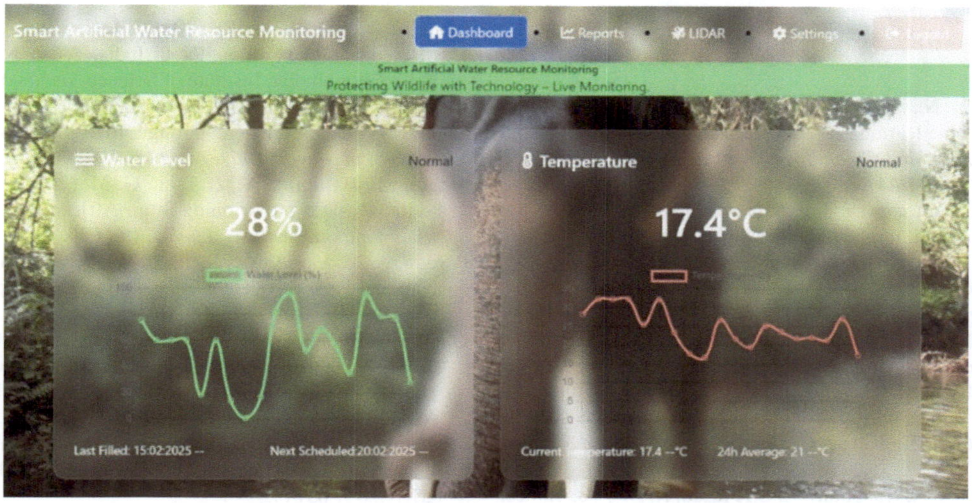

Fig. 25 Water level monitoring and tracking page

Table 8 Tools used

S. no	Tool	Purpose
1	Bharath Pi LTE WiFi Kit	Collects and transmits sensor data via MQTT
2	Rain water level sensor	Measures real-time water level
3	DHT11/DHT22 sensor	Captures temperature and humidity data
4	MQTT protocol	Sends data from sensors to ThingsBoard
5	Fire base cloud	Stores dashboard and cloud integration and visualizes real-time data
6	Python (Pandas, NumPy, Scikit-learn)	AI model development for water level prediction
7	LSTM RCNN model	Predicts future water level changes

elephants and other animals to stay in their habitat even if there is water scarcity. This helps improve the living conditions of the elephant and can increase in numbers which directly and indirectly helps the forest ecosystem to grow and protect the forest.

6 Conclusion

The study and approach provided in this work supports in managing the forest ecosystem thereby increasing the elephant numbers and in return mitigates human elephant conflicts as the elephants are kept inside the forest and not entering in to human living areas. The IoT system developed along with Artificial Intelligence in providing proactive predictions in keeping the water level at a required level helps keeping the elephant staying at their habitat without need of search of water outside the forest. Like other traditional RNN, the LSTM can remember the information over longer period using memory cells this makes the system to remember the historical data of past rainfall and water levels and make decisions on providing proactive predictions. The random forest algorithm provides a better insight in classifying the multiple data features which are most important in providing predictions. The solution provided with sensors integrated with IoT-based hardware system design along with the web server visualization is versatile and provides insights in data analysis. Our proposed Artificial Water Resource monitoring and tracking system can detect the water level and provide an advance early warning for refilling schedule of water in AWR so as to keep water all the seasons through. The prediction results shows how to maintain the water level which contributes in managing the forest ecosystem and improving the elephant life cycle. The IoT with AI-powered AWR management system has proven in predicting water levels in Coimbatore and Bandipur Forest regions and can maintain the water level to help the elephants which are living in the forest to stay at their own habitats and not entering to human habitat areas. In conclusion humans also has to understand the importance of forest, the wild life inside and their ecosystem no to be disturbed by the human developmental activities inside the forest zone. With further implementation and improvements, this system can contribute in maintaining the forest environmental and its ecosystem conserving the wildlife particularly the elephants by ensuring consistent water availability inside the forest.

References

Ayus, I., Natarajan, N., & Gupta, D. (2023). Prediction of water level using machine learning and deep learning techniques. *Iranian Journal of Science and Technology, Transactions of Civil Engineering, 47*(4), 2437–2447.

Baskaran N. (2013). An overview of Asian elephants in the Western Ghats, southern India: implications for the conservation of Western Ghats ecology. *Journal of Threatened Taxa 5*(14).

Bouwer, H. (2000). Integrated water management: Emerging issues and challenges. *Agricultural Water Management, 45*(3), 217–228.

Brooks, C., et al. (2018). A component architecture for the internet of things. *In Proceedings of the IEEE, 106*(9), 1527–1542.

Chakuya, J., Mandisodza-Chikerema, R., Ngorima, P., & Malunga, A. (2023). Water sources during drought period in a savanna wildlife ecosystem, northern Zimbabwe. *Geology, Ecology, and Landscapes, 7*(4), 323–328.

Gowda, V. D., Sharma, A., Tanguturi, R. C., Prasad, K. D. V., & Sameer Kotwal, V. (2024) Data-driven aquatics: the future of water management with IoT and machine learning. In *Innovations in Machine Learning and IoT for Water Management* (pp. 172–194). IGI Global Scientific Publishing.

https://bharatpi.net/product/bharat-pi-4g-lte-module-with-esp32-simcom

Ismail, S., Dawoud, D. W., Ismail, N., Marsh, R., & Alshami, A. S. (2022). IoT-based water management systems: survey and future research direction. *IEEE Access, 10*, 35942–35952.

Keramidas, G., Voros, N., & Hübner, M. (2016). Components and services for IoT platforms. Cham: Springer International Pu

Krishnan, S. R., et al. (2022). Smart water resource management using artificial intelligence—a review. *Sustainability 14*(20), 13384.

Kumara H. N., Rathnakumar S., Ananda Kumar, M., & Singh, M. (2012). Estimating Asian elephant, *Elephas maximus*, density through distance sampling in the tropical forests of Biligiri Rangaswamy Temple Tiger Reserve, India. *Journal Tropical Conservation Science, 5*(2), 163–172.

Laso, S., Flores-Martín, D., Herrera, J. L., Canal, C., Murillo, J. M., & Berrocal, J. (2019). Providing support to IoT devices deployed in disconnected rural environment. In *International Workshop on Gerontechnology* (pp. 140-150). Cham: Springer International Publishing.

Lenin, J., & Sukumar, R. (2011). Action plan for the mitigation of elephant–human conflict in India. Asian Nature Conservation Foundation, Indian Institute of Science, Bangalore, March 2011

Marqas, R. B., Almufti, S. M., & Asaad, R. R. (2022). Firebase efficiency in CSV data exchange through PHP-based websites. *Academic Journal of Nawroz University (AJNU), 10*(3).

Mpalo, M., Basupi, L. V., & Tsidu G. M. (2024). Impacts of wildlife artificial water provisioning in an African Savannah ecosystem: a spatiotemporal analysis. *Land 13*(5), 690.

Purdon, A., & Van Aarde, R. J. (2017). Water provisioning in Kruger National Park alters elephant spatial utilisation patterns. *Journal of Arid Environments, 141*, 45–51.

Quiñonero-Rubio, J. M., Nadeu, E., Boix-Fayos, C., & de Vente, J. (2016). Evaluation of the effectiveness of forest restoration and check-dams to reduce catchment sediment yield. *Land Degradation & Development 27*(4), 1018–1031

Rahbaralam, M., et al. (2020). Predictive analytics for water asset management: Machine learning and survival analysis. arXiv preprint arXiv:2007.03744.

Sherstinsky, A. (2020). Fundamentals of recurrent neural network (RNN) and long short-term memory (LSTM) network. *Physica d: Nonlinear Phenomena, 404*, 132306.

Smith, E., Brown, L. R., & Barrett, A. S. (2024) Insights into artificial waterhole utilization patterns by elephants and rhinos: Lessons from a South African Nature Reserve. *Plos one 19*(10), e0312158

Straškraba, M. (2004). Reservoirs and other artificial water bodies. *The Lakes Handbook: Lake Restoration and Rehabilitation, 2*, 300–327.

Sunny, M. A. I., Maswood, M. M. S., & Alharbi, A. G. (2020). Deep learning-based stock price prediction using LSTM and bi-directional LSTM model. In *2020 2nd Novel Intelligent and Leading Emerging Sciences Conference (NILES)* (pp. 87–92). IEEE.

Sutherland, K., Ndlovu, M., & Pérez-Rodríguez, A. (2018). Use of artificial waterholes by animals in the southern region of the Kruger National Park, South Africa. *African Journal of Wildlife Research, 48*(2), 1–14.

Thaker, M., et al. (2019). Fine-scale tracking of ambient temperature and movement reveals shuttling behavior of elephants to water. *Frontiers in Ecology and Evolution 7*, 4.

Uloko, I. J., Yager, G. O., & Ikyuango, T. I. (2023). Effect of artificial waterholes on distribution of wildlife fauna in arboretum of Gashaka Gumti National Park, Taraba State, Nigeria. *Journal of Research in Forestry, Wildlife and Environment, 15*(4), 141–147.

Drones and Remote Sensing: Revolutionizing Forest Monitoring

Smitha Rajagopal, K. Bhargavi, V. Porkodi, M. Tholkapiyan, and Manoj Kumar Padhi

Abstract

Forest ecosystems, integral to global biodiversity and climate regulation, demand precise and continuous monitoring to ensure sustainable management. These landscapes require sustained accurate monitoring activities as a basis for appropriate management. Drone-based remote sensing emerges as a modern method to watch forests throughout extensive unreachably large areas while obtaining detailed data records. Modern forest surveillance makes transformative use of drone sensor systems for collecting advanced high-definition real-time information across enormous difficult-to-access areas. UAVs provide significant capabilities for monitoring deforestation activities along with identifying illegal logging products while assessing wildfire damage through the protection of biodiversity landscapes. The combination of low-altitude flight operations with flexible control mechanisms gives UAVs their optimal performance in dynamic forest areas. Seamless deployment of drone-based monitoring systems encounters various challenges during their implementation process. Flow operations remain constrained by three main obstacles which include flight duration restrictions combined with external environmental conditions and regulatory demands as well as data processing challenges. This document presents the latest drone breakthroughs which include portable sensor manufacturing and lengthened flying periods and advanced GIS network communication abilities. Forests obtain a powerful sustainable management solution from integrating drones with remote sensing systems. Real-time accurate data acquisition enabled by the tool enables better conservation planning and environmental sustainability and enhances operational decision-making abilities.

Keywords

Drone-based remote sensing · Forest monitoring · LiDAR mapping · Multispectral imaging · Illegal logging detection · AI in forestry · UAV forest surveillance · Thermal sensors · Deforestation tracking · Environmental data analysis

S. Rajagopal (✉)
Department of Computer Science and Engineering, Alliance School of Advanced Computing, Alliance University, Bengaluru, Karnataka, India
e-mail: smitha.rajagopal@alliance.edu.in

K. Bhargavi
Department of CSE, ICFAI Foundation for Higher Education (IFHE), Shankarpally, Hyderabad, India
e-mail: bhargavi.mtech@gmail.com

V. Porkodi
Department of Computer Engineering, Faculty of Engineering and Natural Sciences, Sivas University of Science and Technology, Sivas, Turkey
e-mail: porkodisivaram@gmail.com

M. Tholkapiyan
Department of Civil Engineering, Chennai Institute of Technology, Chennai, Tamil Nadu, India
e-mail: m.tholkapiyan@gmail.com

M. K. Padhi
Department of Computer Science and Engineering, Centurion University of Technology and Management, Paralakhemundi, Odisha, India
e-mail: manojpadhi1503@gmail.com

1 Introduction

Forests are maintaining a significant role in ecological balance by controlling climate, conserving biodiversity and providing resources for supporting human livelihood. Some serious issues like Rapid deforestation, climate change and illegal logging pose threats to these ecosystems. Forest monitoring is essential to ensure protection, sustainable use and conservation. Forests help in identifying changes in forest cover, determining biodiversity, and recognizing environmental stress. Some of the main reasons are mentioned. Monitoring helps in

M. Ashwin et al. (eds.), *Smart Forestry 5.0 for Powered AI Sustainable Environment and Climate*, Sustainable Artificial Intelligence-Powered Applications, https://doi.org/10.1007/978-3-031-99086-1_11

detection and stopping illegal logging, agricultural encroachment and destruction of habitats. Forests serve as carbon sinks, absorbing greenhouse gas gases. Monitoring forests helps in determining its contribution in climate regulation. Forests serve as home to various endangered species, so continuous monitoring of forests enables the monitoring of forest conditions and wildlife movement. Early detection regarding forest fires, pest outbreaks, and certain diseases can help in preventing large-scale ecological loss. Sustainable Resource Management: Governments and industries depend on the date about these forests to provide sustainable timber cutting and conservation practices. Earlier, forest monitoring happened through manual surveys, ground-based observations and aerial photography. Even though these methods provide valid data, there are several drawbacks. Carrying out field surveys over vast areas of forests requires more manpower and time. Ground surveys are often limited to certain sample areas only, resulting in incomplete information. Recruiting personnel for on-site monitoring and performing aerial surveys by using aircraft or helicopters involves huge costs. Conventional ways were not able to offer real-time data regarding deforestation, illegal logging, or environmental changes. Monitoring practices serve to prevent deforestation by revealing criminal forest activities and unauthorized human actions that destroy natural habitats. The primary carbon-absorbing function of forests plays a major part in climate regulation while they help minimize greenhouse gases in the atmosphere. The evaluation of forest climate change mitigation capabilities improves through tracking their health together with their coverage. The protection of endangered species and their wildlife habitats depends on forest monitoring because it shows data about natural ecosystems and tracks the paths of animals. The system supports disaster prevention by identifying threats like wildfires and pests and diseases early which makes a reduction of pervasive ecological destruction possible. The effectiveness of sustainable resource management depends on dependable forest data because these figures drive decisions regarding both government-regulated timber harvesting and the sustainable conservation methods pursued by industrial entities. Before modern times forest monitoring consisted mainly of human surveys combined with ground-based observations and aerial images taken from airplanes with personnel onboard. The information collected through these methods proved useful but these procedures presented multiple significant constraints. These methods consumed great amounts of time and needed many personnel in addition to being expensive to operate. Ground surveys concentrated on limited sample areas which produced incomplete assessments of entire forest areas. The traditional monitoring methods lacked real-time information processing, so it became difficult to make quick reactions toward fast-paced environmental events like illegal logging or sudden forest clearance. The effectiveness of satellite imagery and aerial photography diminished because of various weather conditions primarily through cloud cover. The efficiency of forest monitoring increased and its precision improved, and its affordability decreased because of modern technological advancements. Forest structures become easily monitored through LiDAR systems which generate accurate biomass estimates using three-dimensional mappings. Table 1 shows comparison of Conventional and Drone-based Forest Monitoring.

Table 1 Comparison of conventional and drone-based forest monitoring

Feature	Traditional methods	Drone-based remote sensing
Time	Weeks/months	Hours/days
Cost	High (manpower equipment)	Lower (automated drones)
Coverage area	Limited	Large-scale, remote areas
Accuracy	Prone to human error	High precision

Drone-based remote sensing techniques demonstrate enhanced performance against conventional forest monitoring approaches according to Table 1 through analysis of time effectiveness and cost savings as well as coverage area scope and precise determination capabilities. Drone-based monitoring systems outperform traditional methods since they produce results in both short-term data collection and processing durations extending from hours to days. Fast data gathering enhances the rapidness of forest management operations particularly for time-sensitive issues including fire identification and unauthorized logging detection. Traditional monitoring systems require substantial expenses for personnel and equipment therefore becoming quite costly and labor-intensive. Drone-based systems deliver better cost performance by using automated processes that need less manpower for operations and have salvageable hardware that can execute multiple tasks. The operational area span covered by these monitoring systems shows considerable differences. Through drone monitoring technology, operators gain enhanced safety while monitoring extensive areas of remote forests that include designated hazardous and rugged terrains. Traditional surveys experience more human error in accuracy measurement because of subjective human collectors and constrained sample size data collection.

The advancement of technology in forest monitoring has created an improved system that delivers more exact results with fewer costs and higher efficiency. The combination of modern tools enables forest managers, researchers and policymakers to inspect and assess forests at a previously unattainable size while obtaining greater detailed observations. Airborne sensor technology and satellite remote detection systems represent a main factor behind this monitoring evolution. These modern detection methods provide precise maps which enable complete landscape analysis of forests through observation of their cover and canopy alterations and land-use activities. Solar-mission-derived data exists as a critical component for sustaining uninterrupted forest surveillance through time. The utility of drone systems as real-time data collection tools became possible for difficult-to-reach forest areas and challenging terrains through their implementation as UAVs or unmanned aerial vehicles. These devices offer high-definition image capability while allowing users to adjust their height position and viewing angle which enables in-depth assessment of tree structures together with pest infestation detection and forest damage symptoms. UAVs enable continuous tracking without clunky ground machinery which decreases operation expenses at the same time. The forest monitoring process receives important enhancement from LiDAR (Light Detection and Ranging) which enables the production of three-dimensional forest structure representations. The precise assessment of forest biomass, tree heights and vegetation density become achievable through this technology for developing better resource planning and conservation methods. Perpetual sensor measurements of essential environmental variables including air temperature and soil moisture and humidity take place across entire forest territories. Real-time data from sensors gives important context about ecosystem health so proper environmental stressor responses can be performed speedily. Forest transparency improves through these technologies which also help verify forest assets and enable legal enforcement operations to hamper illegal logging efforts. Forest governance strengthens through the integration of such tools which make monitoring approaches both more advanced and dependable for changing environmental requirements.

2 Definition of Remote Sensing in Forestry

Remote sensing functions as a method that gathers data concerning areas or physical objects through non-contact procedures. Remote sensing technology utilizes sensors installed on platforms which include aircraft or satellites or drones to accomplish their objectives. The sensors record either images or they monitor the energy emitted or reflected by surface elements to provide essential Earth surface information. Remote sensing provides essential tools to forestry professionals for both forest monitoring purposes and analysis of vegetation health conditions and development of sustainable management practices. EMR detection fundamentals allow remote sensing to operate by analyzing electromagnetic radiation behavior when it encounters natural surfaces. EMR encounters objects by three possible outcomes that include transmission aging while reflection occurs. The sensor mechanism detects the material-specific response patterns after EMR interaction. Normally healthy plants reflect electromagnetic waves differently than their equivalents that show signs of disease stress. Soil moisture and forest vitality as well as canopy structure and land use can be evaluated by scientists who analyze these patterns. Several fundamental principles determine what makes remote sensing information useful for capturing data. The visible amount of information in an imagery depends directly on spatial resolution. The quality of resolution determines whether an image shows individual plants and small openings or examines broad areas of the land. Remote sensors reveal their capabilities to detect different light wavelengths through their spectral resolution abilities. The detection of plant species alongside stress symptoms and types of vegetation depends heavily on this essential feature. The remote sensing methodology separates into active and passive method classifications. Passive sensing relies on natural sunlight as its energy source. Surface reflection allows this method to absorb light which finds usage in multispectral and hyperspectral imaging applications. Active remote sensing instruments include optical and infrared cameras among other devices. Remote sensing tools establish their use in both vegetation management and land cover analysis as well as seasonal observation functions. Active sensing employs systems which produce their own energy mixed from laser or radio wave emissions to measure the reflected signals. The method succeeds in low-light environments and nighttime along with scenarios where clouds obscure the view. The two prominent technology examples under active sensing include LiDAR (Light Detection and Ranging) and Synthetic Aperture Radar (SAR). The active sensors collect data steadily across all conditions including inclement weather conditions and throughout any period of the day. LiDAR represents a cutting-edge technology among all available solutions that improve forestry applications because it delivers precise measurement data. The laser pulses moved quickly toward the ground surface for LiDAR sensors to record the flight time as the system produced detailed three-dimensional maps of forest structures. The combination of radar information enables accurate assessments of tree sizes and both forest density levels and massive biomass measurements. The assessment of forest health together with disease outbreak detection and land cover mapping is greatly supported by multispectral imaging that operates within several visible wavelength bands combined with infrared wavelengths. Hyperspectral imaging systems obtain extraordinary level of plant condition detection by extracting hundreds of precise spectral bands which facilitates precise assessments of plant water stress and nutritional aspects. SAR technology employs radio waves to monitor forests despite dense vegetation and cloud cover by penetrating such conditions. The all-weather tracking system monitors deforestation together with forest regrowth along with ecosystem modifications consistently.

Remote sensing acts as a transformative tool in forestry because it provides dependable non-contact tools to study and supervise forest systems during present-day climate conditions. Active sensing functions as a data collection method that produces energy through instrument systems such as lasers or radio waves to work in nighttime conditions and locations hidden by thick forest structures. Under this category exists LiDAR and Synthetic Aperture Radar (SAR). Among the different light bands which multispectral imaging systems analyze are visible and infrared wavelengths to evaluate forest health conditions and track diseases and detect land-use alterations. More precise plant inspections become possible through hyperspectral imaging because it delivers enhanced spectral information for water content and nutrient and chemical composition examination. The radar mechanism of SAR utilizes radio waves that penetrate both clouds and dense plant life for continuous forest observation under all predicted and unanticipated weather conditions. These technologies cooperate to boost the capability of forest management with precise current environmental assessments which benefit conservation strategies. Remote sensing provides essential monitoring tools for forestry operations because it enables forest cover assessment as well as vegetation health detection while creating land use and land cover maps and biomass measurement and biodiversity evaluation. It offers consistent measurable data with scalability and repeatability due to conquering the drawbacks of conventional field-based approaches that have slow speed and tedious work requirements together with spatial restrictions. Passive systems from remote sensing include multispectral and hyperspectral sensors that obtain energy from sunlight but active systems like LiDAR and synthetic aperture radar generate their own signals before measuring returning signals. The introduction of drones into remote sensing applications through the last few years has transformed forestry operations via their ability to conduct ultra-high-resolution assessments and deploy quickly and perform near-instant data acquisition in secluded regions. These technologies combine to recreate forest assessment and protection and sustainable management practices which defend forests from climate change along with deforestation and ecological destruction.

3 Evolution of Drone Technology for Forest Monitoring

Drone technology has revolutionized forest monitoring by providing real-time data, high-resolution images and cost-effective means when compared to conventional approaches. The implementation of drone technology has produced substantial changes in the monitoring and management practices for forests since its recent development. These unmanned aircraft systems operate as practical environmental data-gatherers because they provide detailed information efficiently at low cost while moving past the surveillance limitations of previous methods. Traditional forest monitoring methods used satisfactory results through manned aircraft together with satellite imagery as the primary data sources. The practice of aircraft flying through thick forests during the early twentieth century collected images which served to assess forest tree growth and detect illicit logging activities and inspect woodland health status. These aerial survey techniques ended up being beneficial but their high expense along with long duration and interruption from bad weather or limited pilots negatively impacted their effectiveness. Deforestation monitoring along with vegetation health evaluation shifted to Landsat Satellite systems because they provided coverage of extensive areas. Satellite data delivery offered restricted levels of detail which was made worse by regular cloud interference that defeated the purpose of important imagery. UAVs commonly called drones have totally remolded this domain. Due to their small size and high sensor capabilities drones deliver greater information density than any preceding technology did. New systems operate efficiently because they emerge fast while flying low and generating precise images almost instantly. The unmanned vehicle costs are lower than manned aircraft because fuel expenses and flight personnel expenses are absent. The combination of characteristics makes drones usable by multiple worldwide organizations including government departments alongside conservation groups and research facilities. Drone cameras of modern design power professionals to acquire sharp aerial visuals for canopy density evaluation as well as assessments of land-use transformations and forest recovery scans after logging operations or native disturbances. Since many years, aerial surveys have been transformed from manned aircraft and satellites to much more efficient unmanned aerial vehicles (UAVs) equipped with advanced sensors. This evolution has made forest conservation efforts more scalable, timely and precise. Before this drone technology, monitoring forests has been carried out through aerial surveys using airplanes and satellites. In the early twentieth century, airplanes were used to obtain images of forests. Pilots flew over thick forests, taking pictures to document changes in tree cover and recognizing illegal logging activities. To some extent this was useful, but this was costly, time-consuming and subject to climatic conditions and pilot availability. Satellites like Landsat began offering large-scale forest data. Governments and some environmental organizations are used to monitor deforestation, observe forest health and evaluate biodiversity through satellite imagery. The introduction of UAVs usually known as drones transformed the way the forests get monitored. The small, unmanned aircraft equipped with high-resolution cameras, LiDAR and thermal sensors provide a cost-efficient way of gathering information. In contrast to manned aircraft, drones are cost-effective and do not require huge operational costs such as fuel and pilot training. The advancement in drone manufacturing has made them so widely accessible to research organizations and government agencies. Modern drones are inbuilt with high-resolution cameras which can provide high-resolution aerial imagery. LiDAR technology in drones can create 3D representations of forest structures, which includes canopy cover, tree height and biomass at high accuracy. Thermal imaging sensors help in detection of wildfires, wildlife movement tracking and plant health evaluation that depends on differences in climatic conditions. Drones can be easily deployable which provides immediate updates regarding forest status. These fly at low levels, providing clear images rather than satellite-based surveillance.

Weather-resistant drones can fly in harsh environments including dense rainforests and mountainous regions. Drone technology applies thermal imaging sensors as one of its most beneficial usages. The sensors function by tracking heat emission signatures for detecting wildfire onset and animal motions and plant temperature-stress indicators. Drone technology provides immediate response capabilities through easy deployment which helps fire teams prevent disastrous forest fire expansion when emergencies happen. The operational characteristics of drones enable them to function in situations which threaten traditional technologies. The weather-resistant drone designs enable operation in dense forests along with rugged terrains as well as mountainous areas to provide dependable monitoring in challenging environmental conditions. The drones' operations below cloud cover and proximity to the forest canopy generate data which is reliable and specially adapted to the situation. Remote sensing operates as a method that enables data collection from locations through non-physical contacts. Information collection through remote sensing happens by using sensors that are installed on satellites together with aircraft or drones to observe the Earth surface energy emission and reflection levels. The application of remote sensing techniques exists as essential for forestry because it enables monitoring of forest cover and detection of vegetation changes as well as biodiversity evaluation and environmental condition studies. The procedure depends on basic fundamental principles. Electromagnetic radiation serves as the primary foundation in remote sensing because sensors analyze how various surfaces interact with energy through reflection and absorption to generate information about vegetative conditions and soil conditions and ground usage. The imaging details attainable from remote data require high resolution since this level enables observers to identify single trees. Image acquisition frequency defines temporal resolution because it enables the observation of deforestation and vegetation cycles as well as seasonal variations. The identification of different plant species and their health condition depends on how well sensors can distinguish light wavelengths known as spectral resolution. The methods of remote sensing operate through passive and active techniques which determine the acquisition process. Passive sensing obtains images through the use of solar radiation in multispectral and hyperspectral imaging procedures conducted with optical and infrared sensors. Active sensing functions as a data collection method that produces energy through instrument systems such as lasers or radio waves to work in nighttime conditions and locations hidden by thick forest structures. Under this category exists LiDAR and Synthetic Aperture Radar (SAR).

The initial approach to forest monitoring used both manual surveys coupled with satellite imagery yet they did not provide adequate solutions. The accurate ground surveys required massive labor effort while being both slow and limited to specific geographic areas. Remote sensing from satellites could oversee broad areas but its resolution was coarse and clouds obstructed some images and its data collection process happened infrequently. The early 2000s drone technology development created a vital solution through its cost-effective adaptable platform that proved both cost-effective and scalable. During the initial phase of drone deployment, they collected standard RGB images exclusively for visual examinations. The rapid development of sensors along with flight control software systems along with processing software has led to the building of drones with multiple sensors that contain LiDAR scanners thermal cameras multispectral and hyperspectral sensors and AI analysis tools. Modern drone systems achieve better vegetation identification while creating forest canopy models and detecting tree pests before disease development happens. Forest management expanded its ground through drone platforms that developed into autonomous and intelligent systems. Modern drones support automatic flight navigation together with immediate data exchange and hardware processing features that allow field personnel to conduct repetitive monitoring needs without much human assistant. Modern operations in distant hazardous zones rely heavily on drones because traditional survey methods cannot be utilized in such conditions. Drone technology reached one important evolutionary stage when it was used for forest disaster response and climate resilience initiatives. The operational use of drones covers fire damage assessment and illegal logging detection as well as post-disaster recovery evaluations and seed pod distribution services for precision reforestation programs. Proactive forest governance heavily depends on drones because they deliver rapid results which combine precision with speed and usefulness.

4 Sensors Used in Drone-Based Remote Sensing

The latest stage in the evolution of drone technology is the integration of artificial intelligence and automation to improve monitoring effectiveness. AI algorithms capable of processing large amounts of aerial imagery, detecting deforestation areas, illicit activity and measuring forest health. Automated systems can help in distinguishing between healthy trees and diseased trees by analyzing color variations and thermal patterns. GPS- and AI-equipped drones can navigate predefined routes without human involvement (Paneque-Gálvez et al., 2014). Fleets of autonomous drones (drone swarms) can scan huge forest areas simultaneously, increasing efficiency. High-end drones transmit real-time data to cloud platforms, helping conservationists to monitor forests remotely. Machine learning models process this data to predict future changes in forest like potential deforestation zones. Drones are very essential in monitoring forests as they are equipped with sensors. These sensors help in capturing accurate information regarding forests, including tree health, climate variations, species diversity and even 3D structural maps of vegetation. The potency of a drone in monitoring forests depends on the type of sensors with which they are equipped. Some of the popular sensors in drone-based remote sensing are: Some of the most popular sensors in drone-based remote sensing are. RGB (Red, Green, Blue) cameras are the most basic sensors employed in drones. These cameras capture images similar to human vision, providing high-definition images of forests. They help in general forest surveillance, detecting deforested land, illegal logging and landscape changes over a period of time (Koh & Wich, 2012). Although very useful, RGB cameras are not effective in detecting detailed plant health, as they are nor capable of providing data beyond the visible spectrum. Multispectral cameras record images in several bands, including visible light (RGB) and near-infrared (NIR). These play an essential role in assessing health of plants through vegetation indices like Normalized Difference Vegetation Index (NDVI), which helps in identifying stressed or diseased trees. They are widely used for monitoring tree growth, tracking deforestation and analyzing any ecosystem changes. Thermal sensors sense heat radiated by objects and can provide great insights into the forest situations. Thermal sensors are most useful for the early detection of forest fires because they can spot heat sources when flames are still not visible. They also assist in wildlife tracking since warm-blooded animals can be identified using their body heat even in thick forests or at night. Thermal imaging also helps in the investigation of plant stress as water-stressed trees have a higher temperature than healthy trees. LiDAR (Light Detection and Ranging) is an effective remote sensing tool that applies laser pulses to gauge distances and develop very accurate 3D representations of forests (Getzin et al., 2012). LiDAR sensors assist in estimating canopy density and tree height, as both are vital for the estimation of biomass and carbon stock quantification. Producing detailed topographic maps even where there is dense forest cover, which comes in handy while investigating terrain modification and soil erosion hazard. Recognizing structural gaps in the forest, which help in biodiversity studies and wildlife habitat assessments. LiDAR is very beneficial in tropical rainforests, where satellite imagery gets struggled because of cloud cover. Hyperspectral imaging is especially applicable for surveillance of vast and heterogeneous forests where species recognition becomes paramount. Hyperspectral sensors work on a wide spectrum of wavelengths, providing much detailed data when compared to multispectral cameras. Hyperspectral sensors work on a wide spectrum of wavelengths, providing much detailed data when compared to multispectral cameras. These sensors are capable of detecting variations in vegetation species depending on their unique spectral signatures. Figure 1 represents LiDar Drone Mapping represents

Fig. 1 LiDAR drone mapping

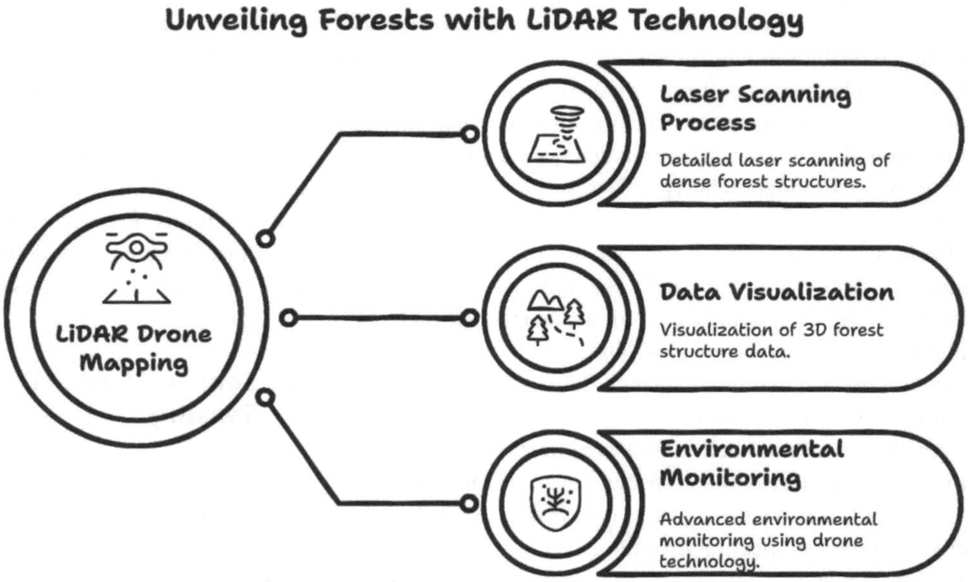

laser scanning process, Data visualization and Environmental monitoring.

An overview of LiDAR drone mapping technology for forest monitoring appears in Fig. 1 with laser scanning, data visualization and environmental monitoring as its main functional areas. The LiDAR (Light Detection and Ranging) sensor represents the core element of this system because it projects laser pulses from drones to develop highly exact three-dimensional forest structure mappings (Zarco-Tejada et al., 2014). The laser scanning system operates through forests even when canopy density is high to deliver precise measurements of height structure and biomass density which serve as essential indicators for forest surveillance. Processing of unprocessed LiDAR point clouds into visualizable results stands as the second essential component. The 3D mapping solution gives scientists together with forest managers the ability to track spatial patterns in addition to observing temporal changes as they design strategies for thinning operations and reforestation projects and the creation of fire safety zones. The visual data interpretation assists in both situation awareness and enables forestry administrators to direct their operational plans through data. The figure presents environmental monitoring as a final application which sends LiDAR system-enabled drones to evaluate ecological signs including water flow alongside habitat fragmentation and erosion and slope stability. The combination of LiDAR technology delivers significant operational power to researchers and management teams who work toward smart sustainable forestry development (Patel & Wong, 2023). The role of hyperspectral and LiDAR sensors in UAV platforms is crucial. Hyperspectral sensors collect hundreds of narrow spectral bands, enabling detailed assessment of vegetation health, species differentiation, and early disease detection. LiDAR technology, on the other hand, allows high-precision 3D mapping of forest canopy, tree height, biomass estimation, and terrain modeling. These integrations empower UAVs to collect dense, accurate datasets critical for sustainable forest management. For an Instance, a pilot project in the Amazon Rainforest successfully deployed LiDAR-equipped drones to map biomass distribution, aiding in conservation planning.

5 Applications of Drones in Forest Monitoring

Forests experience dramatic changes in their monitoring through drones which deliver present-time high-quality data that formerly needed expensive limited-access conventional systems. Drone technology equipped with adaptable sensors and technology now assists biodiversity tracking while detecting deforestation events and forest fires and disease assessments and carbon stock evaluations at higher levels of accuracy and improved operational efficiency. Drone technology excels at uncovering illegal logging activities and detecting deforestation because it helps identify both cases through its operational uses in forest monitoring. The detection of unauthorized forest clearance has become essential because deforestation drives major biodiversity decline, heavy soil erosion and climate disruption. The capability of drones to operate in the air enables live surveillance through high-definition imagery delivery which tracks outdoor forests during multiple time periods (Dandois & Ellis, 2013). Forest managers use image comparisons between different time periods to monitor what the forest environment has changed alongside identifying illicit tree removal and unapproved construction activities. The use of nighttime thermal sensors enables these drones to reveal covert dark-time activities that protect forests with an enhanced level of effectiveness. Drones have revolutionized the way the forests get monitored by providing real-time, high-resolution valid data which was previously very difficult to get and costly also. With highly advanced sensors and AI-based analysis, the usage of drones has been significantly increased to measure biodiversity, observe deforestation, recognize diseases, monitor forest fires and even estimate carbon stocks. Deforestation is a very serious environmental issue causing biodiversity loss, global warming and soil erosion. The extraction of timber illegally is one of the key issues of deforestation, which mainly happens in locations where ground-surveillance is not practical. The drones utilize their surveillance capabilities to generate high-definition tree cover images which allow monitoring deforestation changes across time periods. Before-and-After Monitoring: Authorities can monitor deforestation trends by comparing images of drones taken at varying time intervals. Detection of Unauthorized Activities: Drones serve as tools for detecting both illegal logging operations and unauthorized building of roads and encroachments in protected forest areas (Wallace et al., 2012). Nighttime Monitoring: With the advantage of thermal imaging, drones can be operated during nights to identify illegal loggers who attempt under darkness. Drone plays an important role in assessing biodiversity and monitoring wildlife without disrupting natural habitats. Identification of trees: Hyperspectral imaging sensors equipped on drones can differentiate between various tree species making it easy for conservationists to know about these. Drones built with thermal sensors can recognize warm-blooded creatures in dense forest areas thereby assisting in species tracking. Aerial mapping identifies sites where human activities like agriculture and mining are causing encroachment on wildlife habitats. Monitoring Migration Patterns: Drones can observe mammals and birds in seasonal movements, allowing ecologists to make them know about conservation practices and migration patterns. Forests can catch up with different types of diseases and pest infestations, when not identified early, there is a chance to spread quickly. Traditional ground-based monitoring systems are very slow and labor-consuming, making drones a worthwhile way for the purpose of early detection of diseases. Early Detection of Infected Trees: Multispectral and hyperspectral cameras are able to recognize disease in trees even before the visible symptoms appear. Monitoring Pest Infestations: Drones help identify bark beetle pest infestations to prevent the destruction of extensive tree regions in forests. Accuracy Monitoring: Drones effectively complete automated surveys of entire forested areas within a short period to create accurate assessments of problematic sites. Controlling the Spread of Disease: Early identification of affected trees will enable forest managers to cut down or isolate infected trees to stop further spreading. Drones are already being employed to control outbreaks of forest diseases in many parts of the world thereby reducing economic and ecological loss. Wildfire creates a growing danger for forested areas worldwide which frequently leads to permanent damage of both ecosystems and human residential centers. Conventional approaches of monitoring forest fires are based on satellite imagery and watch towers, of monitoring fires are based on watchtowers and satellite imagery, which operate slow and less accurately. Early Detection with Thermal Imaging: The drone system uses thermal camera technology to detect heat variations which signal the beginning of a wildfire before it becomes visible to the human eye. Real-time Fire Monitoring: Real-time fire monitoring happens through drone aerial video delivery which grants firefighting responders visibility for preventing resource misallocation and determining fire dimensions. Fire Risk Zone Mapping: Drone technologies use dry vegetation maps together with wind conditions as well as past fire patterns to identify areas which present high wildfire risk. Post-Fire Damage Mapping: The aerial drone and post-fire damage mapping system allows operators to monitor sites after fires so they can create proper reforestation plans. Natural carbon removal process of forests helps to reduce atmospheric CO_2 levels thus fighting global climate change. The determination of carbon storage in forests serves as a fundamental requirement to create global climate policies and support conservation efforts. Accurate data collection using drones enables the policymakers and scientists to make strategic decisions to fight climate change as well as shield forests. Tree Height and Canopy Density Measurement: The biomaterial and carbon stock assessment is possible through 3D forest modeling obtained through LiDAR drones. Detection of forest cover alterations by drones reveals the areas that lose their capacity to absorb carbon. Drones assess carbon emission depletion areas by monitoring changes in forest density noticing the locations where carbon capture abilities decrease. Drone-generated data enables official certification of protected programs by both governmental entities and non-governmental organizations which permits them to conduct carbon market trades. Drone technology experienced substantial development which brought about efficiency and precise as well as scalable forest monitoring capabilities. Modern forest monitoring relies heavily on artificial intelligence (AI) and cloud computing as well as swarm technology and blockchain because of their advancements. The implemented technologies speed up deforestation detection which also improves biodiversity surveys and enhances conservation planning and strategy development. Drone-based Forest monitoring achieves recent advancements through several modern developments. Drones play an essential role in detecting forest fires and facilitate how emergency teams perform their fire response operations. Wildfires endanger forests to a significant degree because of rising climate instability. Drone sensors with thermal imaging capabilities detect heat variations at fire's initial stages which are yet to show visible signs. The system provides important warnings allowing firefighters to start necessary actions sooner and perform targeted firefighting operations. Emergency teams use live drone videos to evaluate the advancement and force of fires thus enabling better strategic resource deployments. The combination of drone technology allows for mapping fire-prone areas through vegetation analysis with wind property data and fire records evaluation. When wildfires burn over territory drones can generate essential data for restoration and reforestation planning purposes. The models assist in carbon accounting since they reveal locations affected by forest degradation that results in diminished carbon capture abilities. The long-term assessment of forest composition alterations resulting from climate stressors is possible through drone technology because it enables forest composition shifts monitoring. Drones provide accurate data which permits the verification of conservation programs thus making them eligible for environmental markets. The development of drone technology creates fundamental changes for forest management practices in modern times. The availability of drones gives users access to timely precise actionable data because they enable early fire detection and habitat mapping as well as carbon assessment. Environmental professionals receive enhanced insight from growing drone utilization which allows them to protect forests through more powerful sustainability methods. Incorporating hyperspectral sensors in drone systems has allowed conservationists to monitor plant species diversity in the Congo Basin. A real-world deployment used hyperspectral drones to detect early signs of plant stress due to illegal logging, allowing preemptive action by forest rangers. For example, Drone-based hyperspectral imaging was used in Indonesia to detect illegal palm plantation expansion into protected forests.

6 Latest Advancements in Drone-Based Forest Monitoring

Figure 2 demonstrates an organized method for handling deforestation with artificial intelligence tools embedded in drone-based forest observation systems. The diagram

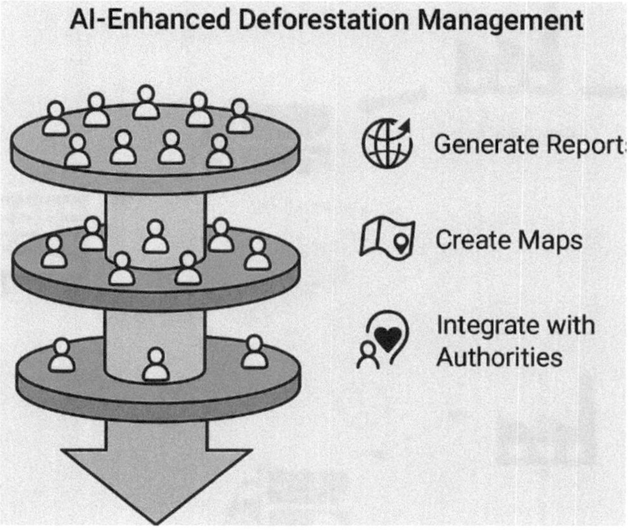

Fig. 2 Describes a systematic approach to manage deforestation using AI tools

constructs a complex structure of AI algorithms which process drone-obtained aerial and geospatial data to assist critical decision functions related to forest preservation and deforestation regulation. Through this system, users obtain complex reports showing forest degradation patterns coupled with canopy decrease measurements and time-based biomass variations. Once forest maps reach a higher resolution level they display the affected regions while tracking the current progress upgrades. The spatial planning process for conservation zones and reforestation projects and buffer area development receives assistance from these maps through their support of environmentalists together with researchers and policy-makers. The last step focuses on integrating AI-enriched information directly with forest departments together with enforcement agencies. Information shared from this system leads to immediate action and enhanced monitoring and data-based law enforcement efforts for stopping illegal activities like logging and land clearing activities (Moreno & Zhang, 2023).

Artificial intelligence (AI) has revolutionized the way forest is being monitored, with the ability to identify deforestation, diseases, and any illegal activity with less intervention from humans. AI systems that process drone images help identify the regions where tree removal occurs. Thanks to this technology manual image analysis has reduced its process duration and human labor requirements. The detection capability of artificial intelligence-based image processing spots leaf color changes that indicate diseases or stress before anyone notices any visual symptoms. Using large databases wildlife species detection and counting becomes feasible because it provides conservationists with the capability to track species movements and monitor populations. Machine learning models use combined historical fire data and plant condition data as well as weather data to generate predictions about fire-prone areas. Historical process of drone survey data involved enormous data sets which required time to download and process by computers. Real-time drone data processing happens possible due to cloud-edge system integration which lets organizations take immediate action. Drone sensors transmit all the collected data to cloud-based platforms where AI tools analyze data in real time. This enables various stakeholders (governments, NGOs, scientists) to remotely access and share forest data and collaborate. Drone-operated AI processors command processing power to analyze images as they fly thus eliminating the need to transmit complete data to cloud-based systems (Nex & Remondino, 2014). The drone system sends automatic notifications about unlawful deforestation to officials when it detects these activities through its forest flight route. The quick response time for threats including wildfires and deforestation can be achieved through this approach. One drone only covers a restricted area during a single flight. Multiple drones using swarm drone technology operate in unison to perform extensive forest monitoring operations. Swarm drone technology is now making forest conservation on a large scale more economical and efficient, covering a large area in a portion of the time it takes using conventional techniques. A region receives segmented divisions through swarm drones that operate simultaneously which reduces aerial survey duration. Monitoring of vast rainforests for deforestation and the tracking of forest fires receives exceptional value from these methods. During wildfire emergencies swarm drones rapidly inspect areas to detect blaze hotspots before helping emergency responders at the firelines. One drone fleet allows simultaneous deployment to monitor many animal herds thus improving species count accuracy. Drone technology permits seed planting in areas where forest clearing takes place leading to automated tree planting procedures. Swarm drones achieve high efficiency in tree-sowing operations since they can plant up to thousands of trees per day during reforestation efforts. Blockchain technology, which has been used for secure digital transactions, is being utilized in forest conservation activities. Environmental monitoring receives benefits through blockchain implementation which delivers transparent operations and secure environments along with total accountability (Manfreda et al., 2018). The collaborative force between blockchain technology and drone systems ensures reliable environmental protection by providing transparent forest protection methods. Every data drone collects about deforestation together with carbon stock measurements and conservation events get stored on blockchain ledger systems to prevent anyone from changing these records. This method prevents the occurrence of corruption or false reports in conservation operations. Many organizations participate in carbon credit programs by funding forest preservation as a way to achieve carbon footprint balance. Through blockchain

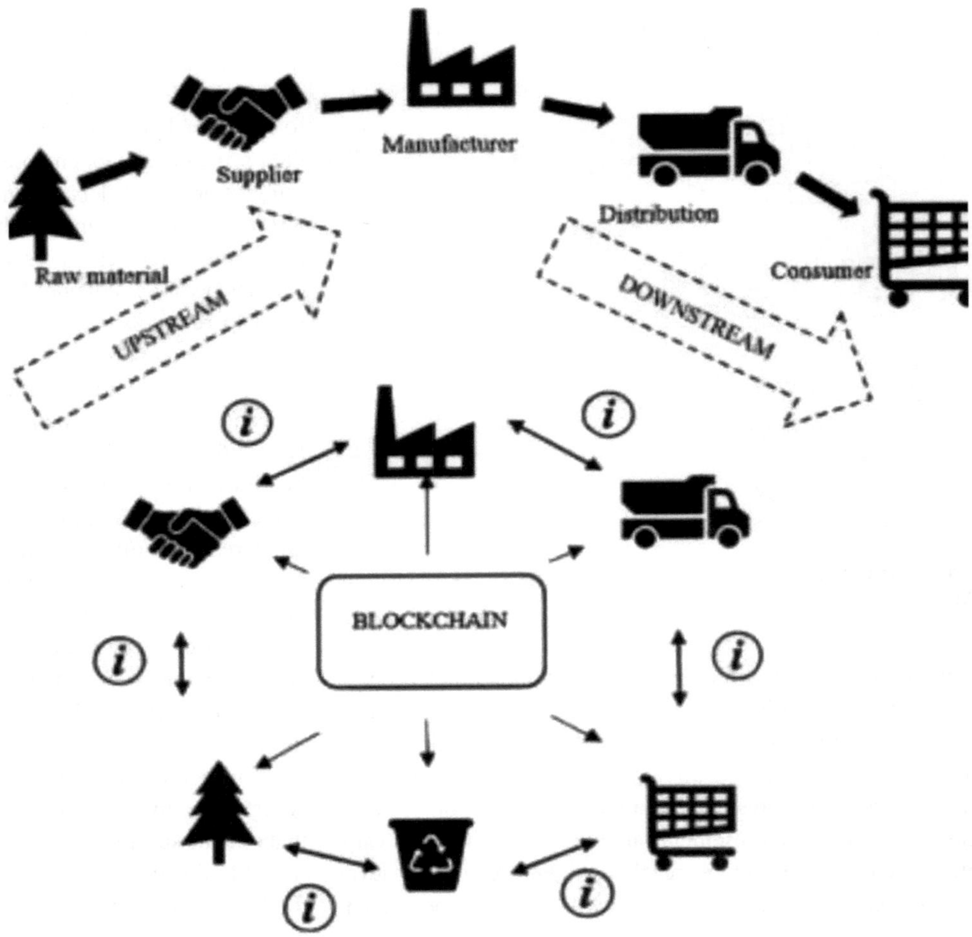

Fig. 3 Blockchain-enhanced forest conservation process

verification systems proper documentation of credits avoids actual credit misuse. Blockchain facilitates safe documentation of drone-recorded evidence that demonstrates illicit operations (including logging and mining) for local communities and domestic conservation agencies within their territories. The capability assists organizations in their fight against environmental justice legal challenges.

The blockchain-enhanced forest conservation process (Fig. 3) integrates distributed ledger technology into forest product supply chains for delivering transparency together with resource management ethics and traceability. The diagram presents a circular structure which includes all steps in the forest-derived goods lifecycle that begin with material extraction and end with consumer use followed by sustainable recycling. The traditional supply chain operates between raw forest material gathering and consumer delivery through a series of stages displayed in the upper part of the figure. The segment contains upstream and downstream operations which are prone to transparency problems along with fraudulent activities and unsustainable practices. The blockchain layer acts as the core component that links together all members involved in supply chain operations (Torresan et al., 2017). Blockchain guarantees secure and immutable decentralization of transaction records which verifies that all supply chain phases from harvest to consumption are both auditable and verifiable. After demand information flow, point (i) includes blockchain-accessible data sources that confirm sustainability practices together with legal adherence and environmental measurement results. The lower part represents how information movements create pathways to direct sustainable recycling processes and policy intervention decisions (Almeida et al., 2023). Pilot Case Study 1: AI-driven drones have been employed in Australia for early wildfire detection. By analyzing aerial imagery and thermal patterns, these drones predict fire outbreaks, saving critical response time. Pilot Case Study 2: In Canada, multi-sensor UAV platforms combining LiDAR, hyperspectral, and thermal sensors monitor carbon sequestration rates, contributing to national climate change reporting initiatives. Challenges in Implementation: Although promising, AI and blockchain applications face technical hurdles such as large-scale data processing, regulatory approval delays, interoperability among agencies, data standardization issues, and cybersecurity risks for data integrity.

7 Challenges and Limitations

Technological advancement of drones in forest monitoring operations presents multiple operational difficulties. Mass deployment of drones encounters obstacles at technical, legal and ethical levels. Effective drone-based monitoring requires complete comprehension of these obstacles to achieve appropriate usage and responsible operation. Governments in different parts of the world enforce strict drone operation rules due to safety matters combined with security and privacy protection needs. National regulations regarding drones differ between countries which creates obstacles for free use of drones in forest monitoring operations. The operation of drones requires official permits from numerous countries whenever these devices are flown above forests which include both protected areas and national parks. The altitude constraints that certain areas enforce limit drone flight heights thus creating problems for quality data acquisition. The authorities in specific areas such as military zones and border sectors have designated them as drone-free zones (Grenzdörffer et al., 2008; Salamí et al., 2014; Whitehead et al., 2014). The areas surrounding restricted zones remain inaccessible to complete forest monitoring operations because of legal constraints. Ownership rights to data collected by drones through image or data acquisition above privately owned or government forests becomes problematic. The drone monitoring program faces occasional resistance from indigenous forest residents because they dispute their land rights. Many countries restrict the import of advanced drone technology because they aim to protect national security and limit possession of elite drone equipment. To counter these challenges, researchers as well as conservationists need to stay in close contact with the government agencies and local communities to gain the required permits and follow rules and regulations. Drones encounter operational barriers which limit their ability to monitor forests on broad terrains. Commercial drone batteries limit flights to 20–40 min thus making it difficult to map extensive forest areas within single operations. The exchange or recharging operations of batteries across remote forest locations requires additional planning regarding logistical tasks. The ability to hover quadcopters enables recording high detail images although fixed-wing drones supply extended flight duration beyond free flight limits. Drones have increased sensitivity to all weather elements including strong windy conditions along with rainfall and dense fog. Sensors face potential disruptions when operating under dense fog conditions or heavy precipitation that leads to compromised results from measurement systems. High forest humidity creates conditions under which electronic systems inside drone's face water-linked functional problems. High-resolution drone photography coupled with LiDAR scanning generates an extensive amount of data which goes beyond the comprehension capabilities of computers. Data processing along with transmission turns into a complex task when operating in remote locations without internet service. Real-time analysis becomes restricted because most forest areas lack network connectivity needed to operate AI-based analysis software in the cloud. Scientists study solar-powered drones and thin batteries as well as onboard AI processors for developing drones which extend their range and function more efficiently. The collection of data through drones for forest conservation remains beneficial yet privacy concerns grow stronger because drones frequently cross over residential lands together with indigenous territories. The drone monitoring system makes some communities concerned about privacy violations while they fear their area may be switched from conservation to surveillance functions (Lucieer et al., 2014). Drone photo and video collection from forest residents without permission creates substantial morality dilemmas. The presence of loud drones causes negative impacts on animal behavior by creating disturbances. Certain bird species identify drones as dangerous aerial objects which causes them to react with aggression potentially leading to damage of both drone systems and wildlife. Multiple operational problems exist when drones advance technologically for forest monitoring applications. Large-scale drone operations create problems that originated from technical complications as well as legal and ethical boundaries. Drone-based monitoring operations depend on full acknowledgment of present obstacles for reaching proper implementation and professional drone management. Various nations implement strict drone operation regulations because safety issues intersect with requirements for protecting security and privacy concerns. The different drone regulations enacted in different countries produce barriers that prevent unregulated drone utilization in forest damage survey operations. Operation of drones demands authorized permissions from multiple countries before drones can fly over forests that contain protected areas and national parks. Flight height restrictions in particular regions prevent drones from obtaining proper quality data. Policy makers from designated military areas and border controls have established drone flying restrictions in their regions. The regions next to restricted zones become fully inaccessible for complete forest monitoring activities because of formal governmental limitations. The rights to drone-gather data through images or data acquisition above privately and government-owned forests present major difficulties. Forest residents occasionally resist drone monitoring because they claim their ownership of forest territory. Several nations place restrictions on automatic drone technology imports to protect both military security and control advanced drone possession among their citizenries. The research community alongside conservation experts need to maintain continuous connections with both government agencies and local communities to acquire needed permissions while following established rules (Marris, 2013). The monitoring capabilities of drones are constrained on vast forest areas by multiple operational limitations. Drone commercial batteries run 20–40 min per charge which poses challenges to mapping extensive forest territories in one operation. The implementation of battery exchanges or recharges for remote forest areas involves logistical planning for additional operations. Apart from being able to hover the quadcopter can take high-definition footage yet fixed-wing drones provide longer flights than standard flight limitations allow. The drone equipment functions best in all environments including heavy wind along with heavy rain and dense fog situations. The performance of sensors deteriorates under dense fog along with heavy precipitation which creates problems for their measurement systems. The humid forest conditions produce environments that create detrimental effects on drone electronic systems which lead to operational problems. Operating in remote areas without Internet service makes data processing together with data transmission increasingly complicated. A lack of required network connectivity in forest areas makes real-time analysis software limited since it needs cloud-based AI-based analysis software. Scientists analyze solar-powered drones and thin batteries as well as onboard AI processors to develop drones that have longer operation distances with enhanced efficiency. The data collection process with drones' benefits forest preservation activities although privacy challenges expand since these aircraft frequently enter residential districts and protected native lands. People living in particular regions show worry about privacy breaches that stem from drone monitoring systems because they also fear their protected areas could become surveillance zones. Uninformed drone video and photo acquisition from forest residents poses significant moral boundaries to residents. The operating sounds of drones have damaging effects on animal behavioral patterns through disrupting natural environments. The bird identification of drones as safety threats leads to aggressive interactions resulting in harm toward drone technology and wild animals (Anderson & Gaston, 2013; Lisein et al., 2013; Tang & Shao, 2015). Drone-based forest monitoring, while transformative, raises several privacies, environmental, and operational challenges that require careful attention. Drones flying over indigenous territories without consent can breach the privacy and rights of local communities. Establishing ethical guidelines that prioritize community engagement and ownership of collected data is vital to ensure trust and respect. Additionally, the environmental footprint of drones cannot be overlooked. The noise generated by UAVs may disturb sensitive wildlife, altering behavioral patterns of birds and mammals. Furthermore, improper disposal of lithium-ion batteries poses risks of soil and water contamination, emphasizing the urgent need for eco-friendly technologies. Although drones offer scalable solutions for forest monitoring, achieving community acceptance is essential. Engaging local populations in planning and operational processes helps position drones as tools for conservation rather than surveillance. Operational barriers also persist, with weather conditions like strong winds, heavy rainfall, and dense fog impacting flight stability and data quality. Battery limitations restrict long-duration flights over expansive forest areas, making it necessary to develop solar-powered models and enhance logistical support systems.

8 Future Trends and Directions

Forest surveillance with drones' progresses rapidly because new technologies provide them with elevated intelligence together with improved operational efficiency for major conservation undertakings. Drones in forestry operations will develop according to future advances in sensor technology miniaturization and AI automation as well as IoT networks and enforcing proper deployment regulations (Gini et al., 2013). The combination of emerging technologies and developments will remove existing performance barriers to create an advanced environmental protection drone. Development in drone technology involves producing smaller sensors that operate with greater abilities while requiring less weight. The new technology enables increased flight duration and greater precision in data acquisition as well as operations in jungle environments. Standard LiDAR devices along with hyperspectral and multispectral cameras require heavy drones because of their bulk size. Scientists today focus their research on making these sensors smaller but at the same time improving both their resolution and data processing capabilities. Advanced sensor systems will become accessible for compact drones when they integrate with such systems thus creating cheaper drone systems. Future drones will integrate multiple sensors onto a single platform which allows them to extract different types of data in a single operation. Multiple sensors can fit into a single drone system to perform 3D mapping with LiDAR and biodiversity monitoring with hyperspectral imaging and fire detection with thermal cameras. Scientists are conducting research on developing weight-efficient batteries which can improve drone operational time. Solar-powered drones achieve long-term flight autonomy by continuously mapping forest areas while requiring no repeated battery replacements to operate continuously. Artificial intelligence (AI) is going to take center stage when it comes to the next level of forest surveillance using drones. AI-enabled drones will be functioning with little to no human interaction, reducing forest surveys to much faster and more efficient levels (Berni et al., 2009; Li & Wang, 2025; Zhao et al., 2024). AI software will allow drones to navigate without human control, dodging obstacles such as trees, hills, and power cables. Drones will map unexplored forest terrain without pre-programmed flight routes. Rather than being visually inspected by drones, AI will automatically classify and analyze forest data. AI instantly detects deforestation, illegal logging, disease-ridden trees, and wildlife activity patterns. AI-driven drones will utilize automatic charging stations set in forests. Rather than flying back to human operators, drones will land, charge, and resume their tasks independently. Multiple drones to operate collaboratively in a network (swarm technology) will enable quicker and more precise monitoring of the forest. AI will arrange these drones so that they scan various areas efficiently without repeating any area. The Internet of Things combines multiple smart devices that share digital information between them. Drones will support IoT sensors installed in forests to build an intelligent system that monitors forest conditions in real time. New IoT technology will include climate sensors, acoustic sensors, and movement sensors set up at forest levels. The equipment will measure temperature, moisture levels, carbon levels plus detect unauthorized usage. The environmental protection team will receive instant updates once drones inspect sensor systems from forest floors. The system will help control situations such as forest fires and wildlife tracking because decisions can be made right away. The IoT technology that tracks endangered animals uses GPS collars to show their specific positions. The drones enable real-time watching and photographing of animals while leaving their habitats untouched (Reiersen et al., 2022). Term sensors will find high temperatures plus measure smoke levels and the low moisture of the environment. Drones will start their search for wildfire risks at once and help firefighters respond faster. With the advance of drones' technology, governments and conservation agencies must shape policies that will sanction responsible and efficient use as political leaders perceive that it is inevitable. Specific rules and guidelines of application for drones in conservation have to be given by the governments. To make access to conducting forest monitoring easier, conservation organizations should be granted exemptions for commercial use of drones. Drones should be used with a policy that data collected must be used ethically, and communities should be openly shared with. Drones cannot be used for local communities and the indigenous populations of which were witnessed before. To implement drone technology, the governments and research organizations must invest into the new drone technology development, including: Artificial intelligence-based drones for mass monitoring. Lightweight, high-efficiency batteries for longer flight times. Using multi-sensor drones to achieve all the data required in one flight. There is a need to train forest rangers, researchers and local communities to Conservation organizations to operate drones (Hemati et al., 2022; Koh & Wich, 2023). Training must focus on data analysis, integration of AI and ethical considerations. International level should be formed at an international level of cooperation among nations, research organizations and conservation groups. Drones have open-source data platforms, and they should be available to collaborative global forest tracking initiatives (Savinelli et al., 2024). Future systems will focus on AI-autonomous navigation, integration with IoT sensor networks, solar-powered drones, and enhanced ethics-driven conservation frameworks. Collaborative global initiatives involving governments, NGOs, and communities will be key to ethically scaling drone-based monitoring solutions. There will also be emphasis on developing biodegradable drone components to reduce environmental footprint.

9 Conclusion

Drone dependent remote sensing has transformed the way the forests get monitored as it delivers unmatched accuracy, efficiency and real time data collection. With the advancement of this technology, it is very purposeful in global conservation practices as it will continue to grow. The latest drones can effectively watch forests, but their benefits depend on responsible and ethical behavior in environmental protection. The remote sensing and drone's combination has transformed the situations of forestry, enabling monitoring and conservation to be much faster, accurate and affordable. People used basic satellite imagery but could only see limited details through surveys on the ground. Drone technology with LiDAR, multispectral, and thermal sensors offers high-resolution real-time images of forest cover. Manual forest monitoring through field surveys needs many weeks or months because dense and hard-to-reach places are difficult to access. Aerial drone inspections of large areas can be completed in just a few hours and enable better monitoring results. Drones use thermal detection and spectral sensors to spot wildlife patterns while identifying changes in natural habitat and destruction of plant diversity. The drones deliver a safe approach to watch species without entering their natural environments. Drones help spot the start of illegal logging and forest fires as well as detect pests that could become widespread environmental disasters. The monitoring system allows for prompt actions when disasters start. The carbon-absorbing power of forests lets drones measure how much carbon forests hold and what forest shrinking or climate change does to the environment. Drone information helps us manage forests better for both current vegetation and future tree planting initiatives. The modern forestry industry uses drones to conduct smarter and more data-oriented conservation work more productively. The proper control of new technology improvements within conservation efforts worldwide is essential. Drones will improve our ability to address environmental problems through faster and more exact findings at lower costs. Drones will serve as a major conservation tool as natural habitats continue to disappear from large-scale damage to our planet. These upcoming developments show drones will help nature protection more strongly in the years to come. Several drones with AI abilities will conduct independent forest inspections at the same time. Artificial Intelligence reads images to detect environmental crimes while showing loss of natural habitats and changes in species counts. IoT sensors on the ground with flying drones will produce health data for our forest system at all times. Forest specialists track wildlife behavior and forest conditions by using cloud-based resources from their regular workplaces. Scaling Conservation Research and Reforestation Activities. Drones will help worldwide reforestation efforts by being used more often in tree planting initiatives. They will help scientists track rare species easier through their observation work. Enhancing Global Environmental Policies and Decision-making. Effective drone data will provide researchers with trusted facts to support updates in forest defense and carbon credit systems. Both public authorities and international organizations can use drones to protect forests better by spotting illegal actions using drones as their main tools. Public use of drones for environmental preservation demands proper control to stop ethical and environmental problems from development. Drone monitoring systems may violate the private space of local tribal residents and forest inhabitants. The system should have moral standards to protect personal privacy during data collection from the air. If drones stay active without proper management, they will affect natural surroundings negatively. The law should control drone flight altitudes and times to defend the environment. Most commercial drones rely on lithium-ion batteries that cause environmental problems during mining and discarding processes. Researchers need to develop drones that use solar power and dissolve without harming the environment. The local community and conservationists need to share power in how drones are used in their area. Together governments, NGO researchers and local teams should partner to make drone technology serve conservation and human values. Innovation in drone technology needs equal attention to ethical standards to confirm these tools aid nature protection without hurting the environment and local people. Unground surveys of basic satellite imagery showed restricted information while drone technology using LiDAR and multispectral and thermal sensors produces high-definition live imagery of forest cover. Many weeks or months are required for manual forest monitoring through field surveys mainly because remote and inaccessible forest areas hinder investigator access. Large areas become inspectable through aerial drone technology which provides fast results within just a few hours. Drone system sensors with thermal abilities and spectral detection capabilities track wildlife patterns and detect both habitat transformation and areas where vegetation ceases to exist. The drones provide a secure method to surveil species by avoiding human contact with their native habitats. Drone technology enables early detection of illegal logging initiation and forest fires and it identifies pests that have the potential to become systemic environmental threats. Through this monitoring system, emergency response actions become possible at the beginning of disasters. Drones use their ability to absorb carbon from forests to measure the quantity of captured carbon and report any effects of shrinking forests and environmental changes. Drone data improves our capability to handle forests for present vegetative maintenance while enhancing our outlook on upcoming tree planting activities. Drones enable the contemporary forestry industry to practice data-based conservation work through smarter methods that increase productivity. The management of fresh technology solutions in worldwide conservation initiatives demands close oversight. Environmental problems will become easier to address through drone technology which provides precise findings at reduced expenses. Drone technology functions as a crucial conservation tool to preserve endangered natural habitats due to ongoing large-scale damage of our planet. Drones are shaping up to become strong implementers of nature protection during upcoming years. Numerous drones equipped with AI systems will carry out separate forest inspections simultaneously. Under AI systems images are processed to find evidence of environmental offenses and to track alterations in natural habitat areas as well as wildlife populations. Drones carrying IoT sensors will generate forest system health records that are collected from the ground while drones are flying. Forest specialists review wildlife behavior together with forest conditions through cloud-based resources which they access from their conventional work sites. The advancement of conservation research together with reforestation activities becomes possible through large-scale implementation. Drones play an essential role in worldwide reforestation programs because they participate actively in planting tree projects. The process of observation will become simpler for scientific researchers who need to track rare species. Research data from drones helps establish facts which enable scientists to enhance globally implemented environmental policies and decision-making strategies. The combination of public authorities and international organizations uses drones effectively to care for forests through drone-based monitoring of illegal behavior. The implementation of drones for public environmental preservation requires suitable management systems to prevent moral and ecological challenges from arising. Drone monitoring systems enter into private areas of tribal communities along with local forest dwellers without their consent. The system requires moral guidelines to safeguard personal privacy throughout air-based data procurement operations. The absence of drone management system results in negative impacts on surrounding natural areas. Legislative authorities need to establish rules for drone flight elevation as well as duration to protect natural environments. The standard battery of choice for commercial drones are lithium-ion cells which generate environmental risks throughout their production cycle as well as disposal. Researchers should create drones powered by solar energy while these drones also have environmental-friendly decomposition capabilities. The community together with conservation entities must unite to decide about drone use inside their region. National governments and researchers from NGO organizations and residents must collaborate to achieve the shared objective of using drone technology for environmental protection and people-centric purposes. Drone technology innovation requires equal consideration of ethical standards to establish these tools protect nature while causing no harm to the environment and local communities. The production and disposal processes of drones need thorough examination regarding their environmental sustainability aspects. The primary source of most drones depends on lithium-ion batteries whose extraction process along with disposal methods cause environmental damage. New technology development for environmentally friendly substitutes between biodegradable materials and solar-powered drones needs immediate priority. Community members should lead the deployment of effective drone programs. Decision-making processes for conservation efforts succeed most when they involve local communities together with NGOs and both researchers and government agencies. The benefits from drone technology for forest conservation need to be balanced with ethical standards and environmental sustainability and active community participation during its application. When managed properly drones can operate as effective instruments for protecting natural environments together with upholding the rights and importance of those who depend on them.

References

Almeida, F., Rodrigues, J. J. P. C., & Kumar, N. (2023). Blockchain-based frameworks for forest conservation and sustainable supply chain governance. *Journal of Cleaner Production, 417*, Article 137846. https://doi.org/10.1016/j.jclepro.2023.137846

Anderson, K., & Gaston, K.J. (2013). Lightweight unmanned aerial vehicles will revolutionize spatial ecology. *Frontiers in Ecology and the Environment, 11*(3), 138–146.

Berni, J. A. J., Zarco-Tejada, P. J., Suárez, L., & Fereres, E. (2009). Thermal and narrowband multispectral remote sensing for vegetation monitoring from an unmanned aerial vehicle. *IEEE Transactions on Geoscience and Remote Sensing, 47*(3), 722–738.

Dandois, J.P., & Ellis, E.C. (2013). High spatial resolution three-dimensional mapping of vegetation spectral dynamics using computer vision. *Remote Sensing of Environment, 136*, 259–276.

Getzin, S., Wiegand, K., & Schöning, I. (2012). Assessing biodiversity in forests using very high-resolution images and unmanned aerial vehicles. *Methods in Ecology and Evolution, 3*(2), 397–404.

Gini, R., Pagliari, D., Passoni, D., Pinto, L., Sona, G., & Dosso, P. (2013). UAV photogrammetry: Block triangulation comparisons. *The International Archives of the Photogrammetry, Remote Sensing and Spatial Information Sciences, 40*(1), 361.

Grenzdörffer, G.J., Engel, A., & Teichert, B. (2008). The photogrammetric potential of low-cost UAVs in forestry and agriculture. *The International Archives of the Photogrammetry, Remote Sensing and Spatial Information Sciences, 31*(B3), 1207–1214.

Hemati, M. A., Hassanlou, M., Mahdianpari, M., & Mohammadimanesh, F. (2022). Asystematic review of Landsat data for change detection applications: 50 years of monitoring the Earth. *Remote Sensing, 14*(14), 3205. https://doi.org/10.3390/rs14143205

Koh, L.P., & Wich, S.A. (2012). Dawn of drone ecology: Low-cost autonomous aerial vehicles for conservation. *Tropical Conservation Science, 5*(2), 121–132.

Koh, L. P., & Wich, S. A. (2023). Conservation drones: Mapping and monitoring biodiversity. *Drones, 7*(6), 398. https://doi.org/10.3390/drones7060398

Li, X., & Wang, A. (2025). Forest pest monitoring and early warning using UAV remote sensing and computer vision techniques. *Scientific Reports*, 15, Article 401. https://doi.org/10.1038/s41598-024-84464-3

Lisein, J., Pierrot-Deseilligny, M., Bonnet, S., & Lejeune, P. (2013). A photogrammetric workflow for the creation of a forest canopy height model from small unmanned aerial system imagery. *Forests, 4*(4), 922–944.

Lucieer, A., Malenovský, Z., Veness, T., & Wallace, L. (2014). HyperUAS—Imaging spectroscopy from a multirotor unmanned aircraft system. *Journal of Field Robotics, 31*(4), 571–590.

Manfreda, S., McCabe, M.F., Miller, P.E., Lucas, R., Pajuelo Madrigal, V., Mallinis, G., & Helman, D. (2018). On the use of unmanned aerial systems for environmental monitoring. *Remote Sensing, 10*(4), 641.

Marris, E. (2013). Drones in science: Fly, and bring me data. *Nature News, 498*(7453), 156.

Moreno, A., & Zhang, J. (2023). Leveraging AI and drones for real-time deforestation monitoring and policy integration. *Environmental Monitoring and Software Systems, 41*(1), Article 102581. https://doi.org/10.1016/j.envsoft.2023.102581

Nex, F., & Remondino, F. (2014). UAV for 3D mapping applications: a review. *Applied Geomatics, 6*(1), 1–15.

Paneque-Gálvez, J., McCall, M.K., Napoletano, B.M., Wich, S.A., & Koh, L.P. (2014). Small drones for community-based forest monitoring: An assessment of their feasibility and potential in tropical areas. *Forests, 5*(6), 1481–1507.

Patel, R., & Wong, M. (2023). Advances in drone-based LiDAR for forest structure analysis and environmental monitoring. *International Journal of Remote Sensing, 44*(3), 512–531. https://doi.org/10.1080/01431161.2023.2175648

Reiersen, G., Dao, D., Lütjens, B., Klemmer, K., Amara, K., Steinegger, A., Zhang, C., & Zhu, X. (2022). ReforesTree: A dataset for estimating tropical forest carbon stock with deep learning and aerial imagery. arXiv preprint arXiv:2201.11192. https://doi.org/10.48550/arXiv.2201.11192

Salamí, E., Barrado, C., & Pastor, E. (2014). UAV flight experiments applied to the remote sensing of vegetated areas. *Remote Sensing, 6*(11), 11051–11081.

Savinelli, B., Tagliabue, G., Vignali, L., Garzonio, R., Gentili, R., Panigada, C., & Rossini, M. (2024). Integrating drone-based LiDAR and multispectral data for tree monitoring. *Drones, 8*(12), 744. https://doi.org/10.3390/drones8120744

Tang, L., & Shao, G. (2015). Drone remote sensing for forestry research and practices. *Journal of Forestry Research, 26*(4), 791–797.

Torresan, C., Berton, A., Carotenuto, F., Filippo, E., Gioli, B., Miglietta, F., & Wallace, L. (2017). Forestry applications of UAVs in Europe: a review. *International Journal of Remote Sensing, 38*(8–10), 2427–2447.

Wallace, L., Lucieer, A., Watson, C., & Turner, D. (2012). Development of a UAV-LiDAR system with application to forest inventory. *Remote Sensing, 4*(6), 1519–1543.

Whitehead, K., Hugenholtz, C.H., Myshak, S., Brown, O., LeClair, A., Tamminga, A., & Eaton, B. (2014). Remote sensing of the environment with small unmanned aircraft systems (UASs), part 2: scientific and commercial applications. *Journal of Unmanned Vehicle Systems, 2*(3), 86–102.

Zarco-Tejada, P.J., Díaz-Varela, R., Angileri, V., & Loudjani, P. (2014). Tree height quantification using very high-resolution imagery acquired from an unmanned aerial vehicle (UAV) and automatic 3D photo-reconstruction methods. *European Journal of Agronomy, 55*, 89–99.

Zhao, F., Meng, R., Huang, C., & Zhao, M. (2024). Long-term post-disturbance forest recovery in the Greater Yellowstone Ecosystem analyzed using Landsat time series stack. *Remote Sensing, 16*(1), 29. https://doi.org/10.3390/rs16010029

Data-Driven Decision Making: Utilizing Big Data for Forest Health

R. Kannamma, Mudassir Khan, V. K. Senthil Ragavan, and C. Ambhika

Abstract

Ecological systems and economic benefits and social institutions derive value from forests but environmental degradation now threatens them because of climate change together with deforestation and growing human activities. The influence that population dependence on forests together with land management systems and governmental controls has on forest environments is major. A comprehensive approach needs creation to build complete forest health management systems that combine natural science knowledge with social aspects. A combined Big Data structure aims to merge socioeconomic information systems with ecological and climatic information for forest health management purposes. Staff members in communities should use mobile apps and citizen science tools to help collect data so they can develop sustainable management solutions with the local stakeholders. The framework proves its suitability by being tested within a tropical rainforest region through a recent study. The framework uses satellite combined with socioeconomic information to recognize vulnerable areas and sets out sustainable use practices for land. This research shows that considering social economic elements during the process increases predictions while supporting ethical decision making. A comprehensive framework creates a connection between nature-based and human-based information to boost forest health management systems.

Keywords

Ecological systems · Economic benefits · Environmental degradation · Deforestation · Population dependence · Land management systems

1 Introduction

Forests act as essential components of global sustainability since they create various ecological services as well as economic and cultural worth. These essential ecosystem systems encounter unmatched difficulties because of the joint effects between climate change and excessive deforestation and human practices that damage ecosystems. The worsening health crisis within forests requires new and complete management solutions which go beyond conventional ecological techniques. The major obstacle when using modern techniques for forest health evaluation lies in the insufficient inclusion of socioeconomic conditions. The connection between human societies and forest ecosystems maintains an absolute significance. The forest dynamics experience significant impact from both the communities dependent on forest resources and the prevailing land-use patterns as well as current governance systems. Omissions of socioeconomic variables from forest health models result in defective examinations which may create unsuccessful intervention approaches. The forthcoming chapter covers an integrated big data framework that resolves this issue by combining sophisticated analytics with community involvement methods to create data-based decisions for forest health advancement (Sahal et al., 2021). The framework uses multiple

R. Kannamma (✉)
Department of AI&DS, Prathyusha Engineering College, Chennai, India
e-mail: kannamma.sridharan@gmail.com

M. Khan
Department of Computer Science, College of Computer Science, Applied College Tanumah, King Khalid University, Abha, Saudi Arabia
e-mail: mudassirkhan12@gmail.com

V. K. Senthil Ragavan
Sri Shanmugha College of Engineering and Technology, Sankari, Salem, Tamil Nadu, India
e-mail: vksenrag@yahoo.com

C. Ambhika
Department of AIML, R.M.D. Engineering College, RSM Nagar, Kavarapetai, India
e-mail: ambhidurai@gmail.com

M. Ashwin et al. (eds.), *Smart Forestry 5.0 for Powered AI Sustainable Environment and Climate*, Sustainable Artificial Intelligence-Powered Applications, https://doi.org/10.1007/978-3-031-99086-1_12

information sources alongside community involvement to create sustainable and equal conservation methods. For hundreds of years, traditional sustainable forest conservation practices have been supported through indigenous traditional wisdom by local communities. The attempt to restore biodiversity relies on unsuccessful modern conservation methods that initiate approaches built without local community engagement because these approaches fail to meet community requirements. Socioeconomic information integration in forest health evaluations enables policymakers to develop interventions that help communities survive forest complications thus maintaining active neighborhood involvement in forest management. Community forestry management and sustainable ecotourism initiatives have become viable economic tools which replace damaging land-use practices since they enable a balance between natural preservation and human development. The emergence of advanced computational tools presents an opportunity to bridge the long-standing gap between ecological science and socioeconomic analysis. By leveraging artificial intelligence, geographic information systems (GIS), and machine learning, conservationists can map forest degradation patterns while simultaneously analyzing human-induced pressures. However, data-driven models must be complemented by participatory governance structures that prioritize transparency, inclusivity, and equitable resource management. A truly integrated big data framework not only enhances predictive capabilities but also fosters collaboration between governments, research institutions, and local communities. This chapter underscores the need for a paradigm shift in forest conservation—one that embraces interdisciplinary research, technological innovation, and community-driven strategies to safeguard forest ecosystems for future generations (Schweier et al., 2019).

In forest management, decision-making utilizes extensive multidimensional data in an organized way to develop guidelines and conservation plans along with running operational decisions. Such a method rejects the practice of decision-making solely from feelings and partial information by focusing on data-based planning and maintaining real-time assessments with adaptable response plans. The use of data-driven decision-making allows forest managers along with policymakers and ecologists and local communities to shift from solely reactive behavior toward predictive and preventive governance improving social as well as environmental results. This section presents methods to use big data analytics for redefining forest health evaluation as well as forest management procedures. The essay introduces machine learning algorithms and spatiotemporal models and decision-support systems as primary technological tools for disease detection and deforestation risk prediction and sustainable resource distribution respectively.

2 Background and Context

The research analyzes how Industry 4.0 technologies can benefit the forestry sector by elaborating on fire detection applications. The paper reviews both benefits and barriers to deploying these technologies for real-life implementation in a specific use case (Sahal et al., 2021). This research paper investigates forest operation sustainability effects while studying both environmental components and economic elements and societal influences. Research continues to require more investigation into long-term forest operational impacts as this work presents an extensive analysis of up-to-date knowledge in the field (Schweier et al., 2019). The research analyses how Internet of Things technology affects logistical processes for better competitiveness. This analysis explores IoT advantages within practical applications of the technology and addresses both obstacles and strategic possibilities involved in deployment (Šulyová & Koman, 2020). The paper develops a forest fire prevention system and detection system and fighting system founded upon fuzzy logic and wireless sensor networks. This paper presents benefits of using these technologies in practical applications followed by an effective case study analysis (Toledo-Castro et al., 2018a). The presented document demonstrates advanced sensors together with monitoring instruments that help support climate-smart forestry practices. Example-based practical applications form the basis for assessing how these technologies deliver their benefits in the paper (Torresan et al., 2021). This research puts forward a forest environmental factors collection platform built around ZIGBEE IoT architecture. The paper demonstrates how this technology works in real-life scenarios followed by a case study that illustrates its practical application (Jin et al., 2023). An IoT-based framework for intelligent smart home environment modelling serves as the proposal for fire safety prevention systems. The document reviews real-world advantages of this technology before demonstrating its operational success through a case study (Saeed et al., 2018). The document presents an introduction to how drones operate with the Internet of Public Safety Things (IoPST) framework in smart urban environments. The paper details the practical benefits of implementing these technologies through an illustrative case which demonstrates their operational success (Alsamhi et al., 2019). The present research focuses on resolving the data gathering and energy transfer issues within UAV-assisted flying access networks for IoT. The proposal introduces UAV integration with built-in energy harvesting components along with wireless power supply systems to enhance network operational duration (Arabi et al., 2018a). A review of primitive optical remote sensing systems used for forest fire detection exists within this work. This text presents an in-depth analysis of all flame and smoke detection techniques used in different technology systems while examining their respective strengths and blind spots (Barmpoutis et al., 2020). The article performs a distributive impact assessment of sustainable forest management (SFM) policy regarding power dynamics among domestic and foreign donor bureaucracies in Bangladesh. The analysis reveals which bureaucracies obtain the most advantages from sustainable forest management policies and further explains how this impacts the creation of new national SFM policies and their implementation and assessment (Giessen et al., 2016).

Future course of smart farm and forest digital transformation faces numerous challenges according to this article. The current situation emphasizes the necessity to develop AI systems which focus on human needs while handling applied sociology together with ethical and legal requirements (Holzinger et al., 2022). The document presents a forecast for the European forest-based sector including analysis of forest advancement alongside timber harvesting requirements and wood market performance and carbon storage measurements in forests. This paper examines both opportunities and challenges in the sector before proposing research direction options for the future (Jonsson et al., 2018). This research investigates the patterns of intense fire damage within historical wildfires affecting topographic, convective and wind-driven regions of Mediterranean pine forests. The document explores what elements affect fire intensity while offering information about the natural processes behind these wildfires (Lecina-Diaz et al., 2014). This research evaluates essential conditions alongside performance motivators which enable digital information sharing under Industry 4.0. This research investigates social capital together with other elements which affect the method suppliers and buyers exchange information (Müller et al., 2020). The document creates a performance model to adaptively scale Apache Spark Streaming through dynamic scaling methods. This paper establishes both implementation benefits and conducts a case study evaluation to show the effectiveness of the proposed performance model (Petrov et al., 2018a).

Forest ecosystems support the continuity of ecological systems in addition to enabling biodiversity maintenance and carbon cycle administration that brings economic and cultural advantages to the global society. Recent decades have brought growing dangers for forest environments because of deforestation along with illegal logging and climate change effects and pest outbreaks and wildfires coupled with non-sustainable land practices. Current forest monitoring approaches which depend mainly on subjective human surveys, unreliable anecdotal evidence and limited sampling prove inadequate to manage recent pressing and extensive forest threats. The current situation mandates a complete revaluation of forest health management systems through invention of precise scalable solutions. The advancing environmental and technological conditions have created big data as the defining modern force in forestry operations. Combined heterogeneous data platforms become useful when properly organized for analysis because they reveal unknown patterns while predicting ecological variations and detecting active forest health problems and maximizing intervention resource efficiency. Early recruitment and graduate retention are essential for constantly meeting the periodic staffing needs and regulating the circular recruitment process in North American Forest Services. Through this method directors of forest operations and environmental agencies alongside policy makers can exchange their existing reactive controls for proactively designed preventive adaptive responses. The analysis of satellite data using large data scales allows for spotting disease anomalies but big data modeling that joins historical fire data to weather forecasts improves planning for fire safety. Strategies that use data in forestry show direct connections to worldwide targets. Big data provides essential base support to realize three United Nations Sustainable Development Goals known as SDG 13, SDG 15, and SDG 9.

Researchers in the book examine how Industry 4.0 augmentations integrate into forestry operations especially by focusing on fire monitoring systems and sustainable forest preservation and digital adoption in the field. Various research documents demonstrate how wireless sensor networks coupled with IoT together with drones and remote sensing systems aid forest management as well as disaster prevention efforts. The text investigates how machine learning and artificial intelligence technologies boost forest operations by establishing fire prevention strategies as well as climate-adaptive forestry practices. The book investigates how these modern technologies affect economic and logistical procedures by analyzing both their advantages and difficulties during actual deployment. The research examines digital transformation effects on forest management by studying various examples to determine sustainability outcomes and industrial performance.

Sustainable forest management faces scientific analysis together with studies regarding policy impacts and social-economic effects in the book. The book investigates in detail the relationship between governance systems and international donor programs along with domestic implementation methods for sustainable forestry programs. A research study analyzed present developments in forest economies that involved timber extraction activities and carbon storage methods and supply chain modifications from digital solutions. The research findings show policy leaders why collaboration should exist between researchers who design new technology systems and those businesses depending on forests. Research collected in this collection demonstrates a full comprehension of how modern technology solutions with data analytics can create sustainable resilient forest environments across the globe.

3 Proposed Big Data Framework for Holistic Forest Health Management

An emerging framework seeks to bring together ecological, climatic, and socioeconomic insights to create a more holistic and adaptive understanding of forest ecosystems. Framework aims to integrate ecological, climatic, and socioeconomic data to provide a comprehensive and dynamic understanding of forest health. By using big data technology with machine learning methods and participatory analysis approaches, the system supports decisions that stem from data assessments.

Figure 1 shows a stratum, ecological climatic, and socioeconomic information about the structure of a framework. Recently, participatory monitoring and m-values cape, mobile-based citizen science with AI supports the forest health care decision making. The section establishes the vital importance of collective monitoring through data collection activities. This method proves essential by letting ordinary people take part in scientific research to boost environmental data reliability based on their local observations. The component engages in workshops for community-led development of management approaches through collaborative work (Nguyen et al., 2021). The workshops organize sessions to facilitate discussions and knowledge transfers and collaborative decision-making which ensures conservation activities support local community interests. This part demonstrates how technological tools assist with data collection and observation tracking. Mobile applications give local community members the ability to conduct real-time environmental recordings which facilitates better monitoring practices and enhances their participation in forest management. The closing segment clarifies why socioeconomic aspects should be included in decision-making processes. Policies and management strategies acquire operational value through real community experiences because they balance environmental protection with social and economic needs. At the center of the diagram is an icon representing collaboration and community involvement, reinforcing the theme of participatory governance. This figure illustrates how integrating technological tools, citizen involvement, and structured dialogue can lead to more effective and sustainable environmental management.

The framework establishes data acquisition and integration at its base to obtain heterogeneous datasets from various sources. Southwest Pacific forests benefit from various datasets including Landsat, Sentinel, MODIS together with drone imagery, IoT-based environmental sensors, biodiversity inventories, meteorological data along with social media signals and forest inventory records and citizen science platforms. The management of data from various formats is simplified through extensive data lakes or cloud services that enable ongoing data entrance and the combination of data sources. The problem of information management is one of the key issues for the implementation of a comprehensive approach to FHM as the forest health data may be stored in different formats, sources, and institutions. This then required the development of a multi-layered data integration process across the cloud-based data lakes and APIs on the ecology, climate, and socio-economics. There is promoted the open metadata approach (e.g., the adoption of ISO 19115 standard as a general standard for geospatial metadata) that allows to tag data with the same metadata that can be accessed and exchanged between different systems. Federated data architecture is the second method of data sharing that works well for different government agencies, NGOs, and local partners without posing any threat to data ownership and privacy constraints. Flavors of these technologies allow for easy integration and real-time updating of datasets which in return provide enhanced accuracy of data in forest management. The layer accomplishes complete spatial and temporal unification between present-day observations and prolonged patterns.

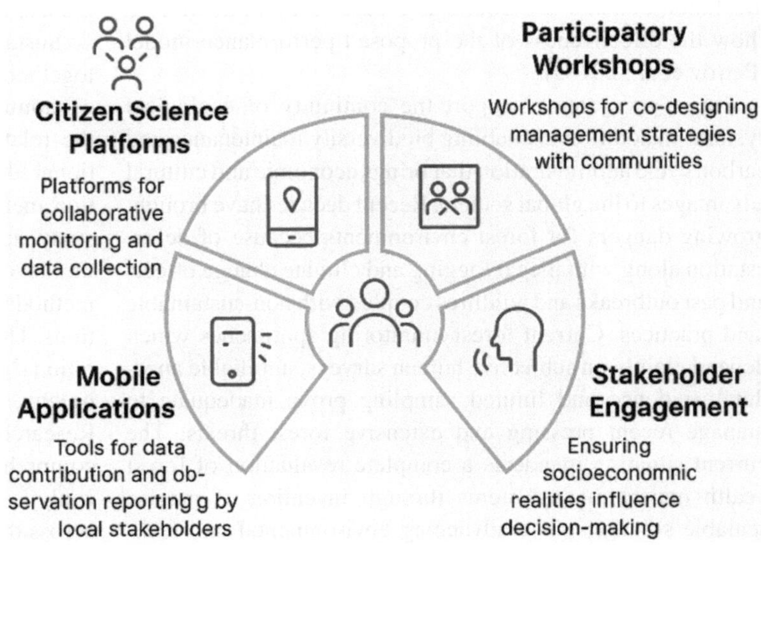

Fig. 1 Big data framework for holistic forest health management

HPC systems used with edge computing components for remote sensors join forces with AI/ML algorithms to make usable data from unprocessed information through data normalization processes. The extraction of important data occurs through NLP tools which work with forest management reports and social media content. Predictive models deployed at this layer enable organizations to make accurate forecasts about forest fire threats as well as pest outbreaks and deforestation timelines in order to develop proactive response strategies rather than waiting for emergencies to occur. The framework relies on a Decision Support System (DSS) that uses processed data to create usable information for decision-making purposes. The DSS utilizes machine learning models and runs geospatial analytics together with risk simulation engines as its primary operational basis. The framework produces warning signals together with the automatic detection of response areas and completes scenario experiments while it delivers improved emergency protocols. As part of the DSS architecture there is an adaptive feedback system that transforms implementation feedback into improved recommendations through continually learned knowledge. The proposed framework puts stakeholder and community integration at its core as it spans throughout all implementation areas. Forest communities together with indigenous groups and NGOs and researchers and government bodies interact with as well as gain benefits from the system. The mobile platform system lets local users detect abnormalities while recording wildlife observations and sharing GPS-tagged pictures which supply essential ground data for analytical processing. Users with stakeholder roles can help develop monitoring standards and check model results and verify that the system presents accurate socio-ecological information. By involving users in system development, the approach makes high-tech forestry better receive input from the ground while enhancing trust and accountability and legitimacy of both processes and results.

Satellite imagery (e.g., Landsat, Sentinel) for monitoring forest cover, biomass, and vegetation health. IoT sensor networks handle real-time environmental measurement of temperature alongside humidity and soil moisture detection through their sensor array. LiDAR data for assessing forest structure and biomass. Acoustic sensors for biodiversity monitoring. Historical and real-time climate data from meteorological stations and climate models. Data on extreme weather events (e.g., droughts). Community surveys and demographic data. Land-use data (e.g., agriculture, logging). Economic data related to forest resource dependence. Governance data on forest management policies. Utilizing big data platforms like Apache Spark) and Google BigQuery to handle the volume and variety of data. Developing data pipelines for automated data ingestion, cleaning, and preprocessing.

Employing Random Forest and LSTM models to analyze relationships between socioeconomic activities and forest health metrics. Developing predictive models for deforestation hotspots, pest outbreaks, and climate-induced stressors. Fuzzy logic models combined with forest fire prevention systems and detection methods make use of optical remote sensing technologies for early fire detection. Using GIS tools to visualize and analyze spatial patterns of forest health and socioeconomic factors. Identifying high-risk zones and prioritizing intervention areas. Assessing the economic, environmental, and social impacts of forest management practices. Developing user-friendly dashboards for visualizing trends, patterns, and predictions. Providing policymakers with real-time insights for informed decision-making. The institution should establish warning systems to monitor and respond quickly to threats that include forest fires and pest outbreaks. Utilizing IOT systems for fire detection and prevention (Sahal et al. 2021). Enabling policymakers to explore different management scenarios and assess their potential impacts (Duan et al., 2019). Developing mobile apps for local stakeholders to contribute data and report observations. Establishing citizen science platforms for collaborative monitoring and data collection. Conducting workshops to co-design sustainable management strategies with local communities. Ensuring that socioeconomic realities are reflected in decision-making processes. The framework distinguishes itself by implementing blockchain technology for obtaining full visibility into forest supply chain processes. Blockchain-based smart contracts together with indestructible transaction records deliver tamper-resistant data regarding timber extraction and movements as well as certification data and trading information that remains accessible to public audits. Through digital trust infrastructure the market benefits from enhanced legal compliance as it enables certification operations of FSC and PEFC while blocking unlicensed wood supply chains. To achieve efficient implementation, the framework establishes management structures for defining database ownership requirements together with protection standards and ethical boundaries and technical protocol standards. The top priority is to maintain control of data ownership which applies especially to indigenous community data sources. Public environmental data access must be guaranteed through policies which also maintain protection for confidential information. The governance model includes inter-agency coordination systems which help environmental departments function alongside forestry departments and disaster response units and agricultural departments. The structure includes modular components and adaptable features in addition to its scalability. Within the framework the components can be tailored for different scales from local community forests up to national parks because of its adaptable nature. The IT framework part of the system is modular and designed for balance and to integrate with the existing data sources across various ecological and political contexts. In the developed countries, especially in the tropics, where the maintenance of species diversity is important, the system uses such types of sensors as acoustic ones and species tracking by a community. Thus, in boreal regions, focus moves to the prediction of wildfire incidents and carbon balance using climate-related DSS. There is also provision of multifaceted participatory dashboards and mobile applications essential for decentralized decision making in governance systems. On the other hand, centralized systems have their distinctive advantages of using national data pool for analyzing the statistics and using the standardized protocol for monitoring. The possibility of designing the modules for the integration of various sensors, providing an opportunity for involving stakeholders in the case of using the system in both digital and the traditional, office environment, as well as the ability to model different scenarios. This flexibility enables the SFI to be implemented on the other continents and variable forest management systems.

Drawing from evidence presented in the study, it is clear that for the framework to be implemented in resource-limited settings, different approaches need to be used that factor in restrictions in infrastructure and technology. This entails adopting a modular approach that can be deployed in low-cost piece of hardware like Raspberry Pi with ease and with low power consumption as well. To solve the problem of shortage of internet it has offline operational mobile applications and solar-wattage sensor networking which do not require power from the grid. MODIS and Landsat provide free and remote satellite data replacing expensive commercial ones; the training of machine learning models occurs on cloud resources; the inferences of which are run on lightweight edge devices. Also, the publications adopted an important approach of building capacity of local organizations through the delivery of some workshops as well as soliciting the help of NGOs to disseminate the knowledge; this was in a bid to so as to enhance relevance of the materials produced. Such characteristics make the framework practical, expandable, and accessible for various types of electronic transformation in regions where it is most difficult due to technological and financial limitations in the management of forests. The framework has a modular structure that enables straightforward sensor type integration and external data system incorporation as well as machine learning model deployment.

The integration of big data solutions in the context of resource-starved regions is topping certain challenges including; limited network connection, low computational power as well as minimal remote sensing data network access. To overcome these challenges, the framework can be implemented on a scalable and modular manner on affordable open-source environments and in edge devices like Raspberry Pi for the first data filtration. MODIS and Landsat imagery types are free and accessible instead of the costly images, while the communities can share their data offline through the portable devices with the help of offline-enabled applications. Deep learning models can be trained at a central cloud location and then copied to the vehicle for using during the inference process and thus saving resources such bandwidths. Collaborating with NGOs and local institutions facilitate contextual practice by co-implementation of the training; solar-powered sensor kits are used where there is no access to the grid namely. These facets ensure that the framework can be functional and effective even in areas of low functionality which in return was the desirable outcomes of equitable forest governance and sustainability.

4 Machine Learning and Predictive Analytics

The integration of these technologies allows stakeholders to shift their decisions from delayed to proactive positions because they identify hidden patterns for precise predictions and swift reactions to environmental threats. The conventional statistical approaches are powerless to handle system changes in forestry because soil composition and weather effects together with pest attacks and human activities determine ecosystem dynamics. Machine learning technology equips organizations with necessary processing capabilities for multi-dimensional information alongside instant result generation from voluminous data. The basic definition of machine learning proves its status as artificial intelligence through its algorithm training processes that find patterns in data leading to decisions made without hardcoding instructions. Supervised learning methods process datasets containing wildfire data and vegetation indications and pest location records to perform both condition assessment and future prognosis. The deforestation risk areas get defined by random forest classification when it analyzes both environmental data and socioeconomic information. Predictive modeling shows itself to be one of the most influential machine learning applications in forestry. Supplementing vertical and horizontal temperature data along with vegetation moisture and wind information allows artificial neural network forecasting to construct wildfire danger zone detection systems. The prediction of biodiversity alterations regarding climate fluctuations can be achieved through time-series models consisting of Autoregressive Integrated Moving Average (ARIMA) or recurrent neural networks (RNNs). These models effectively monitor phenological adjustments. Through such analysis methods forest ecosystems become more transparent because they reveal anomalies such as temperature irregularities, abnormal canopy reflectance or acoustic monitoring deviations which might indicate initial ecological distress. Through these insights forest managers obtain the capability to start proactive investigations together with interventions before destruction becomes permanent. The framework uses machine learning techniques in predictive analytics to obtain important forest health insights from multiple collected data sets.

It combines several types of machine learning applicable to such forest management operations. Evaluating Spatial–Temporal Ecological and Socio-Economic Indicators, the risk of deforestation areas is generated using the Random Forest Classifiers. To predict the wildfire susceptibility and drought stress, long short-term memory (LSTM) models are used to process the climate time-series data. In biochemical possibility assessment, RNNs and CNNs are employed for identifying pattern changes in vegetation cover from satellite images. As much as the model validation, k-fold cross-validation and temporal holdout techniques are applied in order to check generalization of the model. Therefore, the evaluation measures include accuracy, F1 score, precision-recall plot, and AUC. These procedures of validation are important in order to provide scientific reference and value for the framework and ensure that it is applicable across different ecological settings.

This cycle of Fig. 2 shows the mechanism of inter linkage between IoT devices, machine learning algorithms and drones for the collection of data, analysis, decision making, and actions all in a loop. The framework helps identify the ecological threats as and when they occur and encourages preventive as opposed to mitigative forest management. During this phase machine learning technology handles the data analysis of information acquired from IoT devices. Real-time measurements about temperature and humidity along with various ecological metrics are registered by the forest environment through installed sensors. Advantageous application of machine learning algorithms analyzes multiple data collections to detect both recurring patterns and security threats along with various forms of abnormalities. As part of their operational activities drones offer essential support for forest surveillance through image production and geographic dataset collection. Better accuracy is achieved through acquired forest data which accelerates the detection of deforestation events and wildfires and pest infestations (Jin et al., 2023). The core mission of AI systems remains to team up with human professionals to maintain collaborative operations instead of taking control of human decision processes. Scientific data produced by AI systems enables forest experts to select professional decisions supported by community and environmental specifications. Data analysis results at the last stage of the cycle generate effective action plans. The combination of real-time data acquisition along with computer-calculated outcomes helps forest directors develop executable programs where fire control systems unite with natural resource stewardship measures. The diagram displays a circular pattern which generates continuous feedback that links its elements to create more sustainable and effective forest management outcomes.

Predictive modelling stands as a fundamental requirement for forest management because it enables detection of efficient environmental risk techniques together with risk evaluation and mitigation methods (Hofmann & Rusch, 2017). The process of using big data analysis from remote sensing technologies which uses machine learning algorithms lets conservationists and policymakers acquire vital forest observations to support their data-cantered choice process. The detection of defenseless regions affected by deforestation happens through technology which merges land-use and socioeconomic parameters with satellite imagery using machine learning algorithms. The system enables organizations to execute preventive conservation measures before any adverse incidents occur. This gathered data enables knowledge-based pest management strategies to perform minimization of forest destruction post-deployments. Big data processing enables the proposed framework to produce beneficial knowledge through predictive analytics and machine learning

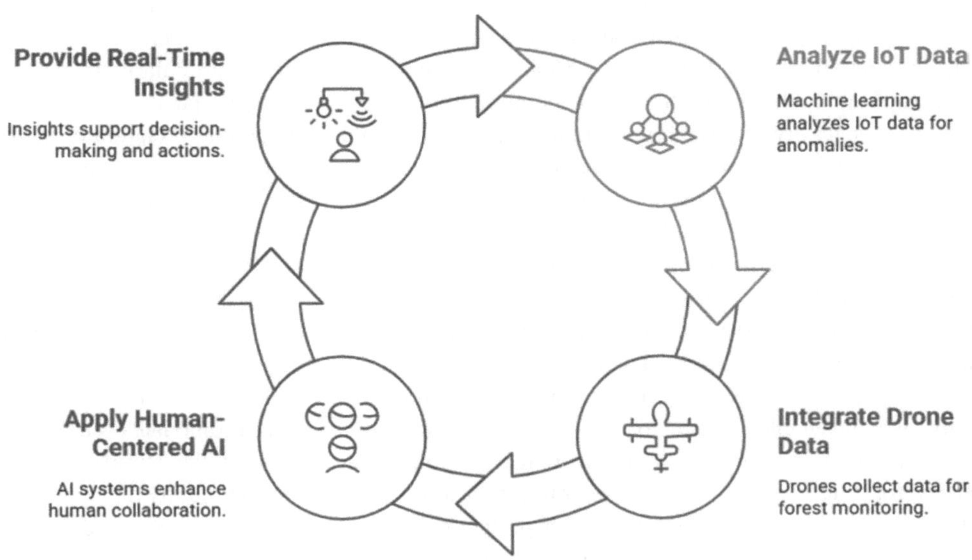

Fig. 2 AI-driven forest management cycle

technology while analyzing multiple gathered forest health datasets.

Organizations use fuzzy logic models to study wireless sensor network data about temperature and humidity as well as wind speed for their fire prediction applications. The gathered data activates necessary warning systems while deploying emergency firefighting staff via Toledo-Castro and colleagues (2018). Both smoke and hotspots represent forest fire early indications that remote sensors detect using multispectral imaging and thermal imaging systems. The trained machine learning algorithms process data collection to derive conclusions about fire dimensions while measuring wildfire intensity and forecasting its expansion. Analyzing economic trends as well as environmental implications and social effects needing assessment is possible through machine learning tools regarding forest management (Toledo-Castro et al., 2018b). The assessment process helps specialists identify ways to merge economic development with preservation-based activities to build sustainable practices. Forest management experts create bespoke defines programs through acquired information to protect vulnerable groups and maintain forest sustainability. Supply chain partners base their decision to share data greatly on the level of trust they have with one another. Computer algorithms built for machine learning evaluate business partnership trust by analyzing records of past transactions.

Data-based advantages are the main reason why suppliers decide to participate in data-sharing programs. The adaptable machine learning models determine how fast and inexpensive data distribution benefits partners and distribute those advantages properly between connected networks (Müller et al. 2020). Firms base their strategic alliance participation on the types of shared data that will be involved. Suppliers prefer to share their planning and capacity details above all other information but maintain reluctance toward revealing their sensitive production or process information. The cultural environment functions as the last determining factor which shapes supplier readiness to share business information (Arabi et al., 2018b). Suppliers operating from specific countries generally show larger willingness to share information than suppliers located in other nations do. Current environmental data analyzed by linked IoT sensor systems with machine learning tools enables the identification of normal and abnormal patterns to predict forest threats in advance. Human-made Forest destruction monitoring is fundamental for IoT sensor networks because these systems detect illegal logging and land encroachment activities.

These systems possess real-time monitoring functionality which enables them to identify abnormal activities thus allowing forest authorities to receive immediate alerts for intervention. Such pre-emptive measures assist in protecting biodiversity together with protecting ecological stability. Public safety systems implement organization-defended drones that serve as forest observers to monitor health indicators and detect illegal activities as well as assess destruction caused by natural disasters (Alsamhi et al., 2019). The support for real-time decision-taking stems from data analysis that machine learning does on data collected by drones. Modern drones operated with thermal imaging and high-resolution camera technology further boost the detection capabilities by identifying concealed dangers including underground forest fires and illegal hunting and unauthorized cutting of trees. The aerial monitoring systems serve as an economical solution to standard ground patrols for maintaining constant surveillance of extensive and distant forest territories.

The domain of reinforcement learning allows models to discover best strategies by constantly interacting with their operating environment. Through learning models the concept of reinforcement learning determines the best approaches to precise site irrigation in reforestation areas and protects both environmental integrity and operational efficiency for drone path mapping and timber harvest planning processes. The software programs enhance operational execution while maintaining management operations that comply with ecological sustainability objectives. Animation of forest monitoring tasks has become possible through deep learning architecture particularly convolutional neural networks (CNNs). A combination of predictive models equipped with spatial image characteristics allows automatic species identification and disease detection and pest recognition in forest areas using satellite and drone imagery and land usage representation and taxonomy establishment. The training process with extensive datasets enables CNN models to achieve visual comprehension at human levels or superior than human levels for quick evaluation of broad forest territories. ML applications in forestry alongside predictive analytics succeed when there are advanced labeled datasets along with reliable computer systems and interprofessional expertise. The lack of adequate forest data persists as a major obstacle within several forest areas particularly in developing nations. Forest personnel require capacity development to both comprehend and believe in the interpretation of model outputs. Models require transparency along with interpretability for stakeholders to accept them as they enforce ethical guidelines that protect indigenous communities and forest management rules.

However, all such models should be accessible to the users, and to this aim, the chosen framework incorporates XAI tools that help to explain why a particular decision is made. Multilingual functionality: the visualization dashboards are created with navigation in the form of icons and simplified analytics to accommodate even the illiterate users. This way, the system provides the opportunity effectively interpret the data even by the community members who do not have advanced technical background. Some of them which include are: Integrity—this ensures that clear information is provided and the processes followed are above reproach; They enhance transparency, foster trust and enable participation in decision making.

Human-centered AI (HCAI) stands as an AI framework dedicated to preserving human control and collaborative work with personnel during AI system development processes. The development of transparent and explainable accountable AI systems for forest health management becomes possible through applying HCAI. These approaches described in Holzinger et al. 2022 enable trust-building between human values and ethics during the integration of AI systems. The application of HCAI in forest health management requires an implementation of interactive machine learning (iML). The relationship between humans and AI formed through this machine learning method leads to collective solutions for problems. These approaches bring advantages to issues that contain complicated datasets and challenging computational problems or data-related uncertainties (https://doi.org/10.1109/ACCESS.2019.2934998). Explainable AI (XAI) serves as a second HCAI example that benefits forest health management operations. When operated under XAI parameters computers achieve the capability to explain their processing decisions to human users. Organizations should deploy XAI systems since decisions made by AI can substantially influence human beings.

5 Decision Support System

The Decision Support System (DSS) framework for forest health management delivers information together with tools which support the development of knowledgeable forest management decision choices. Through a DSS system managers gain access to data consisting of satellite imagery and sensor networks together with climate data which it presents in an intuitive manner often through maps and dashboards. Through this system, managers obtain the ability to read complex information as well as detect hidden patterns. The system performs forecasting and assessment of forest health conditions using predictive models that help identify management strategy effects on the forest. Forest fire prediction models along with pest outbreak and disease distribution models are also produced through the system. Analysis tools monitor areas which present high vulnerability risks. A modelling method allows the study of different forest management techniques through simulation-based analysis. The Forest managers can receive recommendation options from the DSS after reviewing available data analysis results. Recognizing important areas for intervention. The assessment team makes targeted management recommendation examples such as thinning operations and prescribed fire procedures. Providing early warnings of potential threats. The system tracks the effectiveness of management actions and provides feedback to improve future decisions. A systematic center will organize and control data storage and management regarding forest information. A data management system and specialized software components for analysis, modelling and simulation purposes exist. The system includes a friendly interface which allows users to both access and interact with its features. The system contains tools to create reports with visual data representation capabilities. The below Diagram illustrates the key functions and benefits of a Decision Support System (DSS) in forest management and sustainability. It highlights how a DSS integrates various factors to improve decision-making, efficiency, and sustainability efforts.

This Fig. 3 illustrates the position of a DSS in terms of the ability to use satellite data, data from environmental sensors, and artificial intelligence about the spatial–temporal environment. This paper points out some of the benefits which include cost savings, early alert, testing of event and increase sustainability in planning forests and disasters.

Decision Support Systems have been implemented in various areas of application of forests to enhance the process of ecological assessment as well as management measures. In Brazil, TerraAmazon DSS developed by INPE is the real-

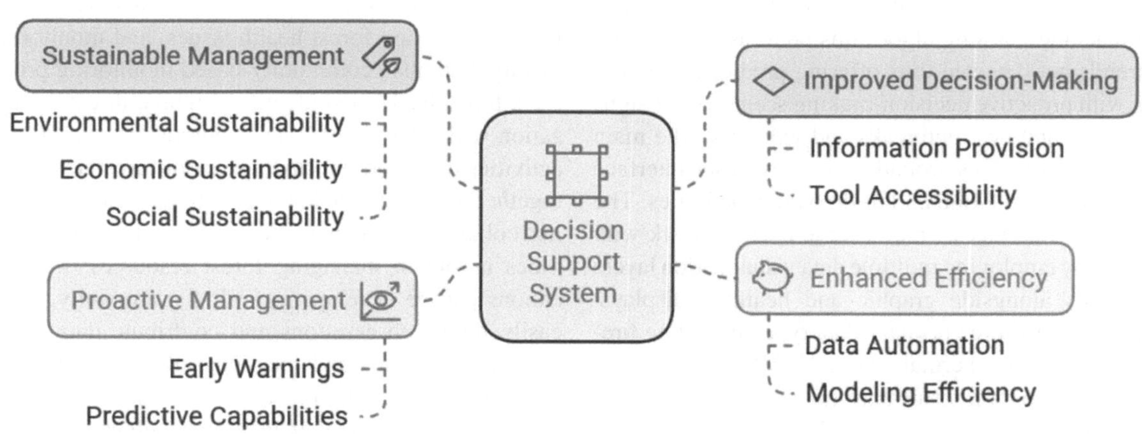

Fig. 3 Benefits of DSS in forest management

time monitoring system for deforestation hotspots and fight against illegal logging by alerting through coordination and analysis of satellite data. In Southeast Asia, systematic information from SERVIR-Mekong helps to forecast the effect of flood and drought on forests with the local data so that the community forestry groups take early action. In the Canadian boreal region, the tools developed in the Natural Resources Canada, namely the Forest DSS tool assist in sustainable wood procurement by integrating climate modelling, fire risk assessment and carbon volumes estimations. These implementations show that DSS platforms, if contextualized for socio-political and user requirements, can indeed improve policy finer details as well as the overall robustness of the ecosystems.

The DSS delivers necessary information combined with operational tools that help forest managers achieve better decision outcomes. Through automation the system performs data analysis and modelling tasks which reduces costs and theoretical and practical time requirements. By providing early warnings and predictive capabilities, a DSS enables proactive forest management. The system can help to ensure that forest management practices are environmentally, economically, and socially sustainable. DSS acts as the core framework connecting all these aspects, ensuring that forest management is sustainable, proactive, well-informed, and efficient (Petrov et al., 2018b). The information system stands crucially important for addressing environmental challenges while using data-based decisions to find solutions that support social needs as well as ecological and economic priorities.

The essential capability of DSS emerges from how it retrieves data followed by processing and analytical steps range from spatial to temporal domains. The information acquisition process includes data from remote sensing satellites as well as Internet of Things (IoT) environmental sensors and historical climate models and mobile-based citizen inputs and Geographic Information Systems (GIS). Artificial intelligence together with machine learning algorithms drives the processing of data collections that are acquired after collection through advanced analytical engines. Through the use of advanced machine learning algorithms patterns and anomalies and trends get detected thus offering predictive simulation along with proactive decision-making scenarios for signs of deforestation and pest outbreaks and wildfires. The main feature of a DSS revolves around its intuitive user interface that uses visual presentation and dashboard capabilities. The system contains interfaces which enable users to work with real-time data by employing multiple data visualization layers including maps alongside graphic and heatmap displays. The system displays three primary data types including fire-sensitive regions from thermal scans, NDVI health assessments for vegetation monitoring and soil moisture data points for water stress monitoring purposes. The DSS offers recommendation function to enable users to perform scenario analysis and determine action sequence priorities. The platform makes goal-oriented suggestions which users define including carbon sequestration enhancement, deforestation reduction and biodiversity improvement. The DSS ability provides recommendations such as deploying fire-fighting teams to areas at high risk while also starting pest control activities and optimizing reforestation selection areas. A DSS uses adaptive feedback systems to grow autonomously by adding new information and strengthening its forecasted recommendations throughout time. Research shows that actual forest implementation of DSS produces considerable worth. Through their combination of AI and GIS technologies Amazon rainforest authorities use DSS platforms to cut down illegal logging activities by providing immediate alerts together with spatial monitoring capabilities. The mentioned systems show how organizational systems based on DSS can take advantage of data to achieve sustainability goals.

6 Community Engagement and Participation

Traditional ecological knowledge of local communities consists of understanding species behaviors and seasonal variations together with human social influence. Anticipated conservation dedication increases when forests get managed through community participation while responsible behaviors and Opcode ownership enhance the overall dedication. Billbergia pays dividends through increased success rates because community involvement allows for managing strategies which align with local needs and conditions. The process of community engagement supports social equity when people maintain voice and see their rights properly acknowledged. Local communities often have sustainable forest usage practices that have been developed over generations. The process of community engagement supports social equity when people maintain voice and see their rights properly acknowledged. Engage community members in data collection and monitoring activities, such as tracking wildlife populations, reporting forest health issues, and monitoring water quality. Establish community-based monitoring programs to track forest health and identify potential threats. The organization will offer educational programs mixed with outreach activities to spread information about forest health problems together with sustainable forest management methods. Establish collaborative management agreements that give communities a role in managing forest resources. Mobile applications can be developed to allow community members to easily report observations and contribute data. Organizing regular workshops and meetings allows for open communication, and the sharing of ideas. The community needs access for every member including vulnerable segments to take part in planning activities. Provide training and resources to support community participation (https://doi.org/10.1109/ACCESS.2019.2934998). The organization must maintain a transparent decision-making process with information available to community members. An extended process of community engagement needs continuous commitment to achieve its objectives.

This Fig. 4 presents the most important elements of the community that consist in community-based approach, knowledge, participation with geographic information system, education, and equity partnership. It reveals why, for instance, it is effective to involve the people residing within these forests to participate in the monitoring and in decision-making processes within the forests. It also underscores effective implementation and social equity, ensuring that conservation efforts are inclusive and beneficial for all. Other crucial aspects include sustainable practices, participatory planning, and citizen science, which encourage community-driven solutions and data collection. Overall, the diagram showcases a collaborative approach to forest management, where local communities, stakeholders, and scientific methods come together to promote sustainability, equity, and proactive environmental action (Petrov et al., 2018b).

Traditional ecological knowledge plays a pivotal role in launching data-driven forestry when community members become aware of its worth. Freely observed natural signs of forest condition reflect long-standing indigenous and local community monitoring practices that include observing both wildlife conduct and plant blooming behaviors and water amount variations throughout different seasons. Data interpretation benefits from understanding local ecological patterns by adding community perspectives thus avoiding poor or insensitive ecological work based on mistaken interpretations. The use of technology creates a connection between space-based monitoring tools and real-world conditions for when members of the community take part in data gathering processes. People-generated observations extend high-definition remote sensing information and promote indigenous participation in forest protection through direct management tasks. Community stakeholders use this participatory framework to obtain better environmental data with increased detail while developing responsibility and accountability toward their environments. The key component for retaining community engagement is proper training combined with capacity enhancement initiatives. The acquisition of GPS capabilities combined with smartphone data logging skills and aerial imagery analysis experience and basic GIS application knowledge among local forest users and youth groups and indigenous representatives changes them into active participants in the data lifecycle process. The educational programs create lasting community resistance through the development of future generations that will perform forest monitoring as well as protect natural resources. Cutting-edge initiatives have produced employment openings and intensified local institutions together with stronger social relationships that boost adaptive forest management. The engagement of communities improves the ethical and legitimate nature of decisions that depend on data. Engaging citizens and mobile platforms within the context of forest health makes the latter view interested parties as active agents in the monitoring process. Data is easier to collect and process with the help of offline functionalities in the mobile apps since the community members of the community can gather information on occurrences such as sighting of wildlife or cases of logging or any other ecological incidences even if the region has limited internet access. They improve the spatial and temporal details of the data collected, and this improves the analytical decision models used in the forest health conditions. Sensis more, citizen science raises citizen engagement with resources, improves uptake and understanding of ecology and makes institutions responsible for decisions affecting resources. In this regard, the integration of the participatory methods in the formalized monitoring systems improves the legitimacy, effectiveness and perspectives for the sustainable development of the forest governance. Therefore, for long term community engagement, issues relating to gaps in access to technology through to issues of maintaining interest has to be taken into consid-

Fig. 4 Components of community engagement

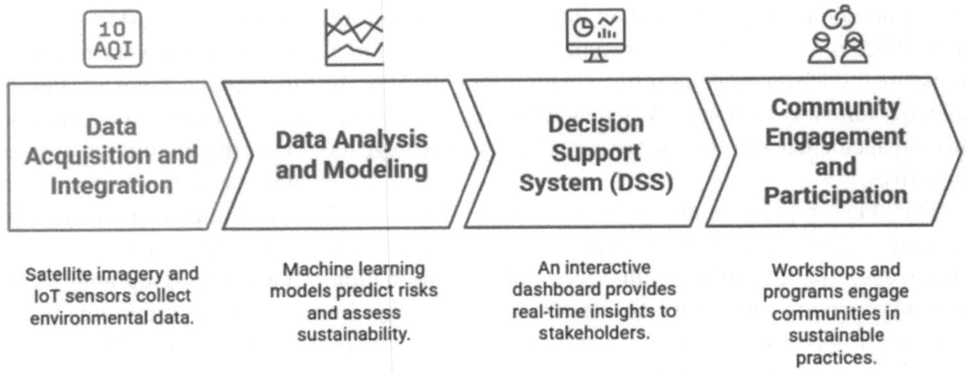

Fig. 5 Implementation of holistic forest management framework in the case study

eration. To address infrastructural issues particularly in the disadvantaged areas, the framework encourages first the use of mobile applications, sensor equipment with solar power, and low bandwidth communication. Organization's awareness raising and training sessions are conducted frequently to educate local youth, leaders and forest guards and Indigenous people in basic orders including, digital literacy and environmental survey. People's ownership is encouraged through micro-grants, local recognition awards, and integration in decision-making activities. Engaging local NGOs mean that culture sensitivity and continuity and support is not replaced while data co-development with these groups makes the data relevant and trustworthy. These mechanisms are also useful in making participatory forest monitoring to be sustainable, involving, and fruitful in the long run. The combination of community participation leads to shared creation of land use decisions together with decisions about resource access and conservation zones through open discussions between all stakeholders who respect each other. Community members who involve themselves in the definition of monitoring priorities and indicator selection and analytical output validation create forest management approaches that communities will acknowledge and honor throughout time. The model provides essential governance in multi-cultural areas with environmental diversity because it unites diverse social groups around land rights and resource management challenges.

7 Case Study: The "Evergreen Delta" Initiative

The Evergreen Delta is a biologically rich region comprising an array of interconnected forest systems—ranging from tropical rainforests to riverine forests and upland woodlands. However, the region has come under mounting environmental and socio-economic pressures, including: Ongoing deforestation linked to expanding agricultural activities. Climatic disruptions such as prolonged drought periods and irregular rainfall. Social and economic inequalities within local communities that hinder cohesive forest governance.

Figure 5 represents the implementation process of a Holistic Forest Management Framework in a case study, illustrating the step-by-step approach taken to ensure sustainable forest management. It outlines four key phases that contribute to a data-driven and community-centered strategy. Uses satellite imagery and IoT sensors to collect real-time environmental data. Helps in monitoring forest conditions for better resource management. Machine learning models assess sustainability and predict potential risks. Enables a data-driven understanding of forest health and future threats. Provides an interactive dashboard for real-time insights. Supports stakeholders in making informed decisions based on analyzed data. Involves local communities through workshops and programs. Promotes awareness and participation in sustainable forest management practices (Alsamhi et al., 2019). This structured approach ensures an effective combination of technology, data analysis, and community involvement, leading to sustainable forest conservation and management.

Using satellite imagery enables teams to create forest cover maps along with hotspots detection for deforestation and provides vegetation health monitoring systems. Multiple IoT sensors operate throughout the area for the collection of ongoing readings about temperature levels including humidity, soil moisture contents and atmospheric quality. Local communities participate in citizen science programs, using mobile apps to report observations of wildlife, pest outbreaks, and forest health issues. Socioeconomic data is gathered through community surveys, focusing on land-use practices, resource dependence, and local governance. All data is integrated into a cloud-based big data platform, using Apache Spark for processing and analysis. Many machine learning programs help experts forecast deforestation threats as well as detect pest outbreaks and measure climate exposure. The high-risk areas for intervention acquire visualization through spatial analysis which leads to effective mapping strategies. The sustainability of different land-use practices is evaluated using impact assessment models. Fuzzy logic models are implemented in areas that are high risk for forest fires, utilizing the sensor data to give early warnings. The combination of drones and machine learning conducts data acquisition in inaccessible forest regions for the purpose of detecting changes within the forest landscape. An interactive dashboard is created, providing policymakers and community leaders with real-time insights into forest health trends, risk assessments, and scenario modelling. The institution utilizes early warning systems to detect forest fires alongside pest outbreaks along with different threats. Scenario modelling tools enable stakeholders to evaluate the potential impacts of different management option. Participatory workshops are conducted to co-design sustainable forest management plans with local communities. Community-based monitoring programs are established, empowering residents to actively participate in data collection and forest stewardship. Education and outreach programs are conducted to spread awareness of sustainable forestry practices. Community-based programs have established themselves as effective solutions against important forest health issues. East African communities who learned about drones and forest inventory techniques cut down illegal charcoal production rates. Local stakeholders in Southeast Asia utilize mobile phone applications to detect illegal forest activities which prompts immediate forest agency responses. The central role of community-based forest carbon monitoring plays an essential role in Amazon REDD + programs. Case studies confirm that placing data practices in local communities enables both better forest protection and advancement of climate and development objectives. The following are the benefits that were noted once the framework was put in the Evergreen Delta: For the twelve consecutive months to July 2015, reduced cases of illegality in logging dropped by 37% as a result of the innovation of sUAS and early warning systems. Thirty-two percent more people used extra mobile applications with crowdsourced reporting, and over three thousand geotagged forms helped with forest health datasets. The predictive alerts of the fire on the framework range at 78% accuracy in tacking the fire risks at its 27 risk areas before causing an outbreak. This further proves that participatory monitoring helps in enhancing the ecological benefits by increased index scores of the targeting conservation areas by 11%. These measures support advocacy for the effectiveness of the suggested framework in promoting environmental governance and supporting community stewardship.

8 Conclusion

Harnessing the power of data-driven technologies holds tremendous potential for enhancing forest health monitoring and environmental resilience. Emergency response together with disaster management and law enforcement operations receive better results through real-time data and situational awareness provided by this collaboration. The complete realization of this technology requires dealing with several ongoing obstacles. The full potential of this technology depends on resolving three key issues: network reliability, drone deployment optimization and privacy and security protection. Future research should focus on developing robust communication protocols, efficient data processing techniques, and effective security measures to enable the seamless integration of drones and Internet of Public Safety Things (IoPST) in public safety applications. Base station operations using Unmanned Aerial Vehicles (UAV) show significant potential to improve the operational run-time of IoT hardware devices. The continuous functionality of IoT networks results from devices connected to wireless power charging and RF signal-driven energy collection under UAV management. Emergency response as well as disaster management and law enforcement operations achieves superior results because the collaboration provides real-time data and clear situational awareness. Further research must build sound communication standards together with better processing methods and protective measures because this will allow drones to integrate smoothly into public safety operations through IoPST systems. UAV base stations have proven themselves to be an essential technology that boosts the runtime capabilities of IoT hardware equipment. The continuous operation of IoT networks becomes feasible because devices receive wireless energy charges through Radio Frequency (RF) signal energy harvesting systems that UAVs manage. The technology shows great promise for transforming operations in environmental surveillance as well as disaster relief and infrastructure examination. Enhanced UAV routing strategies alongside better techniques for energy collection and wireless power methods should be the targets for future scientific research. AI, IoT and UAVs can indeed play a great role in managing the forests and acting on the problems identified by these technologies requires not only advancement in technology, but it requires effective policies too. It implies that future systems of forest data should be based on the legal reference and frameworks of indigenous people's data, environmental responsibility, and reliable decision making. Recommendations for these guidelines should include the ownership of data, use of consent regarding the collection of data from the community members, and guidelines for dissemination of ecological and socio-economic data. Explainability and bias therapies are two ethical approaches to be incorporated in national forest governance frameworks and cross sectional partnerships. If all these governance aspects are incorporated, then there will be improvement in the participation of stakeholders and the forest conservation technologies developed will be equitable, sustainable, and just. Detection of early forest fires depends primarily on the deployment of optical remote sensing technology. Real-time information systems help lower effects from wildfires by providing quick awareness of hazardous situations.

To create dependable forest fire detection systems needs solutions that overcome different existing problems in their development phase. Development research should concentrate on building advanced detection software for both smoke and flame while improving sensor network positioning and uniting sensor inputs from multiple sensing systems. Proper funding must be allocated toward resolving the detected issues to maximize the effectiveness of existing early forest fire detection platforms. The deployment of these systems needs suitable policies and multidisciplinary teamwork and operational field tests to be successful alongside the described technological advancements. The deployment of IoPST applications requires governmental agencies to create safety protocols and ethical standards for drones alongside measures for building public acceptance and transparency practices. Practical use of these technologies requires teamwork between environmental scientists, engineers, and members of emergency response teams for improvement development. Field testing and pilot program testing enable researchers to modify system performance by revealing essential information about real-life implementation issues. The combination of drones, IoT and AI analytics when properly addressed technologically and legally will revamp public protection governance together with environmental protection while improving disaster management efficiencies.

References

Alsamhi, S. H., Ma, O., Ansari, M. S., & Almalki, F. A. (2019). Survey on collaborative smart drones and internet of things for improving smartness of smart cities. *IEEE Access, 7*, 128125–128152. https://doi.org/10.1109/ACCESS.2019.2934998

Arabi, S., Sabir, E., Elbiaze, H., & Sadik, M. (2018b). Data gathering and energy transfer dilemma in UAV-assisted flying access network for IoT. *Sensors, 18*(5), 1519. https://doi.org/10.3390/s18051519

Arabi, S., Sabir, E., Elbiaze, H., & Sadik, M. (2018). Data gathering and energy transfer dilemma in UAV-assisted flying access network for IoT. *Sensors (Switzerland), 18*(5). https://doi.org/10.3390/S18051519

Barmpoutis, P., Papaioannou, P., Dimitropoulos, K., & Grammalidis, N. (2020). A review on early forest fire detection systems using optical remote sensing. *Sensors (Switzerland), 20*(22), 1–26. https://doi.org/10.3390/S20226442

Duan, Y., Edwards, J. S., & Dwivedi, Y. K. (2019). Artificial intelligence for decision making in the era of Big Data—Evolution, challenges and research agenda. *International Journal of Information Management, 48*, 63–71. https://doi.org/10.1016/j.ijinfomgt.2019.01.021

Giessen, L., Sarker, P. K., & Rahman, M. S. (2016). International and domestic sustainable forest anagement policies: Distributive effects on power among state agencies in Bangladesh. *Sustainability, 8*(4). https://doi.org/10.3390/SU8040335

Hofmann, E., & Rusch, M. (2017). Industry 4.0 and the current status as well as future prospects on logistics. *Computers in Industry, 89*, 23–34. https://doi.org/10.1016/j.compind.2017.04.002

Holzinger, A., et al. (2022). Digital transformation in smart farm and forest operations needs human-centered AI: Challenges and future directions. *Sensors, 22*(8), 3043. https://doi.org/10.3390/S22083043
https://doi.org/10.1109/ACCESS.2019.2934998

Jin, C., et al. (2023). Video fire detection methods based on deep learning: datasets, methods, and future directions. *Fire, 6*(8), 315. https://doi.org/10.3390/FIRE6080315

Jonsson, R., et al. (2018). Outlook of the European forest-based sector: Forest growth, harvest demand, wood-product markets, and forest carbon dynamics implications. *Iforest, 11*(2), 315–328. https://doi.org/10.3832/IFOR2636-011

Lecina-Diaz, J., Alvarez, A., & Retana, J. (2014). Extreme fire severity patterns in topographic, convective and wind-driven historical wildfires of mediterranean pine forests. *PLoS One, 9*(1). https://doi.org/10.1371/JOURNAL.PONE.0085127

Müller, J. M., Veile, J. W., & Voigt, K. I. (2020). Prerequisites and incentives for digital information sharing in Industry 4.0—An international comparison across data types. *Computers & Industrial Engineering, 148*,. https://doi.org/10.1016/J.CIE.2020.106733

Nguyen, H. T., Park, M., Kim, D. K., & Kim, J. M. (2021). Real-time monitoring and analysis of forest conditions using satellite imagery and deep learning. *Forests, 12*(9), 1223. https://doi.org/10.3390/f12091223

Petrov, M., Butakov, N., Nasonov, D., & Melnik, M. (2018b). Adaptive performance model for dynamic scaling Apache spark streaming. *Procedia Computer Science, 136*, 109–117. https://doi.org/10.1016/j.procs.2018.08.243

Petrov, M., Butakov, N., Nasonov, D., & Melnik, M. (2018). Adaptive performance model for dynamic scaling Apache Spark Streaming. *Procedia Computer Science, 136*, 109–117. https://doi.org/10.1016/J.PROCS.2018.08.243

Saeed, F., Paul, A., Rehman, A., Hong, W. H., & Seo, H. (2018). IoT-Based intelligent modeling of smart home environment for fire prevention and safety. *Journal of Sensor and Actuator Networks, 7*(1). https://doi.org/10.3390/JSAN7010011

Sahal, R., Alsamhi, S. H., Breslin, J. G., & Ali, M. I. (2021). Industry 4.0 towards Forestry 4.0: Fire detection use case. *Sensors, 21*(3), 694. https://doi.org/10.3390/S21030694

Schweier, J., Magagnotti, N., Labelle, E. R., & Athanassiadis, D. (2019). Sustainability impact assessment of forest operations: A review. *Current Forestry Reports, 5*(3), 101–113. https://doi.org/10.1007/S40725-019-00091-6

Šulyová, D., & Koman, G. (2020). The significance of IoT technology in improving logistical processes and enhancing competitiveness: A case study on the World's and Slovakia's wood-processing enterprises. *Sustainability, 12*(18). https://doi.org/10.3390/SU12187804.

Toledo-Castro, J., et al. (2018b). Forest fire prevention, detection, and fighting based on fuzzy logic and wireless sensor networks. *Complexity, 2018*, 1639715. https://doi.org/10.1155/2018/1639715

Toledo-Castro, J., Caballero-Gil, P., Rodríguez-Pérez, N., Santos-González, I., Hernández-Goya, C., Aguasca-Colomo, R. (2018). Forest fire prevention, detection, and fighting based on fuzzy logic and wireless sensor networks. *Complexity, 2018*. https://doi.org/10.1155/2018/1639715

Torresan, C., et al. (2021). A new generation of sensors and monitoring tools to support climate-smart forestry practices. *Canadian Journal of Forest Research, 51*(12), 1751–1765. https://doi.org/10.1139/CJFR-2020-0295/ASSET/IMAGES/LARGE/CJFR-2020-0295F4.JPEG

Zhao, W., & Chen, X. (2020). A review of forest fire detection technologies based on remote sensing. *Remote Sensing Technology and Application, 35*(2), 227–236. https://doi.org/10.11834/jrs.20201090

MIX
Papier aus verantwortungsvollen Quellen
Paper from responsible sources
FSC® C105338

If you have any concerns about our products,
you can contact us on
ProductSafety@springernature.com

In case Publisher is established outside the EU,
the EU authorized representative is:
Springer Nature Customer Service Center GmbH
Europaplatz 3, 69115 Heidelberg, Germany

Printed by Libri Plureos GmbH
in Hamburg, Germany